Perspectives in Natural Resources Management: Watershed-based Approach

Perspectives in Natural Resources Management: Watershed-based Approach

R.K. Goyal
Mahesh K. Gaur
Editors

CWP

Central West Publishing

This edition has been published by Central West Publishing, Australia

For more information about the books published by Central West Publishing, please visit https://centralwestpublishing.com

Disclaimer

Every effort has been made by the publisher, editors and authors while preparing this book, however, no warranties are made regarding the accuracy and completeness of the content. The publisher, editors and authors disclaim without any limitation all warranties as well as any implied warranties about sales, along with fitness of the content for a particular purpose. Citation of any website and other information sources does not mean any endorsement from the publisher and authors. For ascertaining the suitability of the contents contained herein for a particular lab or commercial use, consultation with the subject expert is needed. In addition, while using the information and methods contained herein, the practitioners and researchers need to be mindful for their own safety, along with the safety of others, including the professional parties and premises for whom they have professional responsibility. To the fullest extent of law, the publisher, editors and authors are not liable in all circumstances (special, incidental, and consequential) for any injury and/or damage to persons and property, along with any potential loss of profit and other commercial damages due to the use of any methods, products, guidelines, procedures contained in the material herein.

NATIONAL LIBRARY OF AUSTRALIA

A catalogue record for this book is available from the National Library of Australia

ISBN (print): 978-1-925823-94-3

Foreword

Land and Water are most important natural resources for the existence and survival of living beings on the earth. There are historical evidences that the prosperity of any civilization or the nation to a great extent depends on availability and sustainable management of these resources. Land and water resources are the backbone for industrial and socio-economic development of any area. Development of land and water together with sustainable production system when confined to natural drainage unit termed as watershed leads to sustainable development. Watershed management has, therefore, emerged as a new paradigm for planning, development and management of land, water and biomass resources with a focus on social and institutional aspects apart from biophysical aspects.

Looking into the vast potential of watershed management techniques for sustainable resources conservation, Government of India started its focused efforts in 1983 with the launching of 47 model watersheds across the country. Over the past three decades, India has addressed the challenges of land degradation and enhancing productivity of rainfed areas head-on and has made major investments in the area of watershed management through an appropriate combination of technical innovations, participatory approaches and an enabling policy environment. Various hydrological models and modern tools like GIS and Remote Sensing were developed and employed for watershed management. Common guidelines were prepared by National Rainfed Area Authority for Watershed Development Projects during 2008. These guidelines broadly indicate a fresh framework for the next generation watershed programmes. Generally, the impact of integrated watershed management projects is analysed through tangible benefits accrued out of the program in terms of productivity components. The environmental externalities and intangible benefits are rarely accounted for. Even by considering tangible benefits, IWM programmes were found to be economically sound with benefit: cost ratio varying from 1.10:1 to 2.94:1.

The present book covers the experiences of watershed management programs from different parts of the country. The book deals with important components like basic concept of watershed to details of various options for soil and water conservation measures, agroforestry, livestock and integrated pest management in watershed development process. Management of aquifers in urban area, conservation of wetlands and irrigation scheduling for efficient water management have also been discussed. Basics of geospatial applications in land degradation and watershed management, impact of climate change on groundwater resources are also included in the book.

The editors/contributors have attempted to provide a comprehensive information under a unified cover. I am sure that this book will be of interest to researchers and practitioners in the field of hydrology, agricultural engineering, watershed management as well as to those engaged in water resource planning and management. It is hoped that the book will be widely referred by all concerned with this important subject.

P.R. Ojasvi
Ex-Director
ICAR-Indian Institute of Soil and Water Conservation
218 Kaulagarh Road,
Dehradun (UK)-248195, INDIA

About Editors

Dr. R. K. Goyal, B.E. in Agricultural Engineering (1987) from College of Technology And Engineering, Udaipur; M. Tech in Water Resources Engineering (1989) from Indian Institute of Technology, New Delhi and Ph.D. in Water Resources Engineering from M.B.M. Engineering College, Jodhpur, is presently working as Principal Scientist (Land and Water Management Engineering) at ICAR-Central Arid Zone Research Institute, Jodhpur. He possesses expertise of more than 27 years in arid lands' hydrology; water/watershed management and climate change studies. Dr. Goyal is recipient of certificate of appreciation "SAVING THE DRYLANDS" from United Nations Environment Program (UNEP) for outstanding contribution in combating desertification and controlling land degradation in dryland environments (1996). Dr. Goyal has more than 125 national and international publications. He has visited The University of Queensland, Brisbane, Australia (1996) under AusAID fellowship of Government of Australia.

Dr. Mahesh K. Gaur is working at the ICAR-Central Arid Zone Research Institute, Jodhpur, India as a Senior Scientist, since September 2012. He specializes in arid lands geography and the application of satellite remote sensing, GIS and digital image processing for natural resources mapping, management and assessment. He also researches drought, desertification, land degradation, indigenous knowledge systems, and the socio-economic milieu of the Thar Desert of India. Prior to this, he has worked as an Associate Professor for more than 13 years with the Department of Higher Education, Government of Rajasthan State, India and has also worked at IIM Ahmedabad, IHS Hyderabad and IN-RIMT in various capacities. He has conducted many research studies sponsored by ISRO, SAC, UGC, and ICARDA. He has published 35 research papers in national and international research journals, 50 conference papers, 35 book chapters, five popular articles, and seven books. He is a member of numerous national and international scientific organizations, such as the Association of American Geographers and the Society for Conservation Biology, and several editorial boards of journals. He has been to South Africa, Namibia and Costa Rica on academic assignments. He is awarded the Citizen Karamveer Award 2011 by iCONGO for working in the field of Higher Education, Environment and Technology Applications for Community Upliftment; the Millennium Award for working in the field of Environment, and prestigious recognitions by the UGC of India and Scientific Assembly of the International Committee on Space Research (COSPAR).

List of Contributors

A.K. Patel
ICAR-Central Arid Zone Research Institute, Jodhpur-342003 (Raj.), India

A.K. Shukla
ICAR- Central Arid Zone Research Institute, Regional Research Station (RRS), Pali Marwar–306401(Raj.), India

Ajita Gupta
Scientist (Land and Water Management Engg.), ICAR-Central Institute of Agricultural Engineering, Bhopal (M.P.), India

Amit Daiman
MPRA Division Madhya Pradesh Council of Science & Technology, Bhopal (M.P.), India

Archana Verma
ICAR-Central Arid Zone Research Institute, Jodhpur-342003 (Raj.), India

Arpat Suparyatna
Senior Engineering GIS and SCADA of PT. Adhya Tirta Batam; & IT and System Department, PT. Adhya Tirta Batam, Indonesia

Ashish Kumar Vishwakarma
Department of Mining Engineering, Indian Institute of Technology (BHU), Varanasi-221005 (U.P.), India

Bilas Singh
Extension Division, Arid Forest Research Institute, New Pali Road, Jodhpur–342005 (Raj.), India

C. Mathuvanthi Nachiyar
Regional Remote Sensing Centre (South), NRSC, ISRO, Dept. of Space, Govt. of India; ISITE Campus, Bangalore–560037 (Karnataka), India

Chetankumar P. Sawant
Agricultural Mechanization Division, ICAR-CIAE, Bhopal-462038 (M.P.), India

D.C. Gupta
Department of Earth Sciences, Barkatullah University, Bhopal (M.P.), India

Dipak Kumar Gupta
ICAR-Indian Agricultural Research Institute, Gauria Karma-825411 (Jharkhand), India

G. Nirmala
ICAR-Central Research Institute for Dryland Agriculture, Hyderabad, India

Gaurav Tiwari
Disaster Management Institute, Bhopal (M.P.), India

H.M. Ravishankar
Regional Remote Sensing Centre (South), NRSC, ISRO, Dept. of Space, Govt. of India; ISITE Campus, Bangalore–560037 (Karnataka), India

Himanshu Singh Mahawar
S.P.C. Government College, Ajmer (Raj.), India

J. P. Shukla
Water Resources Management Group, CSIR-AMPRI, Bhopal (M.P.), India

Jadeja Yogendrasinh J.
Arid Communities and Technologies, Bhuj (Gujarat), India

K. Ganesha Raj
Regional Remote Sensing Centre (South), NRSC, ISRO, Dept. of Space, Govt. of India; ISITE Campus, Bangalore–560037(Karnataka), India

K. Sreenivas Reddy
ICAR-Central Research Institute for Dryland Agriculture, Hyderabad, India

K.S. Ramesh
Regional Remote Sensing Centre (South), NRSC, ISRO, Dept. of Space, Govt. of India; ISITE Campus, Bangalore–560037 (Karnataka), India

Kartar Singh
Department of Earth Sciences, Banasthali Vidyapeeth, Banasthali-304022 (Raj.), India

Keerthika A
ICAR- Central Arid Zone Research Institute, Regional Research Station (RRS), Pali Marwar–306401 (Raj.), India

M.B. Noor Mohamed
ICAR- Central Arid Zone Research Institute, Regional Research Station (RRS), Pali Marwar–306401 (Raj.), India

Mahesh K. Gaur
ICAR-Central Arid Zone Research Institute, Jodhpur-342003 (Raj.), India

Milan Kumar Yadav
S.P.C. Government College, Ajmer (Raj.), India

Mili Ghosh
Department of Remote Sensing, Birla Institute of Technology, Mesra, Ranchi (Jharkhand), India

N.C. Das
Department of Soil and Water Conservation, Bidhan Chandra Krishi Viswavidyalaya (BCKV), Mohanpur, Nadia–741252 (W.B.), India

Narendra Gupta
Maharana Pratap PG College, Chhitorgarh (Raj.), India

Neetu Negi
Department of Remote Sensing & GIS, Kumaun University, Almora (U.K.), India

Nisha Patel
ICAR-Central Arid Zone Research Institute, Jodhpur-342003 (Raj.), India

P.K. Dhara
Department of Soil and Water Conservation, Bidhan Chandra Krishi Viswavidyalaya (BCKV), Mohanpur, Nadia–741252 (W.B.), India

Prabhat Kumar Pankaj
ICAR-Central Research Institute for Dryland Agriculture, Hyderabad, India

Pramod Dubey
School of Planning and Architecture, Bhopal (M.P.), India

R.K. Goyal
ICAR-Central Arid Zone Research Institute, Jodhpur-342003 (Raj.), India

R.S. Mehta
ICAR- Central Arid Zone Research Institute, Regional Research Station (RRS), Pali Marwar–306401 (Raj.), India

Retinder Kour
Department of Remote Sensing, University of Jammu, (J&K), India

S. Panda
AICRP on Agroforestry, Regional Research Station (Red & Laterite Zone), BCKV, Jhargram–721507 (W.B.), India & Department of Soil and Water Conservation, Bidhan Chandra Krishi Viswavidyalaya (BCKV), Mohanpur, Nadia–741252 (W.B.), India

S. Rama Subramoniam
Regional Remote Sensing Centre (South), NRSC, ISRO, Dept. of Space, Govt. of India; ISITE Campus, Bangalore–560037 (Karnataka), India

Saresh N.V.
Agricultural Research Station (AUJ), Keshwana, Jalore (Raj.), India

Sateesh Karwariya
National Institute of Hydrology (NIH), Ministry of Water Resources, Roorkee-247667 (UK), India

Saurabh Shiva
Disaster Management Institute, Bhopal (M.P.), India

Shailesh Chaure
Department of Geology, Govt. Holkar Science College, Indore (M.P.), India

Shruti Kanga
Centre for Climate Change and Water Research, Suresh Gyan Vihar University, Jaipur-302017 (Raj.), India

Shubha R. Sharma
Department of Bioengineering, Birla Institute of Technology, Mesra, Ranchi (Jharkhand), India

Suraj Kumar Singh
Centre for Sustainable Development, Suresh Gyan Vihar University, Jaipur-302017 (Raj.), India

T. Phanindra Kumar
C-GARD, National Institute of Rural Development and Panchayati Raj, Rajendra Nagar, Hyderabad, India

V. Maruthi
ICAR-Central Research Institute for Dryland Agriculture, Hyderabad, India

Varun Narayan Mishra
Centre for Climate Change and Water Research, Suresh Gyan Vihar University, Jaipur-302017 (Raj.), India

Contents

Preface

There are numerous challenges threatening the management of natural resources in the 21st century. With the ever-increasing population, demand for food production is increasing at an exponential rate leading to tremendous pressure on natural resources. Over-exploitation of natural resources beyond its natural regeneration capacity have jeopardized the sustainable production and therefore, provoking uncertainty in local people as they make and sustain their livelihoods. The various issues related to conservation and sustainable utilization of land and water resources can be linked well when extended to a major land resource unit known as "Watershed" in the planning process. Watershed management is an ever-evolving technique involving the management of land, water, vegetation, and other resources in a defined area for the holistic development. The watershed management concept has emerged as an important option for conservation and efficient utilization of available resources. Watershed management has a long history of evolution in India. In the early 1980s, the main focus was on soil conservation and approach was top-down whereas in 1990s focus shifted to water conservation with the participatory approach. From 2003 onwards, there was a major shift and now the main focus is on productivity enhancement, livelihood improvement and sustainable natural resources with community ownership. During this period lot of new tools like remote sensing, GIS and various hydrological models were developed for better and equitable distribution of benefits to stakeholders under watersheds. The application of new tools and technologies for the management of watershed in different parts of the country was extremely beneficial to the stakeholders. Present book offers a collection of all those diversified studies for the benefit of society. Some of these studies include (i) Effect of watershed management both on productivity and natural resources in climate change scenario; (ii) Scope for additional productive potential through agro-ecosystem approach; (iii) Role of integrated remote sensing and geographic information system for watershed monitoring and management; (iv) Effectivity of Bio-Industrial Watershed model for sustainable production; (v) Agroforestry as an effective and potential tool for watershed management; (vi) Utilization of geographic landscape to enhance income; (vii) Trade-offs between livelihoods and natural resources; (viii) Livestock-based diversification for sustainability; and (ix) Engineering measures for the soil and water conservation for a wide range of ecosystem. This book addresses various issues across different disciplines as well as the incorporation of technological advancements, such as remote sensing, GIS, into watershed management strategies.

It is earnestly hoped that this book will be useful for practitioners, academicians, researchers, policymakers, field-level watershed workers, students, water resources managers, environmentalists, etc. We wish to take this opportunity to express our sincere thanks and appreciation to all the contributors and publisher of this book. We owe our great deal of affection to our lovely daughters Ravina, Anukruti and Oshima for their unconditional affection and dedication. The chapters in the book have been compiled on contributor's individual capacity and they do not reflect any view or opinion of the editors and organizations to which they belong.

R.K. Goyal
Mahesh K. Gaur
(Jodhpur, India)

1

Sustainable Watershed Management through Agro-ecosystem Approach in Indian Context

S. Panda, N.C. Das and P.K. Dhara

Introduction

Watershed is a unique piece of land. It owns its characteristic physiography, hydrology, soil and geomorphology. Climate also plays its conspicuous role for a watershed with regard to its vastness in areal extent. Different types of watersheds vary from few hectares (i.e. micro watershed) to thousands of kilometres for river basins and water resource regions (Table 1). For a particular development programme, the workable size of a watershed is considered. That area for a minor irrigation pond may be few hectares to thousands of kilometres for a hydel project (AISLUS, 1984), though ecosystem with respect to watershed hydrology is strongly influenced by rural micro watershed (Sharma *et al.*, 2014; Reddy, 2014; Smyle *et al.*, 2014).Ascertaining agro-ecosystem of an existing farming condition is needed to deal with micro level aspects of situation management through watershed approach.

Table 1. Classification of Watersheds [based on AISLUS (1984)]

Watershed Delineation Order	Category of Watershed	Size of Watershed (sq.km)
I	Water Resource Region	500,000
II	River Basin	50,000
III	River Sub Basin	5,000
IV	Watershed	500
V	Sub Watershed	50
VI	Mini Watershed	5
VII	Micro Watershed	Below 2.5

Delineation and Management of Watersheds

Geomorphologically the big rivers are well recognised in India. Rao (1979) delineated the whole country into five water zones with the aim of interbasin linking of rivers in the country. However, drawing experiences from delineation of the whole country into six Water Resources Regions in 1949 by the Central Water and Power Commission, followed by the delineation of watersheds larger than 1800 km^2 in the preparation of the 'Atlas of River Basins of the United States' in 1963 with its subsequent revision in 1970, the Ministry of Agriculture, Government of India in 1969 initiated a project work supported by the US Agency for International Development (USAID) for delineating watersheds in India (AISLUS, 1984). That work resulted in 'Watershed Atlas of India' which followed Survey of India topo sheets (1:1 million scale for Water Resource Regions, River Basins and River Sub Basins; and 1:50,000 scale for Watersheds and Sub Watersheds) to village cadastral maps for Mini and Micro watersheds (Table 2).

Table 2. Categories of Base maps followed for delineating Watershed Atlas of India*

<table>
<tr><th>Watershed Delineation Order</th><th>Category of Watershed</th><th>Base maps followed</th></tr>
<tr><td>I</td><td>Water Resource Region</td><td rowspan="3">1:1 million scale</td></tr>
<tr><td>II</td><td>River Basin</td></tr>
<tr><td>III</td><td>River Sub Basin</td></tr>
<tr><td>IV</td><td>Watershed</td><td rowspan="2">1:50,000 scale</td></tr>
<tr><td>V</td><td>Sub Watershed</td></tr>
<tr><td>VI</td><td>Mini Watershed</td><td rowspan="2">Village cadastral maps</td></tr>
<tr><td>VII</td><td>Micro Watershed</td></tr>
</table>

*AISLUS (1984)

Requirement of Watershed Approach

During 3rd Plan (1961–66) launching of River Valley Projects (RVPs) through country wide programme of soil and water conservation required micro level delineation of watersheds. That necessity was further strengthened during 4th (1969-74) Plan for comprehensively treating the priority watersheds in 24 RVPs, extended to 36 RVPs in 6th (1980-85) Plan through the Centrally sponsored scheme for Soil

Conservation. The micro level watershed delineation work was assigned to the All India Soil and Land Use Survey (AISLUS), now SLUSI (Soil and Land Use Survey of India) and the methodology developed was based on Survey of India (1:50,000 scale) toposheets (Bali and Karale, 1979).

Adoption of Watershed Approach

Initially neglected idea of watershed approach gained its momentum through its applications in projects under various government departments and NGOs (Table 3). With the effort of J.S. Bali in 1974, all the centrally sponsored programmes of the Ministry of Soil and Water Conservation were permitted to obtain approval through the adoption of watershed approach (Bali, 1998).

Table 3 (I). Watershed Approach followed in various Government of India Programmes*

S. N.	Programme	Action period	Physical Target/Achievement (million ha)
A. Centrally sponsored Watershed Programmes of the Ministry of Agriculture			
1.	State Programme of Soil and Water Conservation on Watershed basis	1st (1951-1956) to 7th (1985-1990) Plans; 1990 to 1993-94. Increasingly Panchayati Raj was taken into involvement in execution of works.	3.3229
B. Ministry of Agriculture			
2.	River Valley Projects (RVPs)	3rd (1961-1966) to 7th (1985-1990) Plans; 1990-91 to 1991-92; and 8th (1992-1997) to 9th (1997-2002) Plans.	3.944 Enhancing productivity of Degraded lands in catchments of River valleys.
3.	Flood Prone Catchment Scheme	Started from 6th (1980-85) and then upto 9th (1997-2002) - As above.	1.156

4.	Watershed Development Project for Shifting Cultivation Areas (for the North East Region)	9th (1997-2002) Plan (on family basis) and from 1998-99 on watershed basis.	0.15
5.	NWDPRA (National Watershed Development Projects for Rainfed Areas)	7th (1985-1990) to 8th (1992-1997) Plans (2554 micro watersheds); 1997-99; 9th (1997-2002) Plan (2500 watersheds).	7.319
6.	ICAR's ORPs (Operational Research Projects)	Since 1978 to 9th (1997-2002) Plan	47 Projects
	C. Ministry of Rural Areas & Employment		
7.	DPAP (Drought Prone Areas Programs)	1973-74 (From 1995 followed Comprehensive Guidelines for Watershed Projects; As 500 ha for each Watershed)	74.6 (In 947 CD Blocks in 155 districts in 13 states)
8.	DDA (Drought Development Programme);(Co-ordinated from CAZRI, Jodhpur)	1977-78 to 1997-98 (227 Blocks in 36 Districts and 7 States)	45.7 (1747 Projects)
9.	Employment Assurance Scheme (EAS)	Started from 2nd October 1993; 1996: merged with Jawahar Rojgar Yojana; 1997-98: followed guidelines on Watershed basis	(5448 CD Blocks all over India)
10.	Integrated Wastelands Development Projects (IWDPs) (National Wasteland Development Board under Deptt. of Wastelands Development)	1993-94 to 1998-99; 1st Aril 1995: the Board and the Deptt. came under the Ministry of Rural Areas and Employment; operated on a cluster of 10 Micro Watersheds each of about 500 ha under Common Watershed Guidelines	0.005

D. Ministry of Environment and Forests (Watershed Development Programmes)			
11.	Integrated Afforestation and Eco-Development Projects (IAEP)	7th (1985-1990) Plan (rejuvenation and eco-development of Degraded forests and adjoining areas on watershed basis; Each Compact area of not less than 20 ha through Joint Forest Management (JFM))	0.299
12.	Area Oriented Fuelwood and Fodder Project (AOFFP) (Micro watershed project planning)	7th (1985-1990) Plan (In 229 fuelwood deficit districts; Each Compact area of not less than 20 ha through Joint Forest Management (JFM))	0.402
13.	Non-Timber Forest Produce Scheme (NTFP) (Watershed oriented programme on raising of non-timber plantations including medicinal plants in tribal regions)	8th (1992-1997) Plan (through JFM)	0.102
E. Planning Commission (Watershed Development Programmes)			
14.	Hill Area Development Programme (HADP)/ Western Ghat Development Programme (WGDP)	1965: areas identified for Hills; 1972: Western Ghats, (Comprised two hill districts of West Bengal; 25 Talukas in Nilgiri dist. In Tamil Nadu;29 Talukas in Kerala and three Talukas in Goa) 9th (1997-2002) Plan: followed programmes on watershed basis.	7.0 in Hills (with 6.6 m population); 16.0 In Western Ghats (with 38.8 m population)

		Later followed NWDPRA Scheme approach	
F. Department of Land Resources, Ministry of Rural Development			
15.	IWMP: Integrated Watershed Management Programme; Integrated: IWDP, DPAP, DDP (**Hariyali** Programme)	2010 1 April 2003: Hariyali Programme	31.593 (from 1995-96 to 2007-2008)**
16.	WDC – PMKSY: Watershed Development Component of Pradhan Mantri Krishi Sinchai Yojana; Supported by **World Bank assisted Neeranchal Project**	2016-17 to 2021-22 (450 sub watersheds, each of about 5,000 ha)	2.25 (Two districts in each participating state)

**Details are presented in Table 5

Table 3 (II). Watershed Approach followed under Internationally Funded Watershed Projects*

S. N.	Programme	Action period	Physical Target/Achievement
A. World Bank			
1.	Integrated Watershed Development (Hills) Project	10th May 1991 to 30th June 1997	Haryana, Punjab, Himachal Pradesh and Jammu & Kashmir
2.	Integrated Watershed Development (Plains) Project	-as above-	Rajasthan, Gujarat and Orissa
B. DANIDA (Danish International Development Assistance)			
3.	Comprehensive Watershed Development Project (CWDP)	Phase I: 1991-92 Phase II: 1997-2001	0.0041082 Mha (Belgaon and Dharwad districts in Karnataka)
4.	CWDP in Degraded areas	Phase I: 1991-92	0.0062 Mha (Tirunelvelli in T. N.)

		Phase II: 1995 – To 6 years	0.0045 Mha (Bijapur) 0.0045 Mha (Gulbarga, Karnataka)
5.	CWDP - Orissa	1992-93 to 6 yrs.	0.0043022 Mha (Koraput, Orissa)
6.	CWDP – Tamil Nadu	-As above-	0.0011 Mha (Ramanathapuram,TN)
	C. UNDP/FAO/Govt. of Netherlands		
7.	UNDP Farmer Centered Agriculture Resource Management (Farm)	1992-93 to 1997-98	Alwar (Rajasthan); Almora (Uttar Pradesh); North Arcot (Tamil Nadu); Dr. Y. S. Parmar University of Horticulture & Forestry, Solan, Himachal Pradesh
8.	Participatory Watershed Asia (PWMTA)	1996-2000	
	D. European Economic Community (EEC)		
9.	IWMP (South Bhagirathi Phase II)	April 1989 to 7 years	0.0048915 M ha (Nainital, U.P.)
10.	Bhimtal IWMP	April 1993 to 7 years	0.0021582 M ha (Dehradun and Tehri districts, U.P.)
	E. Indo German Watershed Projects (0.3 M ha)		
11.	Bilateral Project: Watershed Management	1989-2005	Advisory and financial support for Integrated Watershed Management (IWM); Capacity building; Monitoring of watershed
12.	Watershed Self Help Programme (WOTR)	2001-2004	Capacity building of NGOs and community based organizations
13.	Changar Eco-Development Project	Phase I: 1994-1999 Phase II: 1999-2006	Building capacity of local animators and future institutional development
14.	Reorganization and Strengthening of Watershed Training Institutes in Maharashtra	2002-2005	
15.	Capacity Building and Strengthening of Decentralized Watershed Management	2003-2007	

*Bali (1998), Kotru (2005), Annon. (2012), Annon. (2017), Annon. (2018), Annon. (2019)

Governance and Management of Watersheds

From the review (Table 3: I&II) of Watershed Management Projects and the adoption of watershed approach in various schemes (1951 to 2018) it was found that sixteen Govt. of India Programmes under different states, and various Deptt.s under the Ministries of Agriculture, Rural Areas & Employment, Environment and Forests, Rural Development and Planning Commission and ICAR (Narayana, 1993; Narayana *et al.*, 1997). Apart from those, fifteen internationally funded watershed projects were executed through the Govt. of India. Five international funding authorities, viz. i) World Bank, ii) DANIDA, iii) UNDP/FAO/Govt. of Netherlands, iv) European Economic Community (EEC)and v)Indo German Watershed Programme forwarded their financial support for various watershed projects implemented in India.

From 1951 to 2016 (Table 4) all those projects covered more than160.3 million hectares (Mha) in India which is about half (48.78%) of the geographical area (about 328.7 Mha) of India. Upto 2013-14 a quantum of 120.4 Mha of land still remained as degraded areas (Annon. MOA, 2017). Contemporarily from 1995-96 to 2007-2008 (Table 5) three watershed based projects like DPAP, DDA and IWDP comprehensively covered 31.593 Mha areas under National Watershed Development Programme of Deptt. of Land Resources, Ministry of Rural Development, Govt. of India (Annon., 2012).

Table 4. Targets and achievement for Development of Land Resources through Watershed Approach (1951 to 2016)*

Executed by	Achievement / Total Area Targeted (Mha)
I.GOVERNMENT OF INDIA PROGRAMMES	**159.9999**
II.INTERNATIONALLY FUNDED PROJECTS	**0.33176**
Total	**160.3317**
% of total geographical area of India (328.7 Mha)	**48.78%**
Gross Cropped area (2013-14)	**200.9**
Net Sown area (2013-14)	**141.4**
Total Degraded land (2013-14)	**120.4**

*Table 3; Annon. MOA. (2017)

Table 5. Areas Covered under different watershed based projects sanctioned from 1995-96 to 2007-2008 under National Watershed Development Programme (Deptt. of Land Resources, Ministry of Rural Development)*

Name of Schemes	No. of projects sanctioned	Area covered (Mha)
DPAP	27439	13.02
DDP	15746	7.873
IWDP	1877	10.7
Grand Total	45062	31.593

*Annon. (2012)

In the meantime from 1976 onwards upto 2004, ten different Govt. agencies tried to estimate areal extent of degraded lands in India. Among those Ministry of Agriculture (Soil and Water Conservation Division, 1978) and Ministry of Agriculture (1985) closely estimated the areas of Degraded lands (2 & 5 in Figure 1) of about 175.00 and 173.64 Mha respectively (Trivedi *et al.,* 2010).

Such efforts of delineating and estimating degraded areas were targeted to convert marginal lands for crop production, and delineation of Watershed Atlas of India (1984) and subsequent application of watershed approaches in various projects coincided with the delineation of Degraded & Wastelands of India (Trivedi *et al.,* 2010).

However, dry areas and wastelands were targeted since 1928 by Royal Commission on Agriculture (Annon., 1928) to extract inherent potentiality of those lands for crop production, followed by establishment of Dry Farming Research Stations at Sholapur, Bellary and Rohtak in 1933. That followed sorting out of some simple methods like 'wattband' in Punjab foothills and bunding and then contour bunding based on Bombay Dry Farming Methods. Contour bunding on dry lands then could increase the yield on those lands by about 25% which could take a leading role to ward off famine, to build assets and to provide employment and, thereby, creating congenial environment for provision of food for the poor. All those successes gave foundation of the strong Soil Conservation Department of Maharashtra (Bali, 1998).

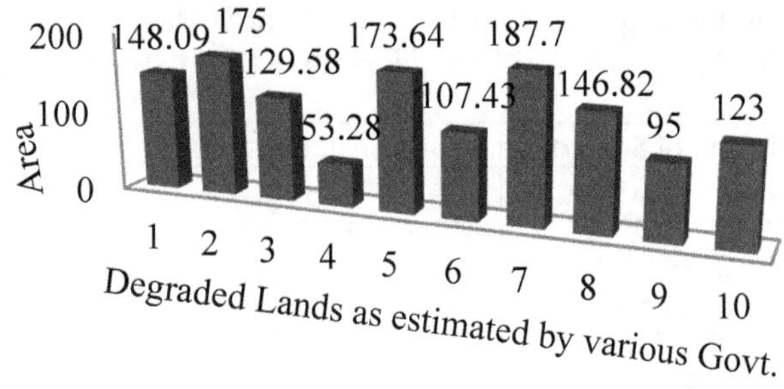

Agency estimated the Degraded Lands in India

1 National Commission on Agriculture (NCA, 1976)
2 Ministry of Agricul ture (Soil and Water Conservation Division) (1978)
3 Society for Promotion of Wastelands Development (SPWD) (Bhumbla and Khare, 1984)
4 NRSA (1985)
5 Ministry of Agriculture (1985)
6 Ministry of Agriculture (1994)
7 NBSS&LUP (1994)
8 NBSS&LUP (2004) (revised)
9 Department of Environment (Vohra, 1980)
10 National Wasteland Development Board (1985)

Figure 1. Areas of Degraded Lands as estimated by various Agencies in India* [*Source of Estimates: Trivedi *et al.,* 2010].

Proper Implementation of Watershed Projects – Agro-Ecosystem Approach

Before 1985 project areas were mostly larger than micro watersheds and most of those projects could not achieve expected results. For example, Hill Area Development Programme (HADP) and Western Ghat Development Programme (WGDP) under the Planning Commission, Govt. of India, initiated from 1965, were also unable to yield satisfactory results. Apart from that Aravallis and other low hilly areas like Eastern Ghats, being denuded and was in need of urgent attention for soil and water conservation, afforestation and all other components of watershed programme, were not considered for inclusion in those projects for administrative reasons of low heights (Bali, 1998). Then NWDPRA (1985) was implemented and those difficulties would no longer arise before application of effective methods of soil and water conservation in all areas as that project was targeted to the micro watersheds involving various livelihoods of farming communities (Table 3). That period was markedly important regarding enactment of the 'The Environment Protection Act, 1986' by the Ministry of Law and Justice, Govt. of India (1987); and adoption of 'National Water Policy' by the Ministry of Water Resources, Govt. of India (1987), with its subsequent explanations as 'Agricultural Scientists' Perceptions on National Water Policy' by the National Academy of Agricultural Sciences (NAAS, 1995).Thus, some reliefs came with regard to achieving successes from the implementation of NWDPRA projects by targeting watersheds at micro level.

Obtaining optimum production potential from existing farming situations was still then away from attaining achievement. During those times the delineation of map of Agro-Ecological Regions in India by NBSS & LUP, Nagpur was done in 1992 (initiated in 1989) with its subsequent revision in 2015 (Mandal *et al.*, 2015).In the mean time during November 1998 the National Agricultural Technology Project (NATP), a World Bank assisted programme, was started jointly by the ICAR (NATP) and the Deptt. of Agriculture and Cooperation, Govt. of India. NATP included three major components of O & M (Organization & Management), Research and ITD (Innovations in Technology Dissemination). Those activities made some significant reforms in management systems, creating some software towards solution of critical problems faced in institutional set-up and farming situations and effective linking of both the institutions and farming community and resourceful persons towards reaping optimum harvest from the farm. NATP also started Production System Research (PSRs) in all Agro-Eco Regions in India, which enabled to delineate a farming situation with respect to Agro-Eco System (AES), which had taken into account of agroclimatic zoning (as was done though National Agricultural Research Project, initiated in 1979). The Agroclimatic Zoning was also yet to yield fruitful results, because of large areal coverage of two to 4 districts of about 40 to 50 thousand km^2. Such disadvantage could be avoided through NWDPRA and in the meantime application of Agro-Ecological Zoning of India with subdivisions in AESs was potentially helpful for situation management of existing farming conditions to obtain targeted yield from farmers' fields (Ghosh, 1991;

Saxena *et al.*, 2001; Annon., 2004).

All those delineations in the background, the methodological approach for Regional Crop Planning (RCP) was formulated by ICAR-NIPA (National Institute of Agricultural Economics and Policy Research). RCP is a multi-dimensional concept for selecting best set of crops with given constraints in a region to achieve some set of goals of maximizing net income, minimizing cost, maximizing total area cultivated and/or minimizing irrigation water. RCP was taken up in a network mode covering almost all the Agro-Eco-Systems in India and largely used the unit-level Cost of Cultivation Survey (CCS) data of Directorate of Economics and Statistics, Ministry of Agriculture, Govt. of India. The overall objective of RCP was aimed at better resource use efficiency and sustainable improvement of natural resource. RCP, thus, included studies of existing land use, cropping pattern and resource use efficiency in all regions sustainably across production environments (Jain *et al.*, 2015). Thus, RCP is naturally matching with watersheds delineated upto micro level. Also, Regional Crop Planning can be equally applicable for any large watersheds, as micro watersheds cumulatively results into large areas in an agro-eco region.

Improving marginal lands either degraded or wastelands are always of national interest in India, because those areas are mostly inhabited not only by the poor but also by maximum population (Annon., 1928). Taking those underdeveloped vast tracts (Mruthyunjaya, 1983; Venkateswarlu (Ed.) *et al.*, 2005; Reddy *et al., 2017*) under watershed development projects would immensely influence our economy towards better livelihoods of all sections, especially backward rural areas (Panda and De, 1989; Panda and Ghosh, 1996; Panda, 2005), and national food security. Consequently, it may be noted that raising the productivity of rainfed agriculture is the most important objective for some watershed projects, particularly NWDPRA, the ICAR Model Watersheds, World Bank Pilot Projects and IWDP (plains), where opportunities for water harvesting are limited and, thus, begins with conserving soil in rainfed plots which, in turn, is helpful for conserving moisture and nutrients in creating congenial situation for high yielding crops for more production than from traditional farming practices (Kerr *et al.*, 2002; Ahmad *et al.*, 2011; GOI, 2011; Wani *et al.*, 2011; Gray and Srinidhi, 2013). In this way watershed management at micro level (Murty, 1985; Singh *et al.*, 1990; Ryszkowski, 1995; Das and Das Gupta, 1996; Honore, 1999; Tideman, 2000) is at par with sustainably managing existing farming situations in any agro-ecosystems aimed at more income generation and better livelihoods of the whole population around healthy production environment.

Conclusion

Watershed management could be successfully implemented through optimisation of micro level aspects of watersheds as in NWDPRA, ICAR Model Watersheds, IWDP(Plains), etc. Agro-Ecological Regional planning considered agroclimatic regional aspects for securing income and livelihoods of the farming community and, thus, the whole population in India.

Watershed management through Agro-Ecosystem approach effectively considers micro level aspects of watershed and, thus, it was able to promote sustainably increase productivity through effective conservation of soil, water, judiciously following climatic conditions and existing farming situations towards better livelihoods of people around healthy production environment.

References

Ahmad, J., Alam, D. and Haseen, S. (2011). Impact of climate change on agriculture and food security in India. *International Journal Agriculture, Environment and Biotechnology*. 4(2), 129-137.

AISLUS (1984). Explanatory Notes on Watershed Atlas of India (1:1 million Scale). New Delhi: *All India Soil and Land Use Survey (AISLUS)*, p 4.

Annon. (1928). Royal Commission on Agriculture in India – Report. Bombay: *The Government Central Press.* pp. 116-117.

Annon. (2004). National Agricultural Technology Project. DARE/ICAR Annual Report 2003-2004. New Delhi: *ICAR,* pp. 189-201 (D:\2004\04-025\04025-01.P65\9 (19.06.04/22.06.04/23.06.04). https://icar.org.in/files/ar0304/14-NATIONAL%20AGRICULTURAL%20TECHNOLOGY%20PROJECT.pdf

Annon. (2012). National Watershed Development Programme. India.

gov.in, NIC: National Portal Content Management Team, Reviewed on 07-11-2011 (last updated: 09 December 2012).

Annon. (2017). Neeranchal – World Bank Assisted National Watershed Management Project. Deptt. of Land Resources, Ministry of Rural Development, Govt. India. Website last updated 09 October 2017. https://dolr.gov.in/programme-schemes/neeranchal-national-watershed-management-programme

Annon. MOA. (2017). Annual Report 2016-17. New Delhi: *Deptt. of Agriculture, Cooperation & Farmers Welfare, Ministry of Agriculture & Farmers Welfare, Govt. of India, p.* 1.

Annon. (2018). Watershed Development Component of Pradhan Mantri Krishi Sinchai Yojana (WDC-PMKSY). Deptt. of Land Resources, Ministry of Rural Development, Govt. India. Website last updated 02 April 2018. https://dolr.gov.in/programme-schemes/pmksy/watershed-development-component-pradhan-mantri-krishi-sinchai-yojana-wdc-pmksy

Annon. (2019). Indo-German Watershed Development Programs (IGWDP). https://india.diplo.de/in-en/themen/igwdp/1992818

Bali, J.S. (1998). Fifty years of watershed development in India – An overall picture of watershed management. *J. Soil and Water Conservation*, 42 (3&4), 127-143.

Bali, Y. P. and Karale, R. L. (1979). Comprehensive system of delineation and codification of watersheds. *J. Ind. Assoc. of Hydrologists,* 1 (Nos. 1 & 2). (as seen in AISLUS, 1984).

Das, S.C. and Gupta, R.K. (1996). Integrated approach to development of small watersheds in West Bengal. Proceedings 'All India Seminar on Small Watershed Development, 15 February, (1996). Auditorium of Indian Institute of Chemical Engineers. Calcutta: *Indian Association of Hydrologists, West Bengal Regional Centre.* pp.1-8.

Ghosh (Ed.), S.P. (1991). Agro-Climatic Zone Specific Research: Indian Perspective under NARP. New Delhi: *ICAR (Indian Council of Agricultural Research),* p. 5.

GOI. (2011). Common Guidelines for Watershed Development Projects - 2008 (Revised edition 2011). New Delhi: *National Rainfed Area Authority, Planning Commission.*

Gray, E. and Srinidhi, A. (2013). Watershed Development in India: Economic Valuation and Adaptation Considerations, Working Paper, Case Study 1 of 3 for the Scaling Up Good Adaptation Practices (SUGAP) Project - a partnership between World Resources Institute (WRI), Watershed Organizing Trust (WOTR) and Swiss Agency for Development and Cooperation (SDC). Washington, DC: *World Resources Institute,* pp. 32.

Honore, G. (1999). Monitoring and Evaluation of Watershed Management Projects, Our Land, Ourselves - a practical guide to watershed management in India. New Delhi: Indo-German Bilateral Project "Watershed Management". pp. 55-115.

Jain, R., Raju, S.S., Immanuel raj, K., Srivastava, S.K., Pal Kaur, A. and Singh, J. (2015). Manual on Methodological Approach for Developing Regional Crop Plan, under the ICAR Social Science Network Project, Regional Crop Planning for Improving Resource Use Efficiency and Sustainability. New Delhi: *ICAR-NIAP (National Institute of Agricultural Economics and Policy Research,* pp.1-2.

Kerr, J., Pangare, G. and Pangare, V.L. (2002). Watershed development projects in India: An evaluation. Washington, DC: *International Food Policy Research Institute,* Chapter 8 (Research Report 127), pp. 62.

Kotru, R. (2005). Watershed Management Projects in India. In: Achouri *et al.* (2005). Proc. Asian Regional Workshop on Watershed Management, Kathmandu, Nepal, 11 - 13 September 2003. Rome: FAO, Chap. 8, pp. 85–91.

Mandal, D.K., Mandal, C. and Singh, S.K. (2015). Delineating Agro-ecological Regions, Nagpur: *e-Publication ICAR-NBSS Technologies.* https://www.nbsslup.in/assets/uploads/clinks/Delineating%20Agro-Ecological%20Regions.pdf

Mruthyunjaya (1983). Economic Evaluation of Arid Zone Technologies, Desertification and its Control. *ICAR.* pp.366-375.

Murty, V.V.N. (1985). Watershed Management. Land and Water Management Engineering: Chapter 24. Ludhiana: *Kalyani Publishers.* pp. 557-570.

NAAS (1995). Agricultural Scientists' Perceptions on National Water Policy. New Delhi: *NAAS (National Academy of Agricultural Sciences),* pp.1-28.

Narayana, V.V.D. (1993). Watershed Management: Concept and Frame Work. Soil and Water Conservation Research in India, Central Soil and Water Conservation Research and Training Institute, Dehradun. New Delhi: *ICAR.* Chap. 19, pp.437-447.

Narayana, V.V., Sastry, G. and Patnaik, U.S. (1997). Watershed Management. New Delhi: *ICAR,* pp.86-130.

Panda, S. (2005). Systems approach for rainwater harvesting with emphasis on quality of surface water in the coastal region of Eastern India. Proceedings 'International Conference on Crisis Management in Water and Environment (ICCMWE-2005), 15-16 July, 2005. Kolkata: *Indian Association of Hydrologists, West Bengal Regional Centre.* 1, pp.92-96.

Panda, S. and De, P. (1989). Hydrological balance approach for planning irrigation in coastal areas of Contai, Midnapore, West Bengal. Proceedings 'All India Seminar on Role of Hydrology in Efficient Management of Irrigation System, 19-20 May, 1989, Birla Industrial & Technological Museum. Calcutta: *Indian Association of Hydrologists, West Bengal Regional Centre.* Sl. No. 4, pp.1-13.

Panda, S. and Ghosh, A.K. (1996). Water harvesting in coastal and deltaic region. Proceedings 'All India Seminar on Small Watershed Development, 15February, 1996, Auditorium of Indian Institute of Chemical Engineers. Calcutta: *Indian Association of Hydrologists, West Bengal Regional Centre.* pp.1-8.

Rao, K. L. (1979). India's Water Wealth. New Delhi: *Orient Longman Limited,* p 230.

Reddy, C.V.K. (2014). Watershed Development Programme in India (with special reference to Prakasam District, Andhra Pradesh). Chennai: *Vijay Nicole Imprints Private Limited.* pp. 20 & 109.

Reddy, V.R., Saharawat, Y.S. and George, B. (2017). Watershed management in South Asia: A synoptic review. *J. Hydrology.* 551(2017), 4-13.

Ryszkowski, L. (1995). Managing ecosystem services in agricultural landscapes. Nature & Resources. 31(3):27-36.

Saxena, R., Pal, S. and Joshi, P.K. (2001). PME Notes: Delineation and Characterization of Agro-Eco Regions. NATP Taskforce on Prioritization, Monitoring & Evaluation. New Delhi: *National Centre for Agricultural Economics and Policy Research,* PME Notes 6. http://ncap.res.in/upload_files/PME_notes/pmenotes6.pdf

Sharma, S.S.P., Kumar, V.H. and Rani, C.R. (2014). Study on Performance of Institutional Set Up at SLNA / WCDCs under IWMP. Hyderabad: *Centre for water and land Res.* pp.1, 4.

Singh, G., Venkataraman, C., Sastry, G. and Joshi, B.P. (1990). Watershed Management, Manual of Soil & Water Conservation Practices. Calcutta: *Oxford & IBH Publishing Co. Pvt. Ltd.* pp. 328-375.

Smyle, J., Lobo, C., Milne, G. and Williams, M. (2014). Watershed Development in India – An Evolving through Experience. Washington DC: *The International Bank for Reconstruction and Development / World Bank,* p. 7.

Tideman, E.M. (2000). Watershed management – Guidelines for Indian Conditions. New Delhi: *Indo-German Bilateral Project 'Watershed Management'.* pp. 6-9.

Trivedi, T.P., Sharma, R.P., Verma, S.A., Bharti, V.K. and Sastri, A. (2010). Degraded Wastelands of India – Status and Spatial Distribution. New Delhi: *Indian Council of Agricultural Research and National Academy of Agricultural Sciences,* p. 4.

Venkateswarlu (Ed.), B., Ramakrishna, Y.S., Rao, C.S., Ramesh, G., Narasamma, P.L. and Kapil, H. (2005). NATP: Rainfed Agro-Ecosystem – Production System Research, Completion Report 1999-2004. Hyderabad: *Agro-Ecosystem Directorate (Rainfed), Central Research Institu-*

te for Dryland Agriculture. pp. 1-8.

Wani, S.P., Rockström, J. and Sharawat, K.L. (2011). Integrated Watershed Management in Rainfed Agriculture. London : *CRC Press.* pp. 81-82.

2

Watershed Management in Hot Arid Zone of India: A Review

R.K. Goyal and Mahesh K. Gaur

Introduction

The idea of conservation of natural resources, land and water on watershed basis is not new. It was first adopted in 1949 by Damodar Valley Corporation for checking siltation in reservoirs, mitigating of floods, etc. Systematic efforts at long-term ameliorative measures to tackle the problems of drought started after planning for economic development was launched in the country. The first effort to tackle the problem of drought and desertification was the establishment of a Research Centre at Jodhpur in 1952 to carry out research on certain core needs of desert areas such as sand-dune stabilization, shelterbelt plantation, afforestation etc. In 1959, the entire responsibility for research on arid areas was entrusted to the Research Centre, which was then designated as Central Arid Zone Research Institute (CAZRI). During the Second and Third Five Year Plans, the problem of drought-affected areas was mainly sought to be solved by launching Dry Farming Projects, which spread over a few areas with emphasis on moisture and water conservation measures.

The origin of the Drought Prone Areas Program (DPAP) can be traced to the Rural Works Program launched in 1970-71 with the object of creating assets designed to reduce the severity of drought in affected areas. The Program spelt out long-term strategy in the context of the conditions and potentials of the drought prone districts. In all, 54 districts in the country together with parts of another 18 districts contiguous to them were identified as drought-prone for purposes of the Program. The Program covered 12% of the country's population and nearly one-fifth of the area in the country. The Mid-Term Appraisal of the Fourth Plan re-designated the Program as the Drought Prone Areas Program.

The Fourth Plan continued to lay emphasis on dryland farming technology. The Program in the Fifth Five Year Plan followed the strategy and approach of integrated area development based on resource endowment analysis and potential for development. In Interim Report of the National Commission on Agriculture (1974) the hot desert areas were identified and it was suggested that a development program consisting mainly of afforestation and livestock development should be taken up. The Desert Development Program (DDP) was started in 1977-78.

In Jan 1982 certain modifications were suggested in Drought Prone Areas Program and the Desert Development Program. While the ongoing approach and strategy were reiterated, the scope and the objectives of the programs were re-defined. Emphasis was laid more on productive agriculture, dryland as well as irrigated. The DPAP was withdrawn from areas covered under DDP as the program has similar objectives and contents. It emphasized the need for planning the program on a watershed basis in an integrated manner.

However, focused efforts on watershed management was started after 1983 when GOI launched model watershed projects with involvement of ICAR. Under this program 47 model watersheds were identified in different agro-climatic zones all over the country. Under this program, Rajasthan state got its first model watershed during 1986-97 as Jhanwar model watershed (Distt. Jodhpur). The planning commission while emphasizing the role of local level planning in VIII Five Year Plan highlighted the role of drought prone and dryland areas in augmenting the food production level of the country by way of adopting the watershed approach. Later to cover major dryland areas, National Watershed Development program for rainfed area (NWDPRA) was launched as centrally sponsored program by GOI in 1990-91. This scheme covered all blocks where less than 30% area was irrigated. 204 watersheds in 190 Panchayat Samities in state have been covered under the scheme. Now the approach has been adopted as a national strategy for integrated and comprehensive development of drought prone and rainfed areas. It is an approach for bringing green revolution in dryland areas. Watershed Development Fund (WDF) has been established at NABARD with the objective of integrated watershed development in 100 priority districts of 18 States through participatory approach. To involve village communities in the implementation of watershed projects under all the area development programs namely, Integrated Wastelands Development Program

(IWDP), Drought Prone Areas Program (DPAP) and Desert Development Program (DDP), the Guidelines for Watershed Development were adopted w.e.f. April, 1995, and subsequently revised in August 2001. To further simplify procedures and involve the Panchayat Raj Institutions (PRIs) more meaningfully in planning, implementation and management of economic development activities in rural areas, these new Guidelines called Hariyali are being issued. During 11th five year plan, various schemes like IWDP, DPAP and DDP were merged into Integrated Watershed Management Programme (IWMP) under Ministry of Rural Development and common guidelines were issued in 2007 by National Rainfed Area Authority (NRAA).

Need for Watershed Approach

Initial approach for conservation programs were based on particular land problem and individual holdings. In this approach often the upper reaches of the fields were left untreated and runoff from uplands used to damage the conservation work laid out in the individual holdings at downlands. The watershed approach helps in an integrated development of different parts of the watershed in accordance with their nature, problems and potential with best amount of interference. The integrated watershed management principles have accordingly been adopted for holistic development of drought prone and rainfed areas. Though no two watersheds are identical, their characteristics have common parameters, which can be judiciously used for obtaining the required results.

Objectives

A watershed generally consists of three physical sectors, (i) arable or cultivated lands which are privately owned; (ii) non-arable lands which include village, pasture and grazing grounds, culturable wastelands and barren and uncultivable lands and (iii) network of natural drainage lines. These three sub-sections are hydrologically interspersed and would have to become as one organic geo-hydrological entity for project planning and implementation in order to ensure sustainable use of natural resources. The main objective of each watershed development project is to achieve sustainable production of biomass and restoration of ecological balance in the drought prone and rainfed areas improving thereby the standard of livelihood of people. The specific objectives are:

1. Conservation, upgradation and utilization of natural endowments namely land, water and vegetation in harmonious and integrated manner.
2. Generation of massive employment during the project period and regular employment after the project completion for enhancing the work opportunities in the backward rainfed areas with a view to ensure livelihood security, particularly for the underprivileged sections of the rural population.
3. Improvement of production environment and restoration of ecological balance through scientific management of land and rainwater, *in-situ* moisture conservation, introduction of scientific production system, creation of network of runoff management structures and devices for recharge of ground water and recycling for drinking purposes, domestic consumption and life-saving irrigation.
4. To promote social forestry and horticulture activity in all suitable lands, and
5. To enhance cash available with farmers and landless agricultural laborers through increased casual employment, marketable surplus of agricultural and dairy produce.

Watershed Management Technology

The interaction between land and water determines the productivity of a given area, which can be made more productive by adequate understanding of physiography, climate, socio-economic conditions etc. and hydrological behavior of watershed. The PRA, land capability, choice of appropriate soil and water conservation measures, improved management practices, soil conservation technologies are unique feature of detailed watershed planning. Optimization of different possible combinations and alternative uses of a specific resource should be tested. Priorities for the maximum benefits in the shortest possible time have to be selected. The inter-relationship between all factors has to be carefully studied and final decision must be made with local participation. Due to multidisciplinary nature of watershed development projects the various improvement activities need to be coordinated between farming community and government agencies. Therefore, success of such projects depends very much on the degree of co-ordination and co-operation between various organizations.

Economic considerations are a major factor in making decisions on action and timing. However, these decisions have to be fit into the sequence of improvements required for effectiveness of all components in a plan. It is also important to note that the area development plan can be executed gradually, on sub-watershed from top to bottom. Thus, the development period can be spread over a number of years according to availability of finance and trained manpower. If and when all sub-watershed have been treated and developed one by one, and main tributaries have also been improved, then the complete plan would have been implemented in a systematic manner and be based on sound technological and socio-economic principles. The subsequent maintenance is essential to ensure against potential natural hazards to the completed works and improvements.

Hot Arid Zone: A Review

The hot arid regions of India, constituting 0.32 million km^2, occurs dominantly in the country's north-western part. The state of Rajasthan accounts for the largest area (61%), followed by Gujarat (20%), Punjab (5%) and Haryana (4%). About 10 percent area occurs in the southern part of the country especially in Andhra Pradesh (7%) and Karnataka (3%). These areas experience an annual rainfall between 100 and 500 mm with a coefficient of variation varying from 40% to 70%. Low and erratic rainfall combined with extreme temperature results in low water availability and frequent crop failures, which continually affect the agriculture economy in the region. Occurrence of frequent droughts lead to wind erosion and creeping of sand dunes covering fertile agriculture land. Also continuous increase in human and livestock population has put pressure on natural resources. In many areas, mining of ground water resources beyond its replenishment capacity has led to fast depletion of this precious resource and deterioration in water quality. Over grazing of pasture lands and indiscriminate cutting of trees have accelerated in the process of erosion and converted productive agriculture land to wasteland. In view to rehabilitation the degraded agriculture land for the sustaining of local population there is an urgent need to adopt an integrated approaches of developmental activities with the available technology. For maximizing the advantage, all developmental activities should be undertaken in a comprehensive way on watershed basis.

Watershed is the geographical unit that drains into a common point.

This natural unit is carved out, through the interaction of rainwater with landmass and typically comprises of arable and non- arable lands and natural drainage line. Watershed is a self-defended area, which does not allow any water from outside the catchment to enter into it, and allow its excess water to discharge to a common point in a stream. In areas where there is no defined drainage system and internal drainage are choked in sand dune complex, the 'Index Catchment Approach' was considered for taking development activities. Index catchment is demarcated by considering the crest line of sand dunes or eroded rocky out crops from the line of water divide (Vangani *et al.,* 1998).

Component of Watershed Management in Hot Arid Areas

A. Resource Conservation

Soil and water are the two basic resources, needed for crop production. Measures for soil and water conservation can be divided in two main categories i.e. physical/engineering and biological. While the physical measures are the first line of defense and are very necessary to prevent immediate damage, biological measures are the offensive attack made from a position of good defensive strength. Physical measures consist of construction of mechanical barriers across the direction of flow of water to retard/retain the runoff and thereby reduce soil and water losses. Contour bunds, contour cultivation, trenching, sub-surface barrier etc. are some of the physical measures, which are found to be useful in arid areas. Biological measures like vegetative barrier, windbreak/shelter belt, stubble mulching are very effective for moisture conservation and controlling soil erosion due to wind and water.

B. Resource Utilization

Efficient utilization of available resources is of paramount importance for successful crop production. Improved agronomic practices i.e. inter cropping, improved crops and seed varieties, crop rotation, minimum tillage etc. plays vital role in ensuring risk distribution. For aberrant weather situation contingency planning like life-saving irrigation, integrated insect-pest management, differential fertilizer application etc. are crucial to minimize loss of seed, crop and fertilizer. The land which is unfit for conventional farming as per land capability classification may be used for alternative land uses, dairying, agro-forestry and agro-horticulture, cut flowers and fisheries, where water is stored for longtime. Value addition should be in built with the program for income generation and livelihood.

C. Technological integration

The success of watershed development projects greatly depend upon the availability of viable and acceptable location specific technology. Besides, availability of technological options, the success of the program also depends upon beneficiaries' resource base and level of income, timely availability of inputs and the skill of the farmers. As the people and location are the two most important factors, which are required for the selection of appropriate technology and for making suitable modification, it would be better to identify the proven traditional technology and skill, for adoption after suitable modifications.

Impact Evaluation

Impact evaluation of soil and water conservation measures adopted in any watershed area is essential to know an overall assessment of the technical results. It also helps in knowing the appropriateness of the method employed in carrying out the project activities and also to estimate the medium and long term social and economic benefits of the activities, efficiencies and impact of the project in the context of its stated objectives (Gupta *et al.* 2002). Keeping in view the importance of the post project evaluation, an attempt has been made to assess the impact of different soil and water conservation measures adopted in hot arid zone of Rajasthan under different watershed projects under taken by CAZRI and State Govt.

A. CAZRI Experience

Central Arid Zone Research Institute, Jodhpur has done several studies on hydrological behavior of different process in the arid zone and developed techniques for development of arid watersheds. CAZRI has effectively developed more than 6000 ha area under Jhanwar, Sar, Baorali-Bambore, Kalyanpur and Kukma watersheds. For rainwater management, institute has designed underground *tanka* of 10 m^3 to 600 m^3 capacities for different rainfall and catchment conditions.

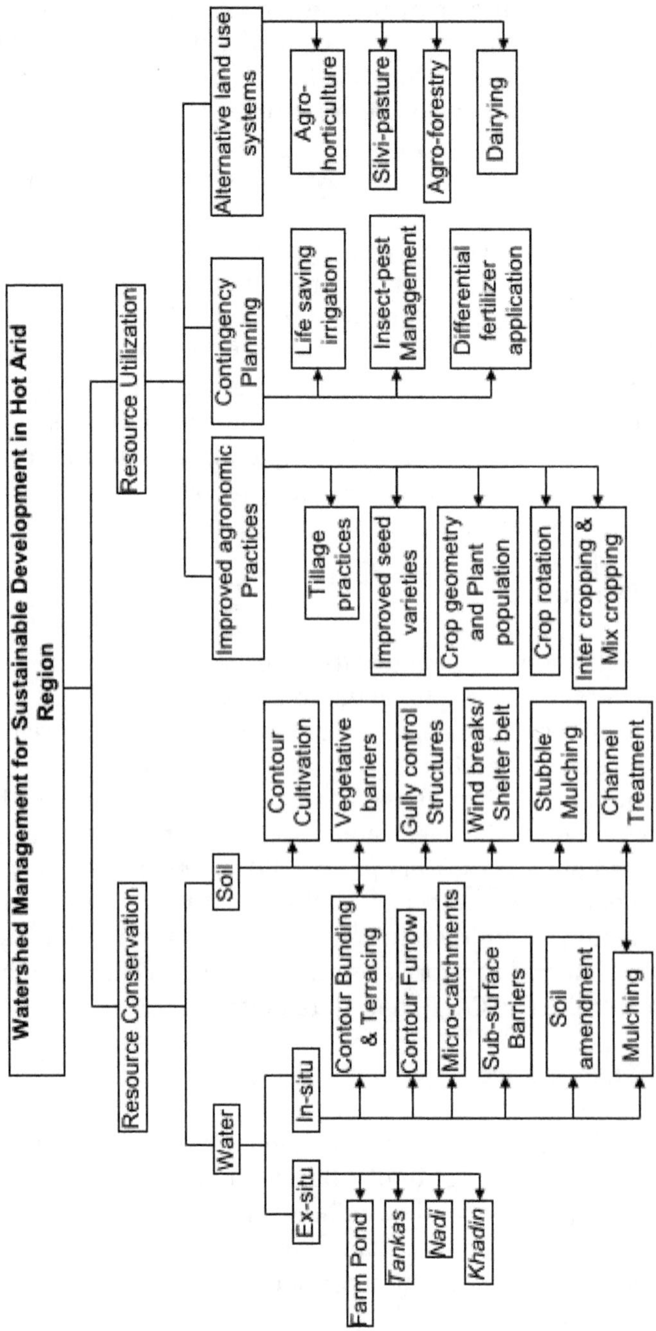

Figure 1. Components of watershed management in hot arid areas.

These *tankas* were successfully constructed in Jhanwar, Sar, and Baorali-Bambore watersheds. Harvested water of these tankas was used to provide life-saving irrigation to plants. The Benefit cost ratio of tanka ranged from 1.25 to 1.40 under different uses (Goyal *et al.*, 1995, 1997; Goyal & Sharma, 2000).

About 60 ha area were covered under contour bunding in Jhanwar, and Baorali-Bambore watersheds. Bunding helped in increase in yield of Pearl millet by 40% over control (3 q ha^{-1}) in Jhanwar watershed. Contour vegetative barriers of perennial grasses or shrubs were constructed for conserving soil and water in sloping lands. Rooted slips of local eight species of perennial grasses (*Cenchrus ciliaris, Cenchrus setigerus, Cymbopogon jwarancusa, Lasiurus sindicus, Panicum antidotale, Panicum turgidum, Saccharum bengalense* and *Vetiveria zizanioides*) and seedling of six species of shrubs (*Agave americana, Aloe barbadensis, Barleria prionitis, Euphorbia antisyphylitica, Ipomoea carnea* and *Leptadenia pyrotechnica*) were transplanted at 1 m vertical interval on contours across the slope. Result indicated that perennial grass species performed the best and formed effective barrier against soil erosion. Runoff volume and specific peak discharge were reduced by 28 to 97% and 22 to 96% respectively (Sharma *et al.*, 1999). In another study conducted at Kalyanpur (Distt. Barmer) during 1998, vegetative barrier of *Lasiurus sindicus, Saccharum munja* and *Cassia angustifolia* were established at horizontal interval of 30 m. The moisture data revealed 36.5%, 72% and 54.2% higher moisture storage as compared to control in *Cassia angustifolia, Lasiurus sindicus* and *Saccharum munja* respectively (Gupta and Rathore, 2002).

Under *ex-situ* rainwater management a *khadin* of 20 ha areas was developed in Baorali-Bambore watershed with surplussing arrangements. Before construction of *khadin*, uncontrolled runoff from upper catchment used to wash away seeds, fertilizers, and standing crops besides loss of valuable water. After construction of *khadin*, farmer could take excellent *kharif* and *rabi* crops.

For farm water management, a farm pond of 20,000 m^3 capacity was constructed at Kukma watershed at Bhuj in Gujarat. Construction of this farm pond resulted in assured availability of 20,000 m^3 water even in as small as 150 mm rainfall region. The collected water was used to provide irrigation to Datepalm, ber, aonla and other fruits plants in nearby area. For individual household roofwater harvesting system was designed and demonstrated at many places in watershed areas.

For *in-situ* rainwater management circular micro-catchment of 5% inward slope with LDPE lining was successfully demonstrated in watershed area for establishment of *ber* and other trees. For severally eroded and gullied catchment loose stone check dams (LSCD) at 1 m V.I. were constructed in Jhanwar watershed area on 17 gullies. Regular observation on these gullies indicate that LSCD proved to be very effective in controlling further extension of gullies and all these gullies got stabilized with adoption of LSCD (Goyal *et al.,* 2007). For channel treatment three masonry anicuts and two loose stone anicuts were constructed on main streams in Jhanwar and Baorali-Bambore watersheds respectively. Construction of these barriers resulted in substantial reduction in velocity of water thereby reduction in erosion at downstream and precipitation of suspended sediments at upstream. Temporary inundation of water at upstream helped in regeneration of vegetation in upstream beside recharge of groundwater. In Sar watershed, artificial recharge of groundwater was superimposed in a 2.8 ha m pond with three infiltration wells to improve water availability for conjunctive uses. For moisture conservation soil, straw and plastic mulch were tried in Baorali-Bambore watershed. The grain yield of Pearl millet was 32.67 and 28.12% higher for plastic and straw mulch respectively over no mulch.

In alternative land use system various systems like agro-horticulture with Pearl millet/Mung/Moth + Ber/Aonla/Pomegranate, Silvi-pastoral system with *Cenchrus ciliaris* + *Prosopis cineraria/Colophospermum mopane/Harwickia binata* were successfully established in watershed areas. *Ditch-cum-mound* fencing and *cut and carry* system was adopted for pasture development in Jhanwar watershed (Bhati and Goyal, 1997). For wastelands alternative crops like *Cassia angustifolia* and *Lawsonia alba* were successfully raised at appropriate locations in the watershed. For arable farming improved varieties of Pearl millet, cluster bean, mung bean, moth bean etc. were introduced in the watershed areas.

B. State Government Experience

As a result of watershed management activities in degraded Aravallis at Siha, Rewari (Haryana), the number of electrified wells have increased from 67 to 205, sprinkler set 2 to 70 and irrigated area 260 to 420 ha. The livestock population increased from 882 to 1396. Consequently, total milk production increased from 2997 to 5724 l day^{-1} and overall income increased by 400 to 500 per cent (Neonand *et al.,* 1996). With proper soil and water conservation measures in watershed in Aravallis, soil loss was reduced from 150 t ha^{-1} to less than 5 t ha^{-1}. Development of water resource resulted in the increase of net irrigated area to 28 ha and cropping intensity from 128 to 210 per cent. Food grain and fodder production also increased substantially (Bhardwaj and Dogra, 1997). Likewise, adoption of graded bunds, gully control structures, contour cultivation, inter cropping, use of cover crop in rotation along with other improved package of practices have been found sustainable in semi-arid region of Rajasthan. Graded bund has reduced the run-off from 20 to 4.8 per cent and soil loss from 24 to 4.12 t ha^{-1} yr^{-1}. Besides other benefits intercropping on contour resulted in 48 per cent higher grain yield (Singh et al, 1997). By adoption of various development activities in Osian index catchment during a five-year period, cropping intensity increased by 31.4 per cent and forage yield by 1.97 t ha^{-1}. Construction of water harvesting structures helped to increase the ground water recharge as indicated by rise in static water level. Sediment deposition against loose stone check dam was 3.86 m ha^{-1} yr^{-1} (Vangani *et al.,* 1998).

References

Bhardwaj, S.P. and Pradeep Dogra (1997). Watershed Management models for Himalayas, Shivaliks and Aravallies *Indian. J. Soil Cons.* 25 (1): 1-8.

Bhati, T.K. and R.K. Goyal (1997). *Conservation of plant genetic resources through watershed approach in Indian arid zone - A case study.* Proc. Symposium on Recent Advances in Management of Arid Ecosystem. (Eds.) S. Kathju, R.K. Abhichandani, Amal Kar and N.L. Joshi, Jodhpur, p. 126

Goyal, R.K., P.R. Ojasvi and T.K. Bhati (1995). Economic evaluation of water harvesting pond under arid conditions. *Indian Journal of Soil Conservation* 23(1): 74-76.

Goyal, R.K., P.R. Ojasvi and J.P. Gupta (1997). Farm ponds: A key to survival in arid areas. *ASIAN WATMANET*, newsletter. No. 11, pp.12-13.

Goyal, R.K. and A.K. Sharma (2000). Farm Pond: A well of wealth for the dryland dwellers. *Intensive Agriculture*, Vol. XXXVIII (July- August 2000), No. 5-6 pp. 12-14 & 26.

Gupta, J.P. and Rathore, S.S. (2002). Biomass production and rehabilitation of degraded lands in arid zone. pp. 20.

Gupta, V.P., R.P. Sharma and R.K. Goyal (2000). *Impact assessment of conservation measures in arid areas: a case study of Osian-Bigmi (Jodhpur) watershed.* Proc. International conference on integrated water resource management for sustainable development, New Delhi, pp. 1012-1019.

Goyal, R.K., T.K. Bhati and P.R. Ojasvi (2007). Performance evaluation of some soil and water conservation measures in hot arid zone of India. *Indian Journal of Soil Conservation*, Vol.35 (1): 58-63.

Neonand, Singh, J., Sharma, A.K. and Yadav, K.S. (1996). All Round Development through Watershed Management - A success story. *Indian Journal of Soil Conservation*, 24 (2): 160-164.

Singh, A.K., Kumar, A.K., Katiyar, V.S., Singh, K.D. and Singh, U.S. (1997). Soil and Water Conservation Measures in Semi-arid Regions of South-Eastern Rajasthan. *Indian Journal of Soil Conservation,* 25 (3): 186-189.

Sharma, K.D., N.L. Joshi, H.P. Singh, D.N. Bohra, A.K. Kalla and P.K. Joshi (1999). Study on the performance of contour vegetative barriers in an arid region using numerical models. *Agricultural Water Management,* 41: 41-56.

Vangani, N.S., Surendra Singh and R.P. Sharma (1998). Index catchment - a new concept for sustainable integrated development. *Annals of Arid Zone,* Vol. 37 (2) 133-137.

3

Integrated Remote Sensing and Geographic Information System for Watershed Monitoring and Management

Mahesh K. Gaur and R.K. Goyal

Introduction

Dryland farming in India accounts for 63% of the cultivated land of 144 million hectares. The crop production on these lands is entirely dependent on the natural precipitation, which is highly erratic in terms of spatial and temporal distribution during the crop-growing season (Chowdary *et al.,* 2001). So, there is a need for national development particularly in developing countries like India, for a comprehensive approach for the management of water resources. Optimal management of water resources on a self-reliant basis requires a long term perspective and it calls for a holistic view in all resource sectors including socio-economic factors. Accurate inventory, assessment, and monitoring of available resources periodically enable the planners to keep an eye on the development process and to take appropriate timely measures.

Remote Sensing in India is notable for its success in implementing its promotion through the unique National Natural Resources Management System (NNRMS). The major achievements of NNRMS are Establishment of facilities for dissemination, services and education; the evolution of appropriate application methodologies dictated by user needs; national level remote sensing application projects on operational basis; development of indigenous capabilities in all areas of the technology including design and development of satellites/ sensors; and encouragement to entrepreneurship. The Indian experience forms an ideal role model for all developing countries to emulate.

Satellite remote sensing has been practiced since 1972, starting with broad channels and moderate ground resolution (Landsat MSS). In the 1980s, Landsat TM and SPOT provided for improved spatial and spectral resolutions. Many satellite images were produced in the last four decades, offering a synoptic view of landscapes that were rich in contrast and that included different terrains. Remote Sensing and Geographic Information System (GIS), now Geospatial technology is being considered a cost-effective source for the management of natural resources most sustainably. The recent advent in technology has led the multifaceted developments in remote sensing technology ranging from less than meter multispectral images to the hyper spectral images to the wide coverage images.

A geographic information system (GIS) captures, stores, analyses, manages and presents data, which is linked to locations or having spatial distribution. It is a computer-based system that provides four sets of capabilities to handle geo-reference data. These are:

- Data capture: graphic data (digitized, converted from existing data) and attribute data (keyed-in, loaded from existing data files),
- Data storage and manipulation: file management and editing,
- Data analysis: database query, spatial analysis and modeling, and
- Data display maps and reports.

GIS provides a valuable tool for information analysis, automated mapping and data integration. The powerful GIS software, tools in problem-oriented systems, provides direct and easy access to large volumes of data. It supports their interactive analysis and helps to display and interpret results in a format directly understandable and useful for decision -making processes. A huge resource of images is available free of cost on the web. Many open-source image processing and GIS software are also available, which can be easily modified/configured tailored to the requirement of the user(s). These free resources on the web have played a catalytic role in the application of remote sensing and GIS in varied areas to generate the various kinds of information related to land use/land cover, geomorphology, water resources, vegetation and specifically biodiversity, agricultural resources, acreage under different crops, crop health, crop vigor, pests and diseases outbreak, etc. Considering the increasing applications of geospatial technology in natural resource management, various training programs have been organized to enhance skills and capabilities of the technical staff mainly by providing them with an understanding on the use of application of remote sensing and GIS for mapping and sustainable management of scarce and depleting natural resources. Participants are expected to acquire hands-on skills in the application of Geographical Information System (GIS) and Remote Sensing (RS) and learn to work with, spatial data, satellite images and GIS tools that are the key to a better understanding of natural resources.

Remote Sensing Technology

Remote sensing is the technique of deriving information about objects on the surface of the earth without physically coming into contact with them. This process involves making observations using sensors (cameras, scanners, radiometer, radar, etc.) mounted on platforms (aircraft and satellites), which are at a considerable height from the earth surface and recording the observations on a suitable medium (images on photographic films and videotapes or digital data on magnetic tapes).

When electromagnetic radiation falls upon a surface, some of its energy is absorbed, some are transmitted through the surface, and the rest is reflected. Surfaces also naturally emit radiation, mostly in the form of heat. It is that reflected and emitted radiation which is recorded either on the photographic film or digital sensor. Since the intensity and wavelengths of this radiation are a function of the surface in question, each surface is described as processing a characteristic "Spectral Signature". If an instrument can be identified and distinguished between different spectral signatures, then it will be possible to map the extent of surfaces using remote sensing. Satellite remote sensing is widely used as a tool in many parts of the world for the management of the resources and activities within the continental shelf containing reefs, islands, mangroves, shoals and nutrient-rich waters associated with major estuaries.

Satellites & Sensors

The mode can be geostationary, permitting continuous sensing of a portion of the earth or sun-synchronous with polar orbit covering the entire earth at the same equator crossing time. The Landsat series of satellites have a repeat period ranging from 16 – 18 days, whereas, in the case of the IRS, it is 22 days. Satellites cover the same area and provide continuous coverage of a fixed area. The sun-synchronous satellites have a synoptic coverage. The ground area covered by the satellite's passes can be obtained by referring to its path and row. Sensors are devices used for making observations. They consist of mechanisms usually sophisticated lenses with filter coatings to focus the area observed onto a plane in which detectors are placed. These detectors are sensitive to a particular region in which the sensor is designed to operate and produce outputs, which are either representative of the observed area as in the case of the camera, or produce electrical signals proportionate to radiation intensity.

Spatial resolution: It is a measure of the smallest angular or linear separation between two objects that can be resolved by the sensor. The greater the sensor's resolution, the greater the data volume and the smaller the area covered. The area coverage and resolution are inter-dependent and these factors determine the scale of the imagery.

Spectral resolution: It refers to the dimension and number of specific wavelength intervals in the electromagnetic spectrum to which a sensor sensitive. Narrow bandwidths in certain regions of the electromagnetic spectrum allow the discrimination of various features more easily.

Temporal resolution: It refers to how often a given sensor obtains imagery of a particular area. Ideally, the sensor obtains data repetitively to capture unique discriminating characteristics of the phenomena of interest. Radiometric sensitivity: It is the capability to differentiate the spectral reflectance/ emittance from various targets. This depends on the number of quantization levels within the spectral band. In other words, the number of bits of digital data in the spectral band will decide the sensitivity of the sensor. Stages of Remote Sensing include:

- A source of electromagnetic radiation or EMR (Sun)
- Transmission of energy from the source to the surface of the earth, through the atmosphere
- Interaction of EMR with earth's surface.
- Transmission of energy from the surface to Remote Sensor mounted on a platform, through the atmosphere
- Detection of energy by the sensor.
- Transmission of sensor data to the ground station
- Processing and analysis of the sensor data
- Final data output for various types of application

Geographical Information Systems

A *Geographical Information System* (GIS) is a system of hardware, software and procedures to facilitate the management, manipulation, analysis, modeling, representation and display of geo-referenced data to solve complex problems regarding planning and management of resources. Functions of GIS include data entry, data display, data management, information retrieval and analysis. The applications of GIS include mapping locations, quantities and densities, finding distances and mapping and monitoring change.

GIS and Remote Sensing Application

- Resource Mapping through Remote Sensing and Photogrammetry
- Image Interpretation and Analysis
- Integration of Remote Sensing and GIS Technology
- Agriculture
- Crop health analysis
- Precision agriculture
- Compliance mapping
- Yield estimation
- Urbanization & Transportation
- Updating road maps
- Asphalt conditions
- Wetland delineation
- Natural Resource Management
- Habitat analysis
- Environmental assessment
- Pest/disease outbreaks
- Impervious surface mapping
- Lake monitoring
- Hydrology
- Landuse-Landcover monitoring
- Mineral province
- Geomorphology
- Geology
- National Security
- Targeting
- Disaster mapping and monitoring
- Damage assessment

- Weapons monitoring
- Homeland security
- Navigation
- Policy
- Database Design and Connectivity for GIS Application
- Surveying, Networking and Demographic Analysis
- Environmental and Forest Management System
- Tourism Development and Database Integration
- GIS-based Application ANALYTICS (Using MS OLAP and Oracle Discoverer)

Benefits of GIS

- Geospatial data are better maintained in a standard format.
- Revision and updating are easier.
- Geospatial data and information are easier to search, analysis and represent.
- More value-added products.
- Geospatial data can be shared and exchanged freely.
- The productivity of the staff improved and more efficient.
- Time and money are saved.
- A better decision can be made.

Watershed Monitoring and Management

Rain-fed agriculture contributes 58 percent to the world's food basket from 80 percent of agriculture lands (Raju *et al.,* 2008). As a consequence of the global population increase, water for food production is becoming an increasingly scarce resource, and the situation is further aggravated by climate change (Molden, 2007). The rain-fed areas are the hotspots of poverty, malnutrition, food insecurity, prone to severe land degradation, water security and poor social and institutional infrastructure (Rockstorm *et al.,* 2007; Wani *et al.,* 2007). Watershed development program is, therefore, considered as an effective tool for addressing many of these problems and recognized as a potential engine for agriculture growth and development in fragile and marginal rain-fed areas (Joshi *et al.,* 2005). Management of natural resources at watershed scale produces multiple benefits in terms of increasing food production, improving livelihoods, protecting the environment, addressing gender and equity issues along with the biodiversity concerns (Sharma, 2002; Wani *et al.,* 2003 a, b; Joshi *et al.,*

2005; and Rockstorm *et al.,* 2007).

Soil, water and vegetation are the most vital natural resources for sustainable development and management, and hence should be handled and managed effectively, collectively and simultaneously. Managing natural resources with a sustainable approach is a rational phenomenon in its natural region. In this approach, the natural regions are invented to be in terms of the flow of water, which influences almost all fields of the environment, where the regions are diversified as a basin, catchment, sub-catchment, macro watershed (50,000 ha), sub-watershed (10,000–50,000 ha), milli-watershed (1,000–10,000 ha), micro-watershed (100–1,000 ha), mini-watershed (1–100 ha) (Nair, 2009). Planning and management of natural resources at the micro-level of the watershed where there is high Spatio-temporal variability in Geo-physical and socio-economic variables, particularly in the fragile arid and semi-arid tropics, is the crucial need of the hour (Aher *et al.*, 2012). The real challenge on water resources planning at a micro-level is to assess the quantum of water demand and availability. Watershed based planning through augmentation of modern techniques such as remote sensing (RS) and Geographic Information System (GIS), for modelling the availability of water resources and sectoral demand is being considered as the most appropriate approach.

Concept of Watershed

The word "watershed" introduced in 1920 was used for the "water parting boundaries". Watershed is that land area that drains or contributes runoff to a common outlet. Watershed is defined as a geo-hydrological unit draining to a common point by a system of drains. All lands on earth are part of one watershed or other. Watershed is thus the land and water area, which contributes runoff to a common point. A watershed is an area of land and water bounded by a drainage divide within which the surface runoff collects and flows out of the watershed through a single outlet into a larger river or lake. Watershed technology is used in rainfed areas.

a. Definition of Watershed

A watershed, also called a drainage basin or catchment area, is defined as an area in which all water flowing into it goes to a common outlet. People and livestock are an integral part of the watershed and their activities affect the productive status of watersheds and vice versa. From the hydrological point of view, the different phases of the hydrological cycle in a watershed are dependent on the various natural features and human activities. Watershed is not simply the hydrological unit but also a socio-ecological entity that plays a crucial role in determining food, social, and economic security and provides life support services to rural people (Wani *et al.*, 2008).

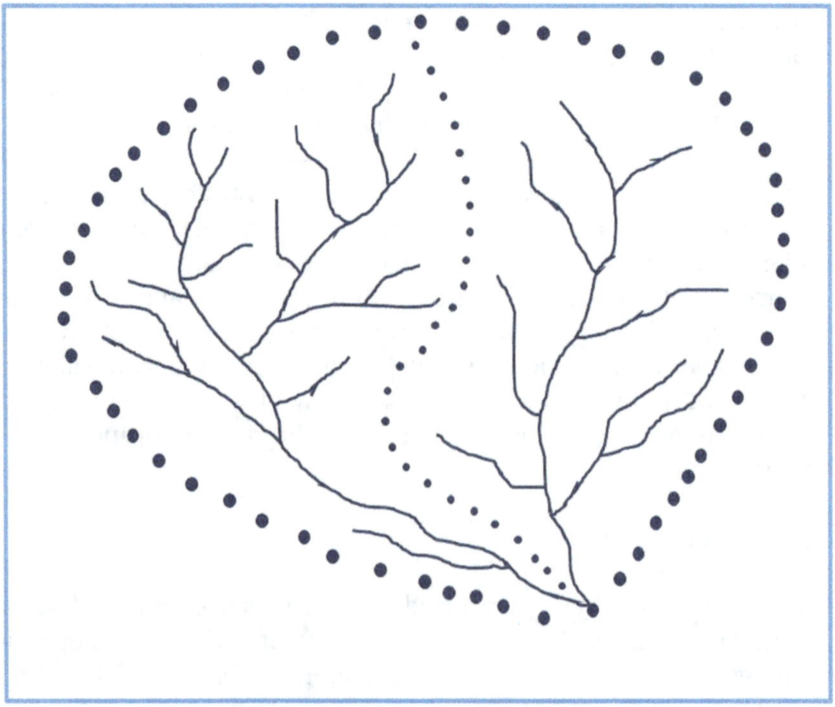

Figure 1. Pictorial representation of a typical fern-leaf shaped watershed.

b. Delineation of Watershed

Hydrologically, the watershed is an area from which the runoff flows to a common point on the drainage system. Every stream, tributary, or river has an associated watershed, and small watersheds aggregate together to become larger watersheds. Water travels from head-water to the downward location and meets with similar strength of the stream, then it forms one order higher stream. The stream order is a measure of the degree of stream branching within a watershed. Each length of the stream is indicated by its order (for example, first-order, second-order, etc.). The start or headwaters of a stream, with no other streams flowing into it, is called the first-order stream. First-order streams flow together to form a second-order stream. Second-order streams flow into a third-order stream and so on (Figure 1). Stream order describes the relative location of the reach in the watershed. Identifying stream order is useful to understand the amount of water availability in reach and its quality, and also used as criteria to divide larger watershed into smaller unit. Moreover, criteria for selecting watershed sizes also depend on the objectives of the development and terrain slope. A large watershed can be managed in plain valley areas or where forest or pasture development is the main objective (Singh, 2000). In hilly areas or where intensive agriculture development is planned, the size of watershed relatively preferred is small. Soil and water are the two most vital natural resources that must be conserved and maintained carefully for the sustainability of agricultural production systems and environmental protection.

Approaches of Watershed Boundary Delineation

There are primarily three approaches for watershed delineation namely, (a) Manual delineation using Survey of India (SOI) toposheet, (b) Field demarcation and (c) Semi-automated delineation using digital elevation model (DEM).

(a) Manual delineation using Survey of India (SOI) toposheet:Any hydrological unit boundary is generally delineated using visual interpretation of drainage map of an area. Survey of India (SOI) toposheet (1:50000 scale) as well as the latest satellite imagery showing the drainage lines can be used for this purpose. Based on visual examination, a rough boundary is first drawn along the drainage divides around a group of drainage lines that join to a common outlet. The rough boundaries of watersheds are then finalized as per the desired size range (Figure 1). For planning on a larger scale, the watershed boundary so obtained are normally enlarged and superimposed over the revenue map (cadastral level).

(b) Field demarcation of the watershed: Proper boundary of a watershed can be demarcated in the field by manual checking with two surveyors. The first surveyor walks along the mainstream in the upstream direction (opposite to the flow direction) until the highest point or the ridgeline is reached. The second surveyor standing at the outlet of the stream, by looking at the skyline, assists the first surveyor to move a little forward or backward to make him stand exactly at the peak point. A landmark post or rock is placed at this highest point. The process is repeated for all the streams to demarcate an approximate boundary in the field. Further fine-tuning of the boundary is possible by using the knowledge of local farmers about general flow directions of surface run-off found after intense rainfall.

(c) Semi-automated delineation using digital elevation model (DEM):Manual delineation method is tedious and time-consuming, but historically was considered to be the most effective and accurate method of delineating watersheds. However, the quality and accuracy of manual delineations are dependent on the scale of the topographic map used, and the delineator's interpretation of the map (Stanton, 2001). Semi-automated watershed delineation using DEM and other ancillary data is fast, less tedious, less time consuming and cost-effective solution than manual delineation. The automated/semi-automated extraction of watershed boundary and other topographic parameters from DEM is recognized as a viable alternative to traditional surveys and manual evaluation of topographic maps, particularly as the quality and coverage of DEM data increases (Qamer et al., 2010).

A DEM is defined as any digital representation of the continuous variation of relief over space, where relief refers to the height of the earth's surface concerning the datum considered. It can also be considered as regularly spaced grids of the elevation information, used for the continuous spatial representation of any terrain. The methodology adopted for hydrological unit delineation using ArcGIS Hydrology Tool, DEM and other ancillary data is presented in Figure 2.

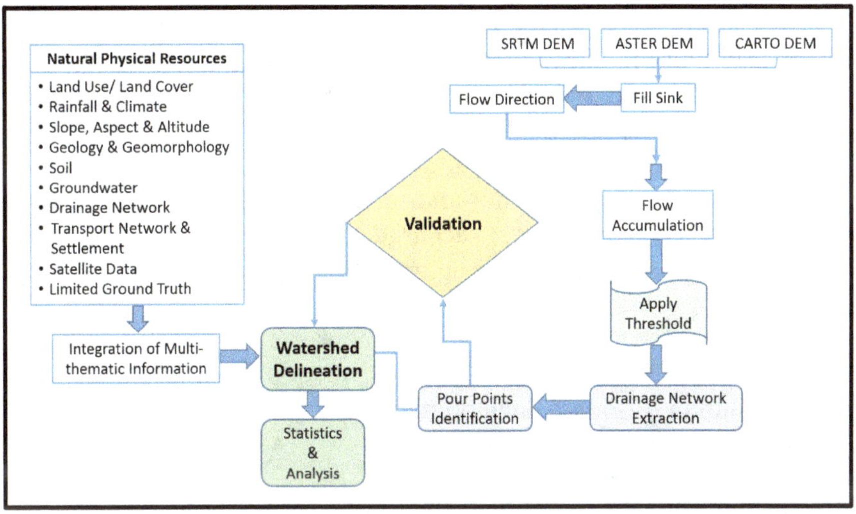

Figure 2. Watershed delineation methodology flow chart.

c. Objectives of Watershed Management:

- Production of food, fodder, fuel.
- Pollution control
- Overexploitation of resources should be minimized
- Water storage, flood control, checking sedimentation.
- Wildlife preservation
- Erosion control and prevention of soil, degradation and conservation of soil and water.
- Employment generation through industrial development dairy fishery production.
- Recharging of groundwater to provide regular water supply for consumption and industry as well as irrigation.
- Recreational facility.

d. Principles of watershed management:

The main principals of watershed management under Mahnot and Singh (1993) are:

- Utilizing the land based on its capability.
- Protecting fertile top layer soil.

- Minimizing setting up of tanks, reservoir and lower fertile land.
- Maintaining and protecting adequate vegetative cover on the soil for controlling soil erosion throughout the year, particularly during the rainy season.
- *In situ* conservation of rainwater.
- Safe diversion of gullies and construction of check dams for increasing groundwater recharge.
- Water harvesting for supplemental irrigation.

e. Components of watershed development

- Human Resource development
- Soil & land management
- Water management
- Crop management
- Afforestation
- Pasture/fodder development
- Livestock management
- Rural energy management
- Form & non – form value addition activities

All these components are interdependent & interactive

f. Categorization of watersheds

Watershed is classified depending upon the size, drainage, shape and land use pattern.

- Based on size: A watershed is termed as macro, micro, and mini watershed depending upon its size. Categories Delineation System are as follows:
  Region (6) .... 1.5-12 lakhs km^2
  Basins (50) ....0.3- 3.0 lakhs km^2
  Catchment ....0.1 - 0.5 lakhs km^2
  Sub-catchment .... .2000 - 10000 km^2
  Watershed ....500 - 2000 km^2
  Sub-watershed .... 50 - 500 km^2
  Mini Watershed .... 10 - 50 km^2
  Micro Watershed ...... 5 - 10 km^2

- Based on drainage: It is related to its drainage pattern, stream density and drainage density.
  a. Stream density of a watershed is the number of streams per unit area (sq. km.). High stream density is an indication of high undulation of the terrain.
  b. Drainage density is the total length of all the stream channels both density indicates the drainage efficiency of the basin. A higher value indicates a well- developed network and high run-off causing an intense flood, while a lower value indicates relatively moderate to low runoff and higher permeability of the terrain.

- Based on shape: A watershed can broadly be described as either fan-shaped (near circular) which provides greater flood intensity of ferm shaped (elongated).

- Based on other criteria: Watershed may also be categorized into several other types. For example, watersheds may be termed as a hill or flat watershed, humid watershed or arid watershed, red soil watershed or black soil watershed, etc., using slope, climate and soil type.

g. Factors affecting watershed management:

- Watershed characters
  i) Size and shape
  ii) Topography
  iii) Soils
- Climatic characteristics
  i) Precipitation
  ii) Amount and intensity of rainfall
- Watershed operation
- Land use pattern
  i) Vegetative cover
  ii) Density
- Water resource and their capabilities
- Geological factor

h. Benefits of the watershed:

- Promotes economic and social development of community

- Employment generation and other income generation
- Ecological balance

Geoinformatics Application in Watershed Studies

Geoinformatics is a core technology of 21stcentury and it is being utilized effectively for various tasks like area prioritization, available resource mapping, DEM preparation, runoff estimation, flow length mapping, geo-database generation, landuse/landcover mapping, action plan preparation, monitoring and evaluation of watershed along with its impact assessment, etc. Medium and high-resolution data with synoptic and repetitive coverage is the gift of Remote Sensing Satellites which producing real-time pictures of the earth's surface for further study. The data produced by remote sensing satellites are used further in GIS software. Generally, the Geographic Information System (GIS) is used to amalgamate the spatial and non-spatial data and further geo-database creation and different kinds of analysis. Global Positioning System (GPS) is supporting effectively capturing ground control points (GCP), locating a particular area on the earth's surface and capturing the exact location of an area. So, geoinformatics play a vital role in geosciences and watershed studies are one of them. The impact assessment of developed watershed can be studied by various points out that some aspects are given below.

- To check the change in water level in the watershed
- To understand the impact assessment on forest canopy and agriculture
- The study of soil erosion and consequent silting up of reservoirs
- The impact on socio-culture aspect of that area
- The overall impact assessment on watershed basin

Watershed Characterization and its importance using geoinformatics

Several watershed characteristics are essential for watershed studies and these can be derived by using remote sensing and GIS. Some of them are discussed below:

a. Size: The size means the areal extent of a watershed which varies from micro to large projects. Each watershed behaves differently from others in terms of runoff and its other characteristics. Hence, in watershed development, size is an important phenomenon.

b. Shape: Generally the watersheds are having their own shapes e.g. it could be square, palm shape, rectangular, oval, etc. The shape of watershed controls the length, width ratio and it also affects the run-off characteristics. In other words, if the watershed is longer then, the greater is the time of concentration. On the contrary, the longer the time of concentration, the greater time available for the water to infiltrate, evaporate and get utilized by vegetation.

c. Slope: This is a very important characteristic of a watershed because it determines runoff volume, soil loss, infiltration, etc. Based on the average slope site selection of the check dam and other treatment plans can be implemented. The contour map of the watershed needs to be prepared for slope analysis. The preparation of the contour map helps to analyze the slope and topography of the region. The contour is the line joining the same height places from mean sea level. The contours are digitized from SOI toposheet. Slope map is now prepared using SRTM/ ASTER/ Carto DEM data.

d. Digital Elevation Model: Any digital elevation is a representation of the continuous variation of relief over space is known as a digital elevation model (DEM). A digital elevation model is an ordered array of numbers that represent the spatial distribution of elevations above some arbitrary Datum's in the landscape (Moore et.al. 1993). DEM describes the elevation of any point in a given area in digital format and contains information on the so-called 'skeleton' lines. Skeleton lines are lines of slope reversals (drainage, crests) and breaks of slopes.

e. Drainage: A large drainage density (length of drainage channels per unit area) creates conducive situation for quick disposal of runoff down the channels. In watershed where drainage density is less overland flow is predominant. Hence it is important aspect which influences the watershed behavior. The drainage is digitized from SOI toposheet using GIS techniques and stream orders are given by Strahler's method (1965). Drainage density depends upon the physical and climatic characteristics of the basin.

f. Geology: The geology is important for study of rocks, faults, extent of outcrops, water recharge areas, etc. The use of construction material, suitable site selection, weathering conditions are associated with geology. Different geomorphic units, their ground water prospect and proneness to erosion and areal extent are calculated at micro-watershed level.

g. Soil: Physical and chemical properties of soil are important aspect in watershed development. It is important for runoff and sedimentation treatment also soil profiles needs to be understood to dug the embankment and small check dams.

h. Climate: It is main aspect for survival of flora and fauna because particular environment is outcome of climate of that region. All climatic aspects like precipitation, temperature, humidity, winds, etc. are essential for understanding of hydrology. Based on it estimation of runoff can be calculated and specific landuse is developed.

i. Landuse: The surface conditions as well as landuse / landcover details can be derived from satellite data. The real picture i.e. ground truth is made available by remote sensing and the planning based on it is near to perfect and more reliable. Land use in the watershed directly affects both quality and quantity of watershed management (Jackson, 1967).The different landuse classes that cause problem in the natural resource management are the existence of wasteland in the area to be taken up for the development purpose. Similarly the depletion of the forest cover, presence of forest blanks in the dense or open forest, presence of scrubs in the larger area, degradation of dense forest into open forest and open into scrubs pose serious problems to the environment. One of the major component of monitoring the impact of watershed development is the change in land use and land cover pattern over a period of time. Keeping in a view, the remote sensing data are used for land use and land cover analysis.

- Habitation: This class of land is identified as mixed type of residential urban or rural cover, which includes low rise and detached buildings and slums.
- Agriculture land: This class of land is broadly defined as the land, which is used primarily for production of field crops. Tobacco, paddy, wheat, etc. are the major crops grown in the study area. Agricultural land is further classified into (a) double cropped (Kharif and Rabi) area, (b) single cropped (Kharif) area. Double cropped area is defined as the area, which is culti-

vated more than once in a year. Single cropped area is defined as the area, which is cultivated during one season in the year, agriculture land is identified mainly on basis of pinkish- red tone, fine texture and its pattern.

- Waste land: Wasteland can be broadly defined as the land which is degraded due to various factor such as erosion, salinity, water –logging, mining, desertification, etc., thereby a posing problem to ecological balance in that area. Wasteland is further classified into: a) Waste land with scrub; and (b) Waste land without scrub. Wasteland with scrub is area with scant scrub vegetation. Wasteland without scrub in the study area is an area without any vegetation.
- Water bodies: This class of land includes ponds, lakes, depression storage, perennial rivers/stream, etc. water bodies are identified mainly on the basis of dark blue/black tone.

j. Ground Water: The level of groundwater is very much important for implementation of watershed. For instance the ground water levels during pre-monsoon and post-monsoon are different. The overall planning of watershed is based on the ground water condition and water level. GIS applications are beneficial in terms of watershed management issues, such as locating possible sites suitable for groundwater recharge, because:

- A large amount of the information required (soils, land-use, and slope maps) to evaluate potential recharge sites currently exists in digital format.
- GIS allows a great number of factors to be viewed on uniform media.
- GIS has the ability to update information on features and corresponding data. This is essential for water resource management projects.
- A GIS database provides decision-makers with a comprehensive visual and tabular means for analyses on which to construct and support decisions.
- Utility of this type of database would be for regional and city planners as well as for water supply and water quality monitoring.
- Groundwater modeling helps to replicate the behaviors of natural groundwater or hydrologic system by defining the essential features of the system in some controlled physical or math-

ematical manner. Modeling plays an extremely important role in the management of hydrologic and groundwater system.

Prioritization of Watersheds

Prioritization is a process of identifying areas of main concern based on single or many parameters and selection of one or few watersheds out of many watersheds by using predefined set of criteria. Watershed prioritization is a prerequisite to operationalize any major scheme as it allows the planners and policy makers to adopt a selective approach considering the vastness of the catchment area, severity of the problems, constraints of funds and man power demands of the local and political system. The prioritization of watersheds varies with the objectives of different schemes but the basic framework of watershed remains the same.

(a) Methods of Watershed Prioritization: It is not feasible to take the whole watershed area at once for its management. Thus the whole basin is divided into several smaller units, as sub watersheds or micro watersheds, by considering its drainage system. Two different methods are listed in this section for prioritization of sub watersheds from a large watershed:

(b)*Sediment Yield Index* (SYI) and *Runoff Potential Index* (RPI) Models: The methods used for determining the priority of the sub-watersheds for treatment from soil erosion and sediment yield point of view are; (i) reconnaissance surveys, (ii) soil and land use surveys, (iii) sediment observations, and (iv) Remote Sensing methods.

(c) A *reconnaissance survey* of the entire watershed gives an idea of the relative erosion status of the sub-watersheds. This procedure is approximate and is to be used when no other method is available.

(d) Detailed *soil and land use surveys* include the erosion information of the sub-watersheds. A careful interpretation of these reports could provide information on the relative erosion status of the various sub-watersheds (Table 1).
(e) Actual *measurement of the silt load* contributed by each of the sub-watersheds will give a clear picture of the extent of erosion in the sub-watersheds. Observations in respect of silt loads recorded over a period of three to five years will indicate the sub-watersheds which are

contributing higher silt loads. The main difficulty with this procedure is that the data has to be collected over a period of years so that reliable conclusions can be drawn.

(f) Remote Sensing Techniques consisting of satellite imagery and its interpretation offer a good scope for determining the priority areas in large watersheds.

Table 1. Criteria selected and data source used for watershed prioritization

Inputs	Weight-age	Source
Silt Yield Index/Soil Erosion Status	20	AIS &LUS
SC/ST population	20	Census of India
Irrigated area	10	Satellite data (NRSC/ISRO)
Wastelands	20	Satellite data (NRSC/ISRO)
Rainfall	10	100 years average, Preliminary Data from DES & analyzed
Agricultural Labourers	10	Census of India
Forest Cover		Satellite data (NRSC/ISRO)

Study of the Survey of India topographical map on 1:50,000 scale helps to have an idea of the catchment area and identification of the major landscape and land use. The methodology of Priority Delineation Survey comprises the following steps.

(i) Preparation of framework of micro-watershed through systematic delineation.
(ii) Codification of different stages of delineation by using Alpha-numeric symbolic code.
(iii) Rapid Reconnaissance Survey on 1:50,000 scale base (SOI Toposheets, aerial photographs and other base material) leading to the generation of a map indicating Erosion Intensity Mapping Units (EIMU).
(iv) Assignment of delivery ratio to various Erosion Intensity Mapping Units.
(v) Computation of Silt Yield Index (SYI)/Run-off Potential Index (RPI) for individual micro-watersheds.
(vi) Based on the descending values of SYI/RPI grading of micro-watersheds into very high, high, medium, low and very low priority cat-

egories is assigned.

(vii) Assignment of weightage values to various Erosion Intensity Mapping Units based on their relative sediment yield/run-off potential.

The EIMU is an assemblage of land and soil characteristics, viz., physiography, slope, land use and land cover with density, surface condition, soil depth, texture and structure of surface and sub-soils, colour, drainage condition, salinity and alkalinity, stoniness and rockiness, erosion condition and existing management practices.

The delivery ratio of an erosion intensity mapping unit indicates the transportability of the soil material detached from the area enclosed by the unit to the site of the dam/reservoir. The maximum values of delivery ratio adjudged for individual EIM unit are based on factors influencing the suspension and mobility of suspended material like texture, mineralogy and pH of the soil, land use/land cover conditions, terrain slope, surface stoniness/rockiness and soil conservation measure adopted. The adjustment delivery ratios are also dependent on the watershed attributes such as drainage pattern/drainage density, watershed gradient, and proximity to active stream resources. The maximum delivery ratio value, assigned to various EIMU ranges from 0.40 to 0.95 (http://ecoursesonline.iasri.res.in/).

Watershed Monitoring

Watershed development is not a new concept, number of watershed development programmes were implemented/are ongoing in our country. Many state governments have separate watershed development departments /units, central Ministry of Rural Development & Department of Agriculture gives lot of importance for watershed development. Accordingly number of projects has been launched by central/state govt. agencies with the participation of NGO's. Number of check dams/*nala* bunds have been constructed, alternate land use practices have been adopted, but how to monitor the impact of implementation of these measures? Remote sensing is the best tool for monitoring the results of the implementation especially the change in land use/cover, water spread/water bodies, etc.

Remote sensing data is found to be useful to assess (i) the status of/change in cultivated, waste land, forest and plantation areas, (ii) new water bodies/ number of water harvesting structures, (iii) change in cropping intensity/cropping system and (iv) soil conservation measures. Number of studies carried out has shown the utility of remote sensing data in watershed monitoring. Implementation of various government schemes are regularly monitored using remote sensing data. The impact of the development measures implemented in watersheds treated under the Watershed Development Programmes of the various State Governments were assessed using IRS data.

Watershed management is not only an anti-erosional and anti-runoff approach but also a comprehensive integrated approach the land water resource management considering the ecological economic and social factors of the locality.

GIS Maps are helpful: Preparation of an Action Plan for watershed is a fairly technical process and requires lot of information in hand, using which the analysis should be performed. The recommendations about the treatment measures have to be based on judicious decisions, otherwise, the impact of the intervention after implementation in the area would hardly have any result. The treatment selection and proper site selection is the crux of the watershed management, and this may be easily achieved with a GIS-based micro-level Watershed. Also such action plans ensure that each and every information is referenced and presented well on village level maps [cadastral maps] where from Khasra specific action plan can be made. It is very essential to understand that it is only the GIS based Watershed Maps which rightly present the picture of current situation and the requirements of the area. Maps may help to show what and where the problem is and where treatment activity be undertaken and what would be impacted area. Once the layers of all kinds of maps - the thematic maps, about the area for basic information and that of the requirements are readied and once these maps are superimposed over each other, only then can somebody state as to what treatment is desired at a site-otherwise not.

The field survey technique is important in geographical study but it is time consuming as well as costly whereas the remote sensing GIS and GPS is a better means of data acquisition (Martin, 2007). The remotely sensed data in addition to collateral data constitute requisite integrated information of watershed development on suitable basis.

In recent years, remote sensing technology made great achievement and contributed significantly in the management of natural resources.

One of the biggest problems generally in the earth is water overexploitation. Watersheds usually are used to irrigate fields, to drill water for various purposes, etc. But water, especially in a basin is not endless. It is very important to monitor the level of the water in every basin with continuous data and keep a record of them. It is an obligation for us to know all the withdrawals and recharges so that to better manage the area.In this case, GIS helps us most to keep the records and represent in story maps and 3D maps the evolution of a watershed. It is not about elaborating data, but to analyze the existed situation in comparison with past data, and make decisions about it (UTZ, 2019).

Conclusions

Remote sensing and GIS are integral to each other. The development of Remote Sensing is of no use without the development of GIS and vice versa. Remote Sensing has the capability of providing large amount of data of the whole earth and also very frequently. GIS has the capabilities of analyzing a large amount of data within no time. The use of Remote Sensing Data for faster assessment of natural resources such as soil, geology, drainage etc. as well as assessment of economic activities through land use and infrastructure of the watershed area is well known. This is also used for monitoring of watershed development at later years. GIS is a very powerful tool for development of the watershed area with all natural and socio-economic facets for better planning, execution and monitoring of the project. It gives a clear perspective for analysis at various levels to different partners of the watersheds. In this case the sustainable improvement of agriculture and related enterprises would be the first challenge for the enhancement of the livelihood of local people, second challenge comes from better utilization of the vegetative covers in terms of forest coverage or horticultural vegetation in the vast agriculturally no-remunerative areas for livelihood promotion. The indigenous rocks and minerals resources can only sustain small-scale enterprises. Proper emphasis on the utilization of the surface run off through better measures suitable for the target region is needed at the planning stage to all partners of watershed. A GIS based model of the terrain with all relevant spatial data related to natural resources, infrastructure and administrative boundaries attached with relevant attribute data will enable the planners, stake holders and funding agencies to develop the watershed keeping in mind to fulfill the stake holders need in the backdrop of natural resources potential as well as limitations. This will also help to plan the infrastructural development needed such as connecting market with local produce. Geo-technical appraisal of all the surface water irrigation projects necessary to avoid unfavorable natural conditions will be easier to develop through participatory method. This model provides a holistic picture to enable to share the natural resources and protect them for the betterment of the watershed community.

Since it is very often impractical or impossible to directly measure soil loss on every piece of land, and the reliable estimates of the various hydrological parameters including runoff and soil loss for remote and inaccessible areas are tedious and time consuming by conventional methods. Therefore it is desirable that some suitable methods and techniques are used/evolved for quantifying the hydrological parameters from all parts of the watersheds. Use of mathematical hydrological models to quantify runoff and soil loss for designing and evaluating alternate land use and best management practices in a watershed is one of the most viable options. The integrated outcome of the hydrological models along with remote sensing and GIS can be helpful to the decision makers to evaluate the best management practices and design the necessary soil and water conservation structures to reduce the soil erosion.

References

Aher, P.D., J. Adinarayana and S.D. Gorantiwar (2012). Use of Morphological Characteristics for Multi-Criteria Evaluation through Fuzzy Analytical Hierarchy Process for Prioritization of Watersheds. Pp. 12-13. In: 21st Century Watershed Technology: Improving Water Quality and Environment Conference Proceedings, May 27-June 1, 2012, Bari, Italy. DOI: 10.13031/2013.41405.

Chowdary, V.M., Paul, Saikat, Srinivas Kumar, T., Sudhakar, S., Adiga, A., Nageswara Rao, P.P. (2001). Remote Sensing and GIS approach for watershed monitoring and evaluation: A case study of Orissa State,

India. Paper presented at the 22nd Asia Conf. Remote Sensing, 5-9 November, 2001, Singapore.

Jackson M. L. (1967). Soil Chemical Analysis. Printice Hall India Pvt. Ltd., New Delhi.p.35.

Joshi, P.K., Jha, A.K., Wani, S.P., Joshi, L. and Shiyani, R.L. (2005). Meta-analysis to assess impact of watershed program and people's participation. Research Report 8, Comprehensive Assessment of watershed management in agriculture. International Crops Research Institute for the Semi-Arid Tropics and Asian Development Bank. 21 pp.

Mahnot, S.C. and Singh, P.K. (1993). Soil and Water Conservation. Inter-Cooperation Coordination Office, Jaipur. P.90.

Martin D and Saha S.K. (2007). Integrated Approach Using Remote Sensing and GIS to study watershed Prioritization and productivity. *Indian Society of Remote Sensing* Vol-35 I –PP-23.

Molden D. (2007). Water for Food, Water for Life: A Comprehensive Assessment of Water Management in Agriculture. Molden D (ed). International Water Management Institute IWMI. Earthscan, London, UK; Colombo, Sri Lanka.

Moore, I.D., Gessler, P.E., Nielsen, G.A., Peterson, G.A., (1993). Soil attribute prediction using terrain analysis. *Soil Sci. Soc. Am. J.* 57, 443–452.

Nair, A.S.K. (2009). A new scientific management approach to water related natural disasters. In:Proceedings of Kerala environment congress, Thiruvanthapuram, pp 143–154, 19–21 Aug 2009.

Qamer, F. M., Saleem R., Hussain N., Akram U., and Raza S. M. (2010). Multi-scale Watershed Database of Pakistan. 10th International Symposium on High Mountain Remote Sensing Cartography. Grazer Schriften der Geographie und Raumforschung Band 45/ 2010.pp.91–96.

Raju, K.V., Aziz, A., Sundaram, M.S.S., Sekher, M., Wani, S.P. and Sreedevi, T.K. (2008). Guildelines for Planning and Implementation of Watershed Development Program in India: A Review. Global

Theme on Agro-ecosystems Report 48. Andhra Pradesh, India: International Crops Research Institute for the Semi-Arid Tropics.

Rockström, J., Hatibu, N., Oweis, T., Wani, S., Barron, J., Bruggeman, A., Qiang, Z., Farahani, J., and Karlberg, L. (2007). Managing Water in Rain-fed Agriculture. In: (Molden D, ed), Water for Food, Water for Life. A Comprehensive Assessment of Water Management in Agriculture. International Water Management Institute. Earth scan, London, UK.

Sharma, R. (2002). Watershed Development Adaptation Strategy for Climate Change. Paper presented in South Asia expert workshop on Adaptation to Climate Change for Agricultural Productivity, organized by the Government of India, UNEP and CGIAR, New Delhi.

Singh R.V. (2000). (Ed.) Watershed planning and management. Yash Publishing House, Bikaner, Rajasthan, India.

Stanton, G. (2001). A Comparison of Manual and Computer-Assisted Drainage Delineation Methods for Hydrologic-Unit Map Development. Pages 1118-1127. In: D.E. Stott, R.H. Mohtar and G.C. Steinhardt (eds). 2001. Sustaining the Global Farm. Selected papers from the 10th International Soil Conservation Organization Meeting held May 24-29, 1999 at Purdue University and the USDA-ARS National Soil Erosion Research Laboratory.

Strahler, Arthur N. (1965). Introduction to Physical Geography. New York: John Wiley & Sons, Inc., Pp. 455.

UTZ. (2019). Watershed Monitoring by GIS. http://uizentrum.de/watershed-monitoring-by-gis/

Wani SP, Sreedevi TK, Reddy TSV, Venkateswarlu Band Prasad CS. (2008). Community watersheds forimproved livelihoods through consortium approach indrought prone rain-fed areas. *Journal of Hydrological Research and Development*, 55-77.

Wani, S.P., Joshi, P.K., Ramakrishna, Y.S., Sreedevi, T.K., Singh, P. and Pathak, P. (2007). A new paradigm in watershed management: A must for development of rainfed areas for inclusive growth. Conservation farming: Enhancing productivity and profitability of rainfed areas. Swarup A., Bhan, S., and Bali, J.S. (eds). Soil Conservation Society of India, New Delhi. pp. 163-178.

Wani, S.P., Singh, H.P., Sreedevi, T.K., Pathak, P., Rego, T.J., Shiferaw, B. and Iyer, S.R. (2003a). Farmer-participatory integrated watershed management: Adarsha Watershed, Kothapally India: An innovative and upscalable approach. Case 7. Pages 123-147. In: Research towards integrated natural resources management: Examples of research problems, approaches and partnerships in action in the CGIAR (Harwood RR and Kassam AH, (eds). Interim Science Council and Centre Directors Committee on Integrated Natural Resources Management, Consultative Group on International Agricultural Research, Washington DC, USA. Rome, Italy: Food and Agriculture Organization.

Wani, S.P., Pathak, P., Jangawad, L.S., Eswaran, H. and Singh, P. (2003b). Improved management of Vertisols in the semi-arid tropics for increased productivity and soil carbon sequestration. *Soil Use and Management*, pp. 217-222.

4

Soil and Water Conservation Measures for Agricultural Sustainability in Arid Lands

R.K. Goyal and Mahesh K. Gaur

Introduction

The United Nations predicts that 1.8 billion people will experience absolute water scarcity in less than 5 years, and worry that by 2025, two out of three persons will be living in water stressed regions. Already every five persons worldwide cannot access their basic everyday water resource. Drylands cover more than 60% of the earth's surface. Arid zones - described as a part of the drylands, & have most severe climatic conditions. Major distinguishing feature -the arid zone has low rainfall (< 500 mm or Aridity Index <0.20) with more than 50% inter-annual variability. This makes - difference in terms of nature of ecosystem, socio-economic environment & challenges for sustainability. High wind and solar regimes further increase the effect of rainfall variability. Whole complexities makes a fragile ecosystem in which small disturbances may cause great loss to the sustainability, which are sometimes irreversible. Hot arid zones of the world are economically and environmentally disadvantaged with unique environmentally disadvantaged, with unique problems. Ecosystems of these zones are highly fragile with large risks that cause. Severe impediments to development programs are cause of concern. Of the total land area in the world arid zones cover 18 8% total land area in the world, arid zones cover 18.8%. These arid zones are diverse in terms of climate, soils, vegetation, animals, and the lifestyles and activities of the people (Gaur & Squires, 2016a).

Arid ecosystem are among the world's most fragile ecosystems made more so by periodic droughts and increasing over-exploitation of meager resources. Arid environments are extremely diverse in terms of their land forms, soils, fauna, flora, water balances, and human activities. Because of this diversity, no practical definition of arid environments can be derived. However, the one binding element to all arid regions is aridity. Aridity is usually expressed as ratio of mean annual precipitation (P) to the mean annual potential evapotranspiration (EPT) where potential evapotranspiration, calculated by method of Penman, taking into account atmospheric humidity, solar radiation, and wind. UNEP (1997) has recognized four main classes of aridity: hyper-arid (P/EPT < 0.03), arid (0.03 < P/EPT < 0.20), semi-arid (0.20 < P/EPT < 0.50), and dry sub humid (0.50 < P/EPT < 0.65). The term "arid zone" is used here to collectively represent the hyper-arid, arid, semi-arid, and sub-humid zones. Of the total land area of the world, the hyper-arid zone covers 4.2 percent, the arid zone 14.6 percent, and the semi-arid zone 12.2 percent (FAO, 1989). Therefore, almost one-third of the total area of the world is arid land and is inhabited by about one billion people, a large proportion of who are among the poorest in the world (Malagnoux *et al.,* 2007).

Soil and water are two essential natural resources for sustaining life and environment. Since advent of human race on the earth these resources are being used for various purposes. The prosperity of mankind on the earth depends to a great extent on these resources and their management. In nature, soil and water resources are often in a fine balance however, with unprecedented increase in human and livestock population; demand for food and fodder have increased tremendously and these resources have been exploited indiscriminately. Though nature has its own process of resource build-up but the present rate of exploitation is much faster than the natural process of resource build-up, thus leading to imbalances in the nature. The imbalances in natural resources had resulted in reduction in productivity of land to sustain human development. Available natural resources of a region form a vital part of ecosystem.

Arid Zone Climate

The arid region of India is spread in 38.7 million hectare (Mha) area out of which 31.7 Mha is under hot arid zone and 7 Mha under cold arid zone. The hot arid region occupies major part of north-western India (28.57 Mha) between 22°30' and 32°05' N latitudes and from 68°05' to 75°45' E longitudes, covering western part of Rajasthan (19.6 Mha, 69%), north-western Gujarat (6.22 Mha, 21%) and 2.75 Mha in south-western part of Haryana and Punjab (Faroda *et al.,* 1999). The climate of Indian hot arid zone is characterized by an abundance of solar energy from cloud-less sky, high diurnal and seasonal temperature variations and annual and inter-annual irregular rainfall with long dry seasons associated with strong winds. Annual rainfall is approximately in the range 100-500 mm, with a coefficient of variation varying from 40 to 80% (Rao and Singh, 1998). The rainfall results largely from convective cloud mechanisms and is characterized by a relatively high intensity, short duration and limited aerial extent. The distribution of rainfall in space is very much influenced by the local terrain. With low ground level and little topographic relief, it is common to see rain evaporating before reaching the ground. The incoming radiation ranges from 15.12 to 26.50 MJ m^{-2} day^{-1} with very little cloud for most of the year. The annual potential evapotranspiration ranges from 1400–2000 mm yr^{-1} leading to a permanent negative water balance (Rao, 2009).

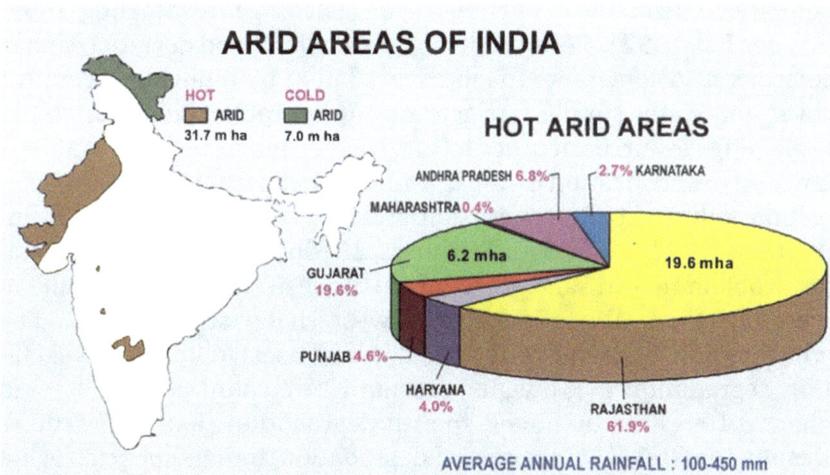

Figure 1. Distribution of arid areas of India.

The soils in arid regions, as a result of these climatic conditions, are characterized by large variations in soil moisture, ground-water storage and high infiltration rates. A dominant feature of these soils is their susceptibility to erosion and degradation by wind, water and human activities. There are frequent dust and sand storms. The surface water resources are meager and directly resulted from rainfall and dry up during the rainless season. Groundwater is deep and often brackish. During the twentieth century, the region has experienced agricultural drought once in three years to every alternate year in one or the other part of the region. The overall probability of drought is 47%. The weather condition, even in average years, for most part of the year remains too dry and inhospitable for successful growth of crops. Under such conditions of uncertainty, conventional cropping is risky and is essentially for sustenance only (Goyal and Vittal, 2009).

Challenges for Arid Lands

Everywhere in the world where people changes a natural ecosystem, the land degrades. The major challenge before arid lands is desertification. The UNCCD defines desertification as "land degradation in arid, semi-arid and dry sub-humid areas resulting from various factors, including climatic variations and human activities". The most commonly cited forms of unsustainable land use are over-cultivation, overgrazing, deforestation, and ill-considered irrigation practices. Soils are being mined of nutrients at an alarming rate (Smaling, 1997; Van der Pol, 1992). Periodic droughts debilitate and destabilize poor societies, and contribute to land degradation by reducing vegetative cover and water supplies, triggering a desperate exploitation of the remaining resources in order to survive. The major physical manifestation of desertification in hot arid zone of India is wind erosion/deposition, followed by water erosion, as well as water logging and salinity (Ghose *et al.,* 1977; Gaur & Squires, 2016b). Wind erosion is a major problem in arid sandy tract of north-western India, especially in west Rajasthan. About 76 % area of western Rajasthan is affected by wind erosion, encompassing all major land uses, followed by vegetation degradation (3%), water erosion (2%), salinization (2%) and about 0.10% area by mined spoiled degradation (Kar *et al.,* 2009). Despite the present state of land degradation, Indian hot arid region supports 22 million human and 29 million livestock population over 19.6 Mha geographical area. The human: animal ratio ranges from 1:1.3 to 1:5 in hyper arid regions against a national average of 1:0.5 (Narain and Goyal, 2009). Acute human and biotic pressure has put tremendous pressure on the fragile water and land resources of this region. Desertification, manifested in different forms of degradation has been a challenging issue to the arid zone of India.

So, the challenges in the arid zone can be summarized as:

- Highly erratic and unpredictable rainfall
- Extremes of temperature (–2 to 48°C), long day sunshine, high wind speed (35–40 km/h) and very high evaporation.
- Soil erosion (wind and water)

Figure 2. Wind erosion is a severe problem in western Rajasthan.

- Low productivity of lands
- Small land holdings
- Lack of water for drinking and irrigation
- Expansion of cultivated fields and diminishing of natural vegetation cover
- Insufficient amounts of organic material and nutrients
- Deforestation
- Burning of organic material (harvest residue, brush fires)
- Free grazing, no clear land use rights
- Poor socio-economic base and poor infrastructure
- Disguised unemployment (Gaur & Squires, 2016a; Gaur & Squires, 2016b).

The Central Arid Zone Research Institute (CAZRI), Jodhpur was established in 1959 to combat desertification and evolve sustainable land management technologies. Over five decades of research the Institute has evolved technologies of soil and water conservation for the management and sustainable development of the arid lands.

Soil and Water Conservation Measures for Management of Arid Lands

There are always strong links between measures for soil conservation and measures for water conservation, and this applies equally to arid lands. Many measures are directed primarily to one or the other, but most contain an element of both. Reduction of surface run-off by structures or by changes in land management will also help to reduce erosion. Similarly, reducing erosion will usually involve preventing splash erosion, or formation of crusts, or breakdown of structure, all of which will increase infiltration, and so help the water conservation (Hudson, 1987). Measures for soil and water conservation can be divided into two main categories i.e. physical/engineering measures and biological measures. While the physical measures are the first line of defense and are very necessary to prevent immediate damage, biological measures are the offensive attack made from a position of good defensive strength. Physical measures consist of construction of mechanical barriers across the direction of flow of water to retard/retain the runoff and thereby reduce soil and water losses. Biological measures like vegetative barrier, windbreak/shelter belt, stubble mulching is very effective for moisture conservation and controlling soil erosion due to wind and water. A proper combination of both mechanical and biological measures is prerequisite for resource conservation and sustained crop production. Conservation measures can also be classified on the basis of their utility for particular land use i.e. arable and non- arable lands. The basic objectives of any conservation measures for arid regions are:

- Increasing the time of concentration for rainwater and thereby allowing more runoff to be absorbed and held in soil profile;
- Intercepting a long slope into several short slopes so as to maintain less than a critical velocity for runoff water
- Protection against damage owing to excessive runoff and conserving moisture.

Watershed Management in the Arid Regions

In India, about 68.4% of the total land area has been degraded by the water erosion process. Intensive agricultural practices accelerate the soil erosion process. Similarly, increased exploitation of groundwater resulted in depletion of groundwater level. Hence, the holistic management of soil and water resources is indispensable for agricultural sustainability as well as for the protection of the natural ecosystem. Development and adoption of improved technologies, judicious use of natural resources, and effective management practices are the need of the hour for protection of soil and water from degradation.

- Mapping of SLM practices
- On-site and off-site and off-site interactions, e.g. dust storms originate on threaten people & livelihoods close by but far away.
- Highland-lowland interactions: People in arid areas are dependent on ecosystem services provided by highland areas, e.g. highlands - give water to surrounding lowlands.
- Regional interactions and rural-urban linkages
- Participation and community involvement:
- Planning for sustainable land management:
- Multi-functional use: helps to reduce risk through diversification, to promote synergies that produce added economic, ecological or social value, & to preserve & strengthen ecosystem services.

Conservation Measures for Arable Lands

Arable lands are those lands which can be used for growing crops. The arable lands include net sown area and current and other fallow lands. Land Capability classes I-IV comes under arable lands. On short term basis these lands have very little limitations for crop production, however, on long term basis these lands need some management practices for conservation soil and water. About 63.67% area of western arid Rajasthan comes under arable land. These land suffers from various degree of soil depth, slope, drainage and erosion limitations, however with proper soil and water conservation measures these lands can be used for crop production on sustainable basis. Agronomic measures are applicable in the landscape of ≤2% slope. Agronomic measures reduce the impact of raindrops through the covering of soil surface and increasing infiltration rate and water absorption capacity of the soil which results in reduced runoff and soil loss through erosion (FAO, 1984). These measures are cheaper, sustainable, and may be more effective than structural measures, sometimes (Yousuf and Singh, 2019). Important agronomic measures are described below.

Contour farming

Contour farming is one of the most commonly used agronomic measures for soil and water conservation in hilly agro-ecosystems and sloppy lands. All the agricultural operations viz. plowing, sowing, inter-culture, etc., are practiced along the contour line. It is beneficial on all slopes where line sown is adopted. All ridges and rows of plants are placed across the slope to form continual series of miniature barriers to water and offer maximum opportunity for infiltration.

Figure 3. Contour farming for soil and water conservation.

Contour operations reduce the power of the water to erode, suspend and carry away soil particles and increase the moisture storage. Increased uniform moisture storage can boost up the yield above 10%. Contour farming alone cannot control runoff volume from higher sloppy lands which may need bunding and grass waterways in natural drainage. For small fields with uniform slopes, one contour guideline is enough and where slopes are irregular two to three guidelines may be enough for farming operation. Contour farming is most effective on moderate slopes of 2 to 7%. Contour farming is reported to reduce soil loss upto 60 % in comparison to up and down farming on 1% slope (Smith and Wischmeier, 1962). The advantages of contour farming include:

- Increase soil organic matter by residue incorporation and deep root system.
- Protection of soil from the erosive impact of raindrops, runoff, and wind.
- Improve water quality and water holding capacity of the soil.
- Improve soil properties, suppress weed growth, and increase crop productivity.

- Act as an obstacle in water flow, reduce flow velocity, and thereby reduce runoff and soil loss.
- Improve nutrients availability to the component crop and succeeding crops through biological nitrogen fixation.

Contour bunding

Contour bunds are narrow base trapezoidal earthen embankment on contour, 1.5 to 2 m wide, constructed across the slope to act as barriers to runoff, to form water storage area on their upslope side and to break up a long slope into short segments. The spacing of bund is dependent on the erosive velocity of runoff, length of the slope, slope steepness, rainfall intensity, type of crops, and conservation practices. Contour bunds are recommended upto 6% slope and rainfall of upto 600 mm. Contour bunds are not suitable for shallow soils having depth less than 7.5 cm. Spacing between contour bunds is usually expressed in terms of vertical interval (VI), which is the difference in elevation between two similar points on two consecutive bunds. The principle involved in fixing the spacing between two bunds is to keep the velocity of water below critical value in order to avoid scour. Vertical interval is generally expressed as a function of percent slope (s)

$$VI = 0.305\left(\frac{s}{a} + b\right)$$

where, V I = vertical interval between consecutive bunds in m
S = land slope in percent
'a' and 'b' are constant specific to particular region
For soils having good infiltration rates value of 'a' and 'b' are 3 and 2 respectively. For soils with low infiltration rate the value of 'a' and 'b' may be taken as 4 and 2 respectively. The horizontal interval (HI) is related to vertical interval (VI) by following equation

$$\text{HI} = \left(\frac{\text{VI}}{\text{S}} \text{x} 100\right)$$

Depth of impounding in calculated on the basis of maximum rainfall of 24 hours for 10 years recurrence interval as follows

$$h = \sqrt{\frac{Re \times VI}{50}}$$

where 'h' is height of impounding (m) Re is 24 hours maximum rainfall (cm) for 10 years recurrence interval. Total height of bund H can be find out by adding 20- 25% freeboard to height of impoundment. For light red loam and sandy loam soils, the side slope of both the sides is taken as 1.5: 1 whereas for sandy soils it is taken as 2:1. The line of seepage should not cross the bottom of the bund while deciding the other dimension of the bund. On average, contour bunds had 27 percent higher soil moisture and 14 to 181 percent higher fodder yield than flat surfaces on grasslands of western Rajasthan (Wasi-Ullah *et al.,* 1972).

Contour Vegetative Barriers

Contour vegetative barriers are hedgerows of perennial grasses or shrubs planted at a regular interval on contours for conserving soil and water in sloping lands. Suitable grass species are grown along contours at suitable vertical interval to intercept part of runoff and to control erosion in agricultural fields having flat to slight undulating topography. The contour vegetative barrier moderates the velocity of overland flow and traps silt at low cost, and augment production of food, fuel and fodder or fiber from lands by growing suitable vegetation species. In recent years, contour vegetative barriers have found acceptability among the farmers as these are cheaper over mechanical measures and are protective while being productive. Contour vegetative barriers can be easily established across a wide spectrum of soil-climatic conditions. Selection of species depends upon purpose of barrier, site-specific conditions, particularly soil and climatic variables. The spacing between plant-to-plant and row-to-row is governed by vegetation species to be planted as barrier. In general the plant-to-plant spacing of 20 to 30 cm at predetermined or 50 to 100 cm vertical interval between the barriers has been found effective for soil and water conservation. Generally, paired row of barrier planted in staggered form across the slope proves more effective.

Generally, dominant grass or shrub species of the region should be preferred for vegetative barrier. Among grasses *Cenchrus ciliaris, Cenchrus setigerus, Saccharum bengalense, Vetiveria zizanioides, Lasiurus sindicus, Panicum antidotale* and *Panicum turgidum* can be effectively used for soil and water conservation in arid areas. Shrubs like *Leptadenia pyrotechnica, Ipomoea carnea* and *Euphorbia antisyphylitica* can also provide good protection against water and wind erosion. Contour vegetative barriers of *Cymbopogan jwarancusa, Cenchrus ciliaris* and *Cenchrus setigerus* transplanted at 0.30 m apart on contours at 0.6 to 1.0 m vertical interval in sandy loam soil of Jodhpur (Rajasthan) have performed well and formed effective barriers in reducing soil erosion and increasing soil moisture storage. In a study conducted during 1992-1994 at 19 farmers field near Jodhpur rooted slips of local eight species of perennial grasses (*Cenchrus ciliaris, Cenchrus setigerus, Cymbopogon jwarancusa, Lasiurus sindicus, Panicum antidotale, Panicum turgidum, Saccharum bengalense* and *Vetiveria zizanioides*) and seedling of six species of shrubs (*Agave americana, Aloe barbadensis, Barleria prionitis, Euphorbia antisyphylitica, Ipomoea carnea* and *Leptadenia pyrotechnica*) were transplanted at 1 m vertical interval on contours across the slope. Result indicated that perennial grass species performed the best and formed effective barrier against soil erosion. Runoff volume and specific peak discharge were reduced by 28 to 97% and 22 to 96% respectively (Sharma *et al.*, 1999; Tiwari and Kurothe, 2006). In another study conducted at Kalyanpur (Distt. Barmer) during 1998, vegetative barrier of *Lasiurus sindicus, Saccharum munja* and *Cassia angustifolia* were established at horizontal interval of 30 m. The moisture data revealed 36.5%, 72% and 54.2% higher moisture storage as compared to control in *Cassia angustifolia, Lasiurus sindicus* and *Saccharum munja* respectively (Gupta and Rathore, 2002).

Graded bunds

Graded bunds are used to dispose-off safely the excess water from the agricultural fields to avoid water stagnation. Graded bunds are suitable in areas where annual rainfall is 500 mm and if the soils are highly impermeable. Graded bunds usually have wide and shallow channels and earthen bund laid along a predetermined longitudinal slope. As graded bunds are essentially means for the safe disposal of excess water from cropped lands, suitable outlets are required to be constructed on graded bund. Draining of excess water from one plot to another through outlets provided in the bund require special attention since considerable amount of soil may be lost through these outlets. Provision should be made to arrest the silt and allow only clear water to flow away. The vegetated watercourse strengthens the system. In areas with high rainfall where the volume of water to be disposed from one plot to the next lower level plot is large and the vertical interval between the plots is reasonably high, a site specific outlet design like a pipe outlet, drop structure etc. must be provided for safe disposal of the excess water. Graded bund is reported to reduce the run-off from 20 to 4.8 per cent and soil loss from 24 to 4.12 t ha^{-1} yr^{-1}. Besides other benefits intercropping on contour resulted in 48 per cent higher grain yield (Singh *et al.,* 1997).

Grass waterways

Grass waterways are developed for safe disposal of excess water from agricultural fields. These may be natural or man-made courses protected against erosion by suitable grass cover. Grass waterways are also used for channelizing and regulating runoff flows for water harvesting purposes. The best location for waterways is a natural depression or along valley line. These may also be constructed along field boundaries for safe disposal of excess rainfall from agricultural fields. Vegetative waterways may be located in all classes of lands except hard rocks, where construction may be difficult. The cross section of waterways may be trapezoidal, triangular or parabolic with shallow depth and flat side slope to facilitate easy movement of man, animal and machinery. The depth of waterways may be kept within 20 to 50 cm and side slope more than 4:1. The channel should have free board of 15 cm. The channel cross-section and bed slope should be such that the computed velocity is within permissible limit (Singh *et al.,* 1990). Cost of construction of grass waterways depends upon type of soil, channel cross-section, length of channel and grass plantation technique.

Shelterbelts

In arid zones, the harsh conditions of climate and the shortage of water are intensified by the strong winds. Living conditions and agricultural production can often be improved by planting trees and shrubs in protective shelterbelts which reduce wind velocity and provide shade. Shelterbelts are barriers of trees or shrubs that are planted to reduce wind velocities and, as a result, reduce evapotranspiration and prevent wind erosion; they frequently provide direct benefits to agricultural crops, resulting in higher yields, and provide shelter to livestock, grazing lands, and farms. Shelterbelt when planted across and on the margins of agricultural fields effectively protect crops and control sand drifting (Ganguli and Kaul, 1969). The effectiveness of shelterbelt in reducing wind velocity depends on wind velocity itself, direction, shape, width, tree height and density etc.

Trees for shelterbelt are planted at right angles to the prevailing wind direction. If the wind direction changes frequently a checkerboard pattern of plantings is required. Otherwise only parallel lines are needed. From 2-5 rows of fast growing trees of different heights should be planted to prevent any possible breaks in single rows, which would create a tunneling action with high and dangerous wind velocities. The shape of shelterbelt should be pyramidal in structure, i.e., tallest trees should be in center followed by flank rows of medium height trees and lateral rows should be of bushy plants or shrubs. Windbreaks should be partly permeable to prevent turbulence. A permeability of 30-40% is considered the optimum. The distance between shelterbelts should be about 20 times the expected height of trees after 5-7 years of growth. No cutting for firewood should be permitted in shelterbelts. In an experiment a three rows shelterbelt in which *P. juliflora* shrubby thicket was used as outer rows with central row of *Albizzia lebbek* reduced wind speed from 12% at 10H (height) distance from shelterbelt to 33% at 2H distance during summer season (Gupta *et al*., 1984). The following trees/shrub species are recommended for shelterbelt plantation in arid areas.

Flank rows: *Prosopis juliflora, Zizyphus nummularia, Acacia bevinosa, calligonum polygonoides*

Lateral rows: *Cassia siamea, Tamarix articulata, Acacia tortilis, Parkinsonia aculeata*

Central rows: *Azadiranchta indica, Albizza lebbek, Acacia nilotica, Ailanthus excela, Hardwickia binnata, Eucalyptus cameldulensis*

Figure 4. Shelterbelt plantation for wind erosion control, CAZRI farm – Jodhpur.

Conservation Measures for Non –Arable Lands

Non-arable lands are those lands, which do not fulfill their life sustaining potential. These can result from inherent / imposed disabilities such as by location, environment, chemical and physical properties of the soil or financial or management constraints. In arid zone of western Rajasthan alone, about 7 million hectares area is classified as wasteland. The area under wasteland is further increasing because of over-exploitation and improper land use, to meet increasing demand for food, fodder and fuel. Also, cultivable area is decreasing because of increasing urbanization, mining and other economic activities. So there is need to develop these wastelands appropriately for sustained and improved fodder production. The adoption of dryland conservation technologies can significantly increase productivity and profitability in these areas (Halvin and Schlegel, 1997).

Contour furrowing

Contour furrowing is the most effective measure to reduce runoff and soil loss, increase in yield and commonly adopted in grasslands and forestlands. However, in very sandy soils or soils with heavy clay pan area, their benefit is limited. Contour furrows varying from 30-60 cm wide and 10-25 cm deep can be used. The shape varies from "V" to square, rectangular, or parabolic. The cross section and depth of furrows mainly depend on soil and equipment used for making them. Furrows spaced 8-10 m apart give a batter distribution of runoff water and higher yield of fodder. The effectiveness of contour furrows to hold water depends upon the degree of slope smoothness of the surface and accuracy in following contours and its life depends upon stability of soil and water storage capacity of the furrow, which can be estimated by following equation

$$Q = \frac{WxD}{100HI}$$

where, Q = Depth of runoff water stored in cm from unit area

W = width of furrow, cm

D = depth of furrow, cm

HI = horizontal spacing, m

Construction of contour furrow is always started from the ridge and progressively extended towards the valley. In a study conducted in arid part of Iran it is found that contour furrow and pitting has significantly helped in controlling soil erosion, increasing water penetration and soil moisture content and promoted propagation of *Hammada saliconica* species, a desirable plant species for both soil conservation and livestock grazing in the region (Jahantigh and Pessarakli, 2009).

Contour Trenches

Trenches are constructed at the contour line to reduce the runoff velocity for soil moisture conservation in the areas having <30% slope. Bunds are formed on the downstream side of trenches for the conservation of rainwater. A contour trench is a useful practice in forestry areas. This practice can be adopted in area which is unsuitable for cultivation but suitable for forestry. Normal standard size of a trench is 60 cm x 30 cm x 60 cm depth with an unexcavated portion 1.5 m after every 50-75 cm. Length spacing or vertical interval depends on the slope of land. Spacing may vary from 30-60 m. After the trenches are excavated to correct size, they are refill partially, and stocking the remaining excavated material as a small bund on the down-stream side. The storage capacity of trench is be estimated by the following equation

$$Q = \frac{WxD}{100HI\left(1 + \frac{X}{L}\right)}$$

where, Q = depth of runoff from area in cm
W = width of trench in cm
d = depth of trench in cm
HI = horizontal interval in meters
X = gap between the trenches in m
L = length of trench in m

Mane *et al.,* (2009) reported effectiveness of continuous contour trench for runoff control and recommended as best soil conservation practice on area having 7 to 8 per cent slope in konkan region of India.

Gradonies

Gradonies are steeply inward-sloping narrow bench terraces constructed on contours. Usually, gradonies are suitable for afforestation in uniformly steep sloping lands. Based on the steepness of slope, vertical interval is kept from 1.0 to 1.5 m. The width of gradonies also varies from 1.0 to 1.5 m. The material dug from the inner side is heaped on the outer edge in order to make a berm of about 20 to 30 cm high with an inward slope of 7.5:1. In the middle of the gradonies, pits of 50 cm x 50 cm x 50 cm are made at spacing of about 3 m. On an average 1000 trees are planted within the gradonies and interspaces in one hectare area (Mahnot and Singh, 1993).

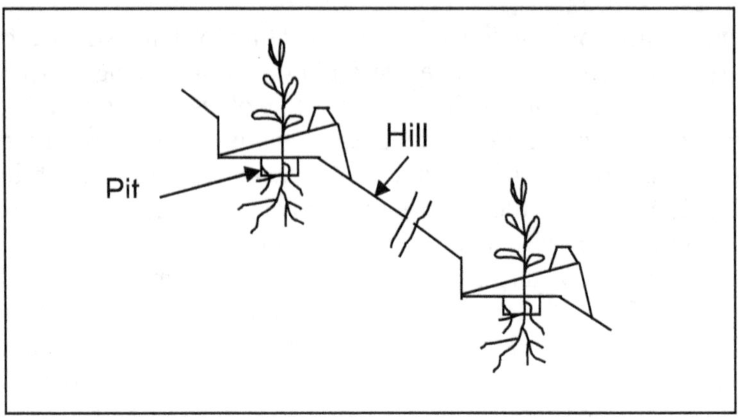

Figure 5. Cross-section of a Gradoni.

Gully Control Measures

About 4 million hectares in India are affected by gullies and ravines that threaten another 4-6 million hectares of productive tablelands. On marginal farmland, gully erosion is major source of soil and associated nitrogen and phosphorous losses (Sharpley *et al.,* 1996). However, with proper management, the gullies can be utilized to store runoff for groundwater recharge and human and livestock consumption. In western countries, construction of gully control works was initiated as far back as 1900. However, in India, such works have been taken up on an extensive scale since 1960 in Gujarat, Maharashtra, Madhya Pradesh and Rajasthan.

Anicut / Check dams

Check dams are masonry overflow barriers (weirs) constructed across seasonal streams. A check dam as such has a relatively limited storage capacity but a large volume of water can still be pumped from such storage as the stream continues to flow and the check dam serves the purpose of an ideal intake structure. A check dam, by storing the base flow, maintains a supply of water for recharge as well as for direct use beyond the monsoon period (Goyal and Narain, 2006). It creates flooding of upstream area, which requires surplusing arrangements at suitable intervals to drain water. Check dam should be avoided in isolation. The number of check dams primarily depends upon the slope of the gully and the quantity of runoff. It may not be advisable to construct check dams on bigger streams with high gradient and where runoff is very high. The bigger streams should be treated with drainage line treatment like gabionic structures and boulder checks with masonry work to curtail the runoff.

Gabionic cheek dams

Gabionic check dams are useful in a locality where stones are readily available and their irregular shape makes them unsuitable for making loose stone cheek dams. If the expected water velocity is very high, gabion is recommended in place of loose rock dams. A gabion is rectangular shaped cage made of galvanized wire, which is filled with locally available boulders, rocks or stones. The gabion may be conveyed flat and are folded to shape at the construction site. Usually, gabions are 1 m wide and 0.75-12 m high with varying lengths ranging between 2-10 m. The gabionic check dams are constructed by connecting several gabions in horizontal and vertical direction. The gabionic check dams are very stable and semi-permanent in nature. These structures are flexible; they may even change shape automatically according to the streambed, even when the bed shape changes due to erosion, without losing stability.

Figure 6. Check-dam at Beriganga research farm of CAZRI – Jodhpur.

Loose stone/dry stone masonry check dams (LSCD)

These structures are effective for checking runoff velocity in steep and broad gullies. These are suitable at upper reaches of the catchment. They have a relatively longer life and, usually require less maintenance. The bed of the gully is excavated to a uniform depth of about 0.3m. Stones are then hand packed from the foundation level. Flat stones of size 20-30 cm are the best for construction and laid in such a way that all the stones are keyed together. Large size stones are placed at the center of the dam and gaps between stones may be filled with small piece stones. The dam should go up to 0.3 to 0.6 m into the stable portion of the sides of the gully to prevent end cutting. In the center of the dam, sufficient spillway is provided to allow maximum runoff to discharge. LSCD constructed at 1 m V.L. in Jhanwar watershed area on 17 gullies proved to be very effective in controlling further extension of gullies (Goyal *et al.*, 2007). Vangani *et al.*,

(1998) reported sediment deposition of 3.86 t ha^{-1} yr^{-1} against loose stone check dam in Osian-Bigmi watershed (Distt. Jodhpur).

Figure 7. A series of Loose Stone Check-Dams (LSCD) at gully in Jhanwar Watershed (Jodhpur).

Brushwood check dams

Branches of tree and shrub species are staked in two rows parallel to each other filled with brushwood and laid across the gully or way of the flow. These are usually built to regulate the overflow in small and medium gullies which are supplemented with vegetative barriers for long term effectiveness. These check dams are constructed by using locally available brushwood and supported by wooden stakes and used in the small gully heads not deeper than 1 m. These check dam are of two types; single row post brush dam and double row post brush dam. Brushwood check dams are constructed in areas where wooden posts, brushwood etc., are available in plenty. These check dams can only be used in the small gully heads not deeper than 1m. Single row post brush dam are made of single row of wood stakes to which long branches of trees are tied length wise along gully with their butt ends facing upstream while in double row post brush dam, the straw and brushwood are laid across the gully between two rows of wooden posts, the distance between the rows being not more than 0.9 m. The longest branches are laid at the bottom and the shorter length branches are laid above it till the required dam height (0.3 to 0.7 m) is obtained.

Rainwater Harvesting for Arid Lands

Since primary source of water in arid zone is rainwater, so for any improvement in availability of water, catchment conditions become crucial. Catchment is the base for harvesting rainwater in form of runoff. Runoff is highly dependent on catchment's shape, size, slope, and type etc. beside rainfall characteristics. Goyal and Issac (2009) has estimated runoff coefficients for different catchments characteristics for hot arid zone of India. Some general modifications can greatly help in enhancing the runoff percentage from catchments.

Techniques for Enhancing Runoff from Catchments

1. Simple earth smoothing and compaction helps increasing runoff from catchment areas. Success is generally greater on loam or clay loam soils. Care must be taken to reduce the slope and/or the length of slope to lessen runoff velocity and thereby reducing runoff.
2. Small amounts of sodium salts - particularly NaCl, $NaHCO_3$ applied to desert soils where vegetation has been removed-causes dispersion of the surface soil, reducing infiltration and increases runoff. However, this type of treatment requires a minimum amount of expanding clays in the soil.
3. Removal of stones and boulder and unproductive vegetation from catchment helps in uninterrupted flow, enhances runoff to collection site.
4. Land shaping into roads and collection of water in channels.
5. Sandy soils have low water holding capacity. Spreading of clay blanket to the soil surface reduces the infiltration and consequently accelerates runoff.
6. Chemical treatments like wax, asphalt, bitumen and bentonite prevent downward movement of water, which augments runoff.

Rainwater harvesting through Tanka

Tankas (small tank) are underground tanks, found traditionally in most part of western Rajasthan. They are generally built in the main house or in the courtyard. *Tanka* is a circular hole made in the ground, lined with fine polished lime for collection of rainwater primarily from rooftop of individual house. The water collected in small *Tankas* is generally used for drinking purposes, however bigger *tanka* can be used for providing supplemental/life-saving irrigation to horticultural plants. During subnormal rainfall when tanka did not get adequate rainwater, water is hauled in camel/bullock cart from nearby wells/*nadis* to fill the household *tankas*. For rainwater management, CAZRI has designed underground *tanka* of 10 m^3 to 600 m^3 capacities for different rainfall and catchment conditions. These *tankas* were successfully constructed in Jhanwar, Sar, Baorali-Bambore (Distt. Jodhpur) and Kalyanpur (Distt. Barmer) villages. Harvested water of these tankas was used to provide lifesaving irrigation to plants. The Benefit cost ratio of *tanka* ranged from 1.25 to 1.40 under different uses (Goyal *et al*., 1995, 1997; Goyal & Sharma, 2000). The improved design of tanka has provision for inlet and outlet with silt-trap for control of silt inflow with runoff. The catchment of a *tanka* is made by spreading the excavated material around the structure. With a view to inducing the runoff, the catchment may be treated either with asphalt or sodium carbonate spray on soil surface. A uniform slope of 2-3 per cent towards *tanka* is provided for harvesting maximum possible runoff. The improved design of tankas have a lifespan of more than 20 years. The improved *tanka* design developed at CAZRI has wide acceptability in the region, which has been widely replicated in large numbers under Rajeev Gandhi National drinking Water Mission.

Figure 8. Improved *tanka* of 21 m^3 capacity at Baorali village (District Jodhpur).

Rainwater harvesting through Nadi

Nadis are village ponds used to store runoff water from adjoining natural catchments during the rainy season. In arid Rajasthan *nadi* system of water harvesting is the oldest practice and still the principal source of water supplies for human and livestock consumption. Across Rajasthan, most *nadis* have a capacity of between 1,200 to 15,000 m^3. Water availability in *nadi* ranged from 2-12 months after the rains. Since *nadis* received runoff from sandy and eroded rocky basins, large amounts of sediments used to deposit regularly in them, resulting in quick siltation. High evaporation and seepage losses through porous sides and bottom, heavy sedimentation due to biotic interference in the catchment and contamination are major bottlenecks. Evaporation losses ranged from 55 to 80% of the total losses in various environments. Seepage losses are greatest during the rainy season (July-September) when *nadi* is completely filled. To overcome these problems CAZRI has developed design for improved *Nadis* with LDPE lining on sides and bottom keeping surface to volume ratio 0.28 and provision of silt trap at inlet (Khan, 1989).

Figure 9. Improved nadi with inlet and outlet.

The site selection of *nadi* is based on availability of natural catchment and its runoff potential. The location of the *nadi* had a strong bearing on its storage capacity due to catchment and runoff characteristics. *Nadis* are 1.5 to 4.0 meters deep in dune areas and those in sandy plains varied from 3 to 12 meters. In addition, planting suitable tree species around the *nadi* creates an oasis in the desert and improves the local environment.

Rainwater harvesting through Khadin for crop production

Khadin is a unique practice of water harvesting, moisture conservation and utilization in hyper arid region of Rajasthan. This system was designed & developed by the Paliwal Brahmins of Jaisalmer (Rajasthan) in the 15th century. This system has great similarity with the irrigation methods of the people of Iraq around 4500 BC and later of the Nabateans in the Middle East. A similar system is also reported to have been practiced 4,000 years ago in the Negev desert, and in southwestern Colorado 500 years ago. The main feature of *khadin* is a very long (100-300 m) earthen embankment built across the lower hill slopes lying below gravelly uplands. Sluices and spillways allow excess water to drain off. The *khadin* system is based on the principle of harvesting rainwater on farmland and subsequent use of this water-saturated land for crop production.

The ratio of farmland and catchment areas is regulated to be about 1:10 so that a suitable moisture supply is uniformly maintained. It is suitable for deep soil surrounded by some natural rock outcrops constituting catchment area. CAZRI has developed *Khadin* of 20 ha areas in Baorali-Bambore watershed with surplussing arrangements. Before construction of *Khadin*, uncontrolled runoff from upper catchment used to wash away seeds, fertilizers, and standing crops besides loss of valuable water. After construction of Khadin, farmer could take excellent *Kharif* and *Rabi* crops (Narain and Goyal, 2005). Collecting water in a *khadin* aids the continuous recharge of groundwater aquifers. Studies of groundwater recharge through *khadins* in different morphological settings suggest that 11 to 48 per cent of the stored water contributed to groundwater in a single season.

This replenishment of aquifers means that subsurface water can be extracted through bore wells dug downstream from the *khadin.* The average water-level rise in wells bored into sandstone and deep alluvium was 0.8 metres and 2.2 metres, respectively (Khan, 1996).

Figure 10. *Khadin* with wastewier at Baorali-Bambore watershed (Jodhpur).

Rainwater harvesting thorough micro-catchment for tree establishment

Micro-catchment technique is particularly suitable for establishment of trees. In this technique a circular catchment of 1 to 1.5 meter radius is constructed around the tree. The catchment is compacted by roller or any other heavy machine. A slope of 5-10% is provided in catchment towards tree for directing flow of water. The catchment can also be lined with locally available materials such as polythene sheets, lime mortar, stone pieces, grass etc. for higher runoff generation. It is reported that the plants with micro-catchment have better chances of establishment in rainfed conditions as compared to conventional plantation technique (Ojasvi *et al.,* 1999).

In another study, Sharma *et al.* (1986) suggested that conversion of canopy area into runoff catchment may be just sufficient for improving its soil moisture profile.

Figure 11. Circular micro-catchment for tree establishment.

Agronomic Practices

All operations carried out in the field, from land preparation to crop harvesting, with the aim of increasing the crop yield are included under agronomical practices. Certain simple agronomical practices like optimum tillage, administration of organic manure, suitable cropping pattern, and strip cropping have been found to be effective in retaining soil fertility as well as giving satisfactory crop yield.

Tillage operation

Tillage (Ploughing) is the practice of breaking and working the soil to the desired depth prior to sowing. In this practice at least 30% of soil surface should remain covered with crop residue before and after planting the next crop to reduce soil erosion and runoff, as well as other benefits such as C-sequestration. Tillage makes soil loose and hence prone to erosion. Timing and depth of tillage are the two important factors, which need special attention. Tillage should be done immediately before the crop season to take advantage of one or two early showers for land preparation. In arid region, land tilled into ridges and furrows across the wind direction has been found to reduce the effects of wind erosion during the summer months. However excessive tillage before the monsoon lowers the percentage of clods and accelerated the wind erosion (Gupta, 1993). Therefore proper tillage is very important to take advantage of moisture conservation during rain and at same time should avoid soil erosion by wind.

Crop management

Row crops that are widely spaced are generally erosion-permitting. To reduce the erosion, the plants should be spaced in such a way as to obstruct the flow of water downhill by crop itself. A proper combination of raw to raw and plant to plant spacing goes in long in reducing the soil erosion beside enhancement in yield. For same plant population increase in raw spacing and corresponding decrease in plant to plant spacing help in creating mini barrier to control erosion by water and wind without adversely affecting the yield. Weeds consume water at much faster rate than crop because of smaller life cycle. Elimination of weed at proper time greatly reduces the competitions for water for the crop. Crops of moth and gaur when kept weed free till maturity produces higher seed yield whereas weed infestation after 30 to 40 days of sowing, resulted significant reduction in yield (Singh, 1984). Optimum plant population is important aspect of crop production in arid areas. Higher plant density enhances evapotranspiration losses relative to non-growth parameter of plant. Higher plant densities do not allow deep percolation of soil moisture as observed in vegetables grown under drip irrigation (Singh, 1978). In arid region, particularly rainfed condition, larger canopy growth may be disadvantageous as it may exhaust the available soil moisture from root zone during drought (Singh, 1977). So in such areas, one has to be more cautious in deciding optimal plant population and row spacing for sustainable crop production. Crop rotation is another practice to maintain the fertility of soil. Crop rotation not only helps to increase the crop productivity and soil fertility, but also improves the water and nutrients use efficiency by reducing weeds, providing conductive micro-climate for plant growth and development.

Strip cropping

Strip cropping for wind erosion control consists of alternate plantation of erosion susceptible and erosion resistant crop against prevailing wind direction preferably across the slope. The erosion resisting crops protects soil from beating action of raindrops, reduces runoff velocity, and thereby increased time of concentration which results in a higher volume of soil moisture and increased crop production

(FAO, 1989) In this system soil eroded from one strip is retained by the next strip and the overall fertility of the land is maintained. Narrow strips are more effective in reducing wind erosion in lighter soils. The width of the strip varies from 6 m in sand to 30 m in sandy loam. Establishment of strips of perennials like *Lasiurus sindicus* and *Ricinus communis* at right angle to the prevailing wind direction at CAZRI Farm - Jodhpur, reduced the impact and threshold velocity of wind to the minimum and checked the erosion. Consequently, crop grown in between the protective strips recorded increased production (Mishra, 1971). Reduction in sand drift due to protective strips of grass at Bikaner and Hingoli (Jodhpur) was also reported by Singh (1989). In another study the productivity and quality of fodder was increased by strip cropping of *C. ciliaris* or *L. sindicus* with *L. purpureus* in association with *C. mopane* or *H. binata* under silivi-pastoral system in arid zone (Patidar *et al.,* 2008). Another advantage of strip cropping is that it helps in the prevention of pest attack on the crops. Since pests are mostly crop specific, one particular strip affected by one particular pest remains confined within that strip itself and does not spread to the next strip, thus preventing the spreading of the diseases to the entire field.

Mulching

Mulching of open land surface is achieved by spreading stubble, trash or any other vegetation. The objectives of mulching are to minimize splash influence of rain drops on base surface; reduce evaporation; increase absorption of the rainfall; obstruct surface flow thereby retarding erosion and allow microbiological changes to occur at optimum temperature. Mulching prevents the formation of hard crust after each rain. Sometimes, spreading of organic residues, instead of mixing can help in reduction of soil and water loss to a considerable extent. Mulching also reduces the weed infestation along with the benefits of moisture conservation and soil fertility improvement. Polyethylene mulches have also been utilized for water harvesting and control of seepage. Trash farming, in which crop remains are cut, chopped and partly mixed in ground and partly left on land surface, is also a form of mulching. Studies conducted at CAZRI have shown that use of organic mulches reduced the maximum soil temperature at 10 cm depth by 1 to 6°C with suppressed weed growth and increase in soil moisture status. Similarly, the application of grass mulch at the rate of 6 t ha^{-1} resulted in the reduced mean maximum soil temperature, reduced evapotranspiration and consequently increase of 40% yield of green gram (Gupta, 1978, 1980).

Figure 12. Grass mulch for evaporation control.

Conclusions

Worldwide arid zones have witnessed increase in human and livestock populations like any other region as a result of natural increase. The increased population pressure has stressed the limited natural resources of this fragile eco-system. Since good lands are already under intense cultivation so focus has shifted towards arid zone all over the world for the crop production. The scientific approach towards understanding the problems and constraints of arid region and adoption of proper soil and water conservation technology can greatly help in achieving the goal of meeting the aspiration of its dwellers on sustainable basis. Healthy soil and availability of water are vital for productivity in all kinds of terrestrial ecosystems because plants require fertile soil with improved bio-physico-chemical properties and good quality of water for their growth and development. Use of soil and water conservation measures including biological (agroforestry and agricultural) and mechanical measures (terracing, bunding, trenching, check dams, etc.) is imperative to reduce runoff, soil erosion and to improve soil quality, water quality, moisture conservation, and overall crop productivity in a sustainable way. Biological measures are economically feasible and environmental friendly; also improve soil properties along with the conservation of soil and water resources.

A successful application of any soil and water conservation measures for improving production requires an integrated approach. In arid areas, conditions vary too much soil, climate, social factors and the list is endless so no two areas are identical, and therefore proper selection of technology for area specific is must. While designing any structure for area specific its cost-effectiveness should be kept in mind. It is generally cheap to repair a structure rather than designing a robust costly structure for extreme events of 100 years. All the technologies discussed above are essentially site specific and different components need to be integrated as a holistic approach to maximize production on sustainable basis. These technologies are time tested and are of proven soundness for extreme conditions such as of arid.

Watershed Management & Strategic Planning for Resource Conservation

The overall goal is alleviation of poverty & upgrading of living standards by means of sustainable development of water & water resources and conservation of the environment. It also incorporates the three following important aspects: (i). Manage, develop and protect water and related resources to meet needs of current and future generations. (ii) Operate, maintain and rehabilitate facilities safely, reliably efficiently to protect the public investment. (iii) Enhance the organizational effectiveness of the water resources coordination system, & promote capacity building. So, the burgeoning population, food insecurity and natural resource degradation are the major issues in the present era of climate change. It has been projected that the world population will be ~10 billion in 2050 (UN, 2017). Therefore, a paradigm shift in soil and water conservation, and its management is needed for agricultural sustainability where watershed management can play a major role. The some of the future concern for soil and water conservation and sustainable agriculture are the following:

- Development of in-situ rainwater harvesting techniques that require less labor and maintenance.
- Development of the cost-effective conservation practices to restore degraded lands and to sustain agricultural productivity.

- Improvement in traditional water harvesting systems.
- The efficient technologies for soil and water conservation should be demonstrated on farmers' fields with their active participation.
- Exploring the potential of arid zone agro forestry as a tool for solving environmental problems. (i) Organic farming in barren and culturable arid lands, and (ii) Promoting eco-tourism.
- Formulation of new policies and development of new technologies based on cultural, social and economic aspects of arid zone.
- Implementation and adoption of effective conservation measures for sustaining agricultural productivity.
- Adoption of efficient management practices and judicious use of soil and water resources.
- Existing soil and water conservation practices should be improved and developed based on the level of natural resources degradation.
- Greater emphasis should be given on participatory approach for effective soil and water conservation.
- Post-impact assessment and monitoring of soil and water conservation measures should be done to evaluate their efficacy in increasing productivity, monetary returns, and livelihood of the stakeholders.
- Emphasis on research, education and extension of soil and water conservation effective technologies to the stakeholders.

References

FAO (1984). Guidelines: Land Evaluation for Rainfed Agriculture. Rome: FAO Soils Bulletin; p. 52

FAO (1989). Arid zone forestry: A guide for field technicians. FAO Conservation Guide – 20.

Faroda, A.S., D.C. Joshi and Balak Ram (1999). Agro-ecological zones of North-western hot arid region of India. Central Arid Zone Research Institute- Jodhpur, pp 24.

Ganguli, J.K. and R.N.Kaul (1969). Wind erosion control. ICAR, New Delhi. *Techn. Bull.* (*Agric.*) 20:1-53

Gaur, Mahesh K. & Squires, Victor R. (Eds.) (2016a). Climate Variability, Land Use change and Impact on Livelihoods in the Arid Lands. Springer.

Gaur, Mahesh K. & R. Squires, Victor (2016b). Geographic extent and characteristics of the world's arid zones and their peoples (Eds. Mahesh K. Gaur and Victor R. Squires: Climate Variability Impacts on Land Use and Livelihoods in Drylands). Springer International Publishing. Pp. 3-20

Ghose, B., B. Singh and Amal Kar (1977). Desertification around the Thar – A Geomorphological Interpretation. *Annals of Arid Zone* 16:290-301.

Goyal, R.K. and A.K. Sharma (2000). Farm Pond: A well of wealth for the dryland dwellers. *Intensive Agriculture*, Vol. XXXVIII, No. 5-6, pp. 12-14 & 26.

Goyal, R.K. and K.P.R. Vittal (2008). Water Resource Management in Hot Arid Zone of India. *Journal of Hydrological Research and Development.* Volume 23, pp. 37-54.

Goyal, R.K. and Pratap Narain (2006). Impact evaluation of watershed conservation measures in hot arid zone of India. *In:* Conference on Natural resources management for sustainable development in Western India (NRMSD-2006) Organized by Indian Association of Soil & Water Conservationist, Dehradun –Pune 11-13 Oct. 2006. pp.77-79.

Goyal, R.K. and V.C. Issac (2009). Rainwater Harvesting Through Tanka in Hot Arid Zone of India. Central Arid Zone Research Institute, Jodhpur, CAZRI Bulletin, pp. 33.

Goyal, R.K., P.R. Ojasvi and J.P. Gupta (1997). Farm ponds: A key to survival in arid areas. *ASIAN WATMANET*, newsletter. No. 11, pp.12-13.

Goyal, R.K., P.R. Ojasvi and T.K. Bhati (1995). Economic evaluation of water harvesting pond under arid conditions. *Indian Journal of Soil Conservation* 23(1): 74-76.

Goyal, R.K., T.K. Bhati and P.R. Ojasvi (2007). Performance evaluation of some soil and water conservation measures in hot arid zone of India. *Indian Journal of Soil Conservation* 35 (1): 58-63.

Gupta, J. P., G. G. S. N. Rao, Y.S. Ramakrishana and Rammana Rao (1984). Role of shelterbelt in arid zone. *Indian Farming* 34(7): 29-30. Gupta, J.P. (1978). Evaporation from a sandy soil under mulches. *Annals of Arid Zone.* 17(3):287-290.

Gupta, J.P. (1980). Effect of mulches on moisture and thermal regimes of soil and yield of pearl millet. *Annals of Arid Zone.* 19:132-138.

Gupta, J.P. (1993). Wind erosion of soil in drought prone areas. (In)Desertification and its control in the Thar, Sahara and Sahel Regions. (Eds. A.K. Sen and Amal Kar), Scientific Publishers, Jodhpur; pp. 91-105.

Gupta, J.P. and S.S. Rathore (2002). Biomass production and rehabilitation of degraded lands in arid zone. pp. 20.

Halvin, J.L. and A.J. Schlegel (1997). Dryland conservation technologies: Enhancing agricultural profitability and sustainability. *Annals of Arid Zone* 36(3): 291-303.

Hudson, Norman W. (1987) Soil and water conservation in semi-arid areas. FAO Soils Bulletin-57

Jahantigh Mansour and Mohammad Pessarakli (2009). Utilization of Contour Furrow and Pitting Techniques on Desert Rangelands: Evaluation of Runoff, Sediment, Soil Water Content and Vegetation Cover. *Journal of Food, Agriculture & Environment* Vol.: 7, No. 2, pp. 736-739.

Kar, Amal, P.C. Moharana, P. Raina, Mahesh Kumar, M.L. Soni, P. Santra, Ajai, A.S. Arya and P.S. Dhinwa (2009). Desertification and its Control Measures. *In:* Trends in Arid Zone Research in India (Eds. Amal Kar, B.K. Garg, M.P. Singh and S. Kathju), Central Arid Zone Research Institute, Jodhpur, pp.1-47.

Khan, M. A. (1996). Inducement of groundwater recharge for sustainable development. In 28th Annual Convention, Indian Water Works Association, Jodhpur, India. pp.147-150.

Khan, M.A. (1989). Upgraded village pond-nadi to ensure improved water supplies in arid zone. *Water and Irrigation Review*, Israel, 19:20-23.

Mahnot, S.C. and P.K. Singh (1993). Soil and Water Conservation. Intercoperation Coordination Office. Jaipur. p .90.

Malagnoux, M., E.H. Sène and N. Atzmon (2007). Forests, trees and water in arid lands: a delicate balance. *Unasylva* (An international journal of forestry and forest industries) 229, Vol. 58, pp.24-29.

Mane, M.S., U.V. Mahadkar; B.L. Ayare, and T.N. Thorat (2009). Performance of Mechanical Soil Conservation Measures in Cashew Plantation Grown on Steep Slopes of Konkan. *Indian Journal of Soil Conservation* Vol.: 37, No.: 3, pp.181-184.

Mishra, D.K. (1971). Ministry of Education, Govt. of India. New Delhi. Agronomic Investigations in Arid Zone. *In:* Proceedings of Symposium on Problems of Indian Arid Zone, pp.165-169.

Narain, P., and R.K. Goyal (2005). Lead paper on 'Rainwater harvesting for increasing productivity in arid zones'. National Symposium on "Efficient water management for eco-friendly sustainable and profitable agriculture". Symposium Abstract. Organized by Indian society of water management and Indian agriculture research institute from Dec. 1-3, 2005 at Water Technology Center, New Delhi. pp.141-142.

Narain P., and R.K. Goyal (2009). Arid zone management strategies in India with special reference to resource mobilization and outsourcing. In: Converting Deserts into Oasis (Eds. J.S.P. Yadav, R.K. Singh and V.P. Gupta). Pub. Stadium Press LLC, USA. pp.176-185.

Ojasvi, P.R., R.K. Goyal and J.P. Gupta (1999). The Micro-catchment Water Harvesting Techniques for the plantation of Jujube (*Zizyphus Mauritania*) in an agroforestry system under arid conditions. *Agricultural Water Management* 41 (1999), pp. 139-147.

Patidar, M., M.P. Rajora and Raj Singh (2008). Forage Production Potential of Different Silvi-pastoral Systems under Arid Conditions of Rajasthan. *Indian Journal of Agronomy* Vol. 53, No. 3

Rao, A.S. (2009). Climatic variability and crop production in arid western Rajasthan. In: Trends in arid zone research in India (Eds. Amal Kar, B.K. Garg, M.P. Singh and S. Kathju). Published by Central Arid Zone Research Institute- Jodhpur. pp.41-61.

Rao, A.S. and R.S. Singh (1998). Climatic features and crop production. *In:* Fifty years of arid zone research in India (Eds. A.S. Faroda and Manjit Singh). Published by Central Arid Zone Research Institute-Jodhpur. pp.17-38.

Sharma, K.D., N.L. Joshi, H.P. Singh, D.N. Bohra, A.K. Kalla and P.K. Joshi (1999). Study on the performance of contour vegetative barriers in an arid region using numerical models. *Agricultural water management.*41: 41-56.

Sharma, K.D., O.P. Pareek, and H.P. Singh (1986). Micro-catchment water harvesting for raising jujube orchards in arid climate. *Transaction of American Society of Agriculture Engineering*. 29(1):112-118

Sharpley, A., S.J. Smith, J.A. Zollweg, and G.A. Coleman (1996). Gully treatment and water quality in Southern plains. *J. Soil and Water Conserv.* 51:498-503.

Singh, A.K., A.K. Kumar, V.S. Katiyar, K.D. Singh, and U.S. Singh (1997). Soil and Water Conservation Measures in Semi-arid Regions of South-Eastern Rajasthan. *Indian Journal of Soil Conservaton,* 25 (3): 186-189.

Singh, G, C. Venkatramanan, G. Sastry and B.P. Joshi (1990). Manual of Soil and Water Conservation Practices. Oxford and IBH Publishing Co. Pvt. Ltd. New Delhi, p385.

Singh, K.C. (1984). Conservation farming in arid areas. *In:* Agro forestry in arid and semi-arid zones (Ed. K.A. Shankarnarayan). CAZRI Publication No. 24, pp. 222-226.

Singh, K.C. (1989). Importance of grasses in management of sandy soil-strip cropping of grasses. International symposium on managing sandy soils, CAZRI, Jodhpur pp. 561-564.

Singh, S.D. (1977). Optimum utilization of limited water supply. *In:*

Desertification and its Control, CAZRI, Jodhpur pp. 237-255.

Singh, S.D. (1978). Effects of planting configuration on water use and economics of drip irrigation system. *Agronomy Journal* 70: 951-954.

Smaling E. M. A., S. M. Nandwa, and B. H. Janssen (1997). Soil fertility in Africa is at stake. In R. J. Buresh, P. A. Sanchez and F. Calhoun (eds.), Replenishing Soil Fertility in Africa. SSSA Spec. Publ. 51. Madison, Wisconsin: Soil Science Society of America.

Smith, D.D. and W.H. Wischmeier (1962). Rainfall erosion. *Advances in agronomy*.14:109-148.

Tiwari, S.P. and R.S. Kurothe (2006). Effect of vegetative barriers on soil and nutrient losses at 2% slope on agricultural lands of reclaimed Mahi ravines. *Indian Journal of Soil Conservation,* 34(1): 37-41.

United Nations. (2017). World Population Prospects: The 2017 Revision, Key Findings and Advance Tables. New York: United Nations; 2017

UNEP (United Nations Environment Program) (1997). World atlas of desertification 2ED. UNEP, London.

Van der Pol, F. (1992). Soil mining: An unseen contributor to farm income in southern Mali. Bulletin no. 325. Amsterdam: Royal Tropical Institute.

Vangani, N.S., Surendra Singh and R.P. Sharma (1998). Index catchment - a new concept for sustainable integrated development. *Annals of Arid Zone* 37 (2):133-137.

Wasi-Ullah, A. K. Chakravarty, C. P. Mathur, N. S. Vangani (1972). Effect of contour furrows and contour bunds on water conservation in grasslands of western Rajasthan. *Annals of Arid Zone* 11(3/4):1709-182.

Yousuf, A, Singh, M. (2019). Watershed Hydrology, Management and Modeling. Florida: CRC Press.

5

Bio-Industrial Watershed Model: A Climate Resilient Unit for Sustainable Production and Profitability in Semi Arid Tropical Regions of India

K. Sreenivas Reddy, V. Maruthi and Prabhat Kumar Pankaj

Introduction

Food security refers to the availability of food and one's access to it. Food security exists when all people, at all times, have physical, social and economic access to sufficient, safe and nutritious food to meet their dietary needs and food preferences for an active and healthy life. Food security is built on three pillars: Food availability, food access and food use. India's current population of 1.3 billion (130 crores) is likely to grow to 1.35 billion (135 crores) by 2023 and demands 23% increase in agricultural production. Land and water are the two important natural resources to achieve demand for food production. India losses more than Rs. 58000 crores worth the agricultural produce due to lack of post harvest and food processing infrastructure (Sharma and Sharma, 2009). Loss in food grain production in India due to contemplated diversification has to be compensated elsewhere to maintain integrity of national food security. Rainfed agriculture being practiced on 55% of cultivated land, supporting 40% of population, 60% of livestock and 40% of food is having significantly low productivity as compared to assured irrigated farming and has some unexploited potentials (NRAA, 2011). In-situ conservation, recharging of ground water and runoff harvesting especially in undulating topographies should be the main driver for improving productivity and cropping intensity. The food security in rural villages can be achieved through bio industrial watershed management. The economic and social benefits of watershed management could be upscaled further by bringing industry to the watersheds and transforming them to the industrial watershed (Bali, 2008). The bio-industrial watershed management involves mobilizing the power of partnership among local communities, government departments, financial institutions and technological experts (Singh and Khera, 2011). Various components of bio-industrial watershed development include, participatory technology development in the management of land, water, cropping system, energy and value addition and processing, inducing diffusion through horizontal learning of technologies by farmers and promoting micro-credit for micro-enterprise by exploiting opportunities arising from farm and non-farm sectors. Basic principle of agriculture and processing is protection of ecology which needs to me matched with country's profile (Table 1). It means agro-forestry, biodiversity, soil and water conservation, and management. Only on this foundation, sustainable and balanced systems of production, based upon plants and animals, can be built. Once a balance has been restored to ecology and production systems, a stage is set for bio-industries, for value addition and profitable marketing.

Table 1. India: Country profile

Geographical area	329 Mha
Flood prone area	40 Mha
Utilizable surface water	690 km^3
Extractable Groundwater	432 km^3
Current per capita water availability	1700 m^3
Projected 2025 per capita water availability	1000 m^3

Source: ACARA Challenge, 2011

Food Security

It is estimated that the total food production in India in 2010-11 is 232.07 Mt. with wheat (81.47 Mt), rice (94.01 Mt), pulses (16.51 Mt) and cotton 33.9 Mt. The Indian population is going to increase 1.35 billion by 2023. So, the present food grain production will not be sufficient to meet the food demand with present rate of production and productivity growth rate (Table 2). The per capita nutritional food is declining due to severe drought or disease destroys the harvest crop and poor infrastructure keeps food from reaching remote areas. Poor economy in the rural level has left people out of work and unable to buy good food and most of the Indian farmer's plot is not large enough to feed his family with the tools he has. The majority of the village farmers receiving food aid in refugee camps are unfamiliar with cooking methods, so they don't eat it and the cooking water is contaminated in many parts of the country due to sever flood and industrial activities in and around the country. All these problems will directly reduce the per capita food availability in the country.

Table 2. Three-year averages of area, production, productivity and growth rate of food grain 1984-86 to 2009-10

Year	Area (Mha)	Production (Mt)	Productivity (t/ha)	Compd. Growth Rate %		
				Area	Production	Productivity
1984-86	128.62	149.45	1.16	- 1.20	-0.64	0.43
1987-89	124.85	151.23	1.21	0.18	8.85	8.97
1990-92	125.49	171.94	1.37	- 1.95	-0.78	1.48
1993-95	123.25	185.08	1.50	0.29	3.29	3.06
1996-98	122.81	190.71	1.56	1.17	3.23	1.99
1999-01	123.11	203.41	1.65	- 1.66	-1.68	0.00
2002-04	120.03	200.27	1.66	0.27	0.08	-0.29
2005-07	121.77	208.08	1.70	1.53	4.66	2.99
2007-09	123.67	227.31	1.83	- 0.20	3.75	3.92
2009-10	-	218.20	-	-	-	-

Source: Agricultural Statistics at a glance. 2009, DAC, Min of Agriculture, Govt. of India

Food Production Challenges: The food production mainly depends on energy and water intensive. In the country wide, the fertile crop land is declining day by day which influence on food declining productivity. Still many farmers in the country are following the mono culture cropping system and another important fact is that, mechanization is not reaching the small level farmers. Feeding the huge population which is manifested with the annual demand of food crops (Table 3), it's a big challenge before the researchers and farmers to meet the expectations of the society. Apart from grains, we need to have matching growth in milk production, meat production and other sub-sectors too.

Table 3. Annual demand for food grains as household food and other uses (in million tonnes)

Food Item/Type of Demand	**Base Year 2004-05**	**By 2011-12**	**By 2020-21**
Cereals (i). Direct Demand as household food	151.7	159.1	166.6
(ii). Indirect food demand and other uses	41.1	59.8	94.9
Total Cereals Demand	192.8	218.9	261.5
Pulses (i). Direct Demand as household food	9.8	11.8	12.5
(ii). Indirect food demand and other uses	4.4	4.3	6.6
Total Cereals Demand	14.2	16.1	19.1
Food Grains (i). Direct Demand as household food	161.5	172.5	187.4
(ii). Indirect food demand and other uses	45.5	64.1	101.5
Total Food Grain Demand	207.0	235.0	280.6

Source: Agricultural Statistics at a glance. 2009, DAC, Min of Agriculture, GOI

Requirement of cereals and pulses for household consumption and other uses at the end of 11th Plan (2011-12) and 2020-21 is expected to be 235 and 280.6 million tones, respectively. Apart from this, India lags far behind in terms of quality protein intake at 20% which needs to be taken care by making available protein rich food products such as eggs, meat, fish, chicken, etc. at affordable prices. India needs to bring out the methods to improve the availability and affordability of protein rich food and food products using the latest environmentally friendly technology without creating further pressure on land and water.

Importance of Food Security: The food insecurity is closely tied to other issues and chronic undernourishment can lead to stunted mental and physical development. This can lead to diminished capacity for work, which impacts not only individuals and families but entire economies. Environment: low-yield production methods increase the need to turn more wild spaces into farmland.

Water Security: Today ground system is the largest, covering - 160 million ha of cultivated land in India with 39 million ha irrigated by ground water, 22 million ha by irrigated canals and about two third of cultivation in India is still depending on monsoon. The water security exists in the country when all the people have access to enough clean water and potable water for their daily life. Threats to water security include drought, flooding, pollution, and overuse of water for agricultural or industrial purposes. In India, proper sanitation is a component of water security; improper removal of waste can contaminate drinking water. There is an inefficient and inequitable use of and distribution of water. Nearly 90% of the Indian population lives in areas with some form of water stress or food production deficit. Ground water has been relatively abundant in most parts of India. However, in some regions, it is becoming one of the most serious resource issues.

Table 4. Projected water demand (BCM)

Sector	Projected water demand (BCM)	
	2010	2025
Irrigation	688	910
Drinking	56	73
Industry	12	23
Energy	5	15

Source: ACARA Challenge, 2011 (BCM-Billion Cubic Meter)

On an average, India receives an annual rainfall of 1167 mm which adds up to an estimated 4000 billion cubic meter (BCM) of water in the country (Majumdar, 2002). Of the total, nearly 100 BCM is contributed by snowfall (Kumar et al., 2005). Annual utilizable surface water in the country is about 690 BCM which can be enhanced considerably through inter-basin transfers from Brahmaputra to Ganga basin (Menon, 2002). The quantum of blue water used from reservoirs and tanks and supplied through canal networks is 300 BCM, and 130 BCM water is made available through pumping from groundwater aquifers. Thus, the total availability of blue water to agriculture is just about 10% of the total precipitation (Majumdar, 2002; Kumar *et al.*, 2005; Gandhi and Bhamoriya, 2011). Average supply of green water, estimated at 107 BCM annually, is about 26% of the total precipitation received over the mainland (Gandhi and Bhamoriya, 2011).

India currently stores only about 6% of its total annual rainfall in large and medium reservoirs (253 BCM) to provide irrigation support for Rabi season crops and meet exigencies of monsoon aberrations during *kharif* season. Currently, *kharif* food grain crops occupy more area (about 14 Mha) than the *rabi* food grain crops, indicating strong dependence of agriculture on monsoons. Yet, the total food production in the two seasons is almost equal. Rabi food production boost is consistent with the irrigated area expansion of about 30 Mha. In most Indian states, 50–80% of the irrigated areas rely on use of groundwater (Milesi *et al.*, 2010; Naraynamoorthy, 2002; Shah *et al.*, 2006), even though one-third of them have issues related to water quality management. Monsoonal pattern of rains has permitted the use of relatively high-salinity groundwater in the post rainy season crops in many arid and semi-arid regions of the country (Minhas and Gupta, 1992; Gupta *et al.*, 1994). With current crop-water management practices, freshwater needs of Indian agriculture by 2030 have been projected around 1500 BCM as against the current water availability of 740 BCM (Addams *et al.*, 2011). This projected water demand is close to total availability of blue (430 BCM) and green water (1070 BCM).

Clean water is not only basic necessity for drinking, but also important for industrial, agricultural, and energy input. Water-borne diseases are responsible for millions of deaths and illnesses every year 1.4 million children die yearly. Disputes over water rights and water usage may escalate into conflicts between neighboring nations. The food and water security are the two major issues for the developmental country. The present practices of agricultural production will not meet the future food demand. This requires the proper usage of available natural resources in rural level where 70 percent of the people are living. There is a need to educate the people in village level on water and food issues and it is very important to reach the small-scale industries to rural farmers. The food and water security can be achieved by adapting the bio industrial watersheds in village level.

Bio-industrial Watershed

In simple term bio-industrial management is defined as bringing the small-scale industries to village level for proper production and processing. In rainfed agriculture, it is very important to achieve food and water security within available natural resources. In the country, many of the watershed programmes are running it is urgent need to convert to bio-industrial watersheds to achieve per capita food and water security in watershed level. It would be prudent to have all the central and state government's watershed management programmes converted in to bio industrial watersheds. There are a number of watershed level activities related to agriculture, horticulture, agro forestry, livestock, fisheries, poultry, sericulture, mushroom farming, honey bee keeping and vermicomposting etc. that could form agri-enterprises for bio industrial watershed management. We need to choose such enterprises suiting to varied socio-economic environmental conditions of different rural areas for greater livelihood, nutritional and environmental security. The new system would have the bio produce processed and value added at the watershed level, thereby, bringing urban profits of industry and commerce to the rural poor.

Components of a Bio-industrial Watershed

The bio-industrial watershed includes three main components namely protection, production and processing.

Protection of natural resources

Natural resources (soil, water, plant and animal diversity, vegetation cover, renewable energy sources, climate, and ecosystem services) are the fundamental for the structure and function of agricultural systems and social and environmental sustainability. However, the basic problem of dryland areas is one of a vicious cycle that starts with degradation of the natural resource base through poor management leading to low productivity and low income (Sharma *et al.*, 2012).

In watershed, a natural resource such as land and water must be protected from extreme events for agricultural production trough agronomic and engineering measures. The agronomic measures would improve the green water for crop production whiles other hand engineering structures would improve availability of blue water for supplemental irrigation in critical stages. The conjunctive use of both green and blue water would improve the water productivity in watershed areas and it is very important to use the available water in judiciously and scientifically. The input banks for crop production in watershed areas is very important ensure sustainability in bio industrial watershed.

Green water management

Agriculture depends on irrigation water supplies provided either through rainwater management (green water) or by withdrawing water from surface or the groundwater (blue water). Green water has two components namely (i) the green water storage (soil moisture storage) and (ii) the green water flow (evaporation plus transpiration). Since moisture retained in the soil is the reservoir of green water supplies, it is a manageable water resource with management control often resting with the farmers. Conservation tillage (CT) covers a spectrum of non-inversion practices from zero-tillage to reduced tillage; it aims to maximize soil infiltration and soil productivity and minimize water losses while conserving energy and labour. In general, though, increasing the residence time of runoff flow in a watershed, e.g. through practices such as water harvesting and CT, may have positive environmental, as well as hydrological, implications downstream. The hydrological implications at watershed and river-basin level of scaling up system innovations, such as water harvesting, are still unknown and require further research (John Rockström *et al.*, 2003). Contour faming is a traditional method of plowing, planting and cultivating the crop across the slope instead of up and down the slope. It is most effective moisture conservation practice used by farmers from around the world. Mulching is the process or practice of covering the soil/ground to conserve the soil moisture for longer duration and make more favorable conditions for plant growth, development and efficient crop production. While natural mulches such as leaf, straw, dead leaves and compost have been used for centuries, during the last 60 years the advent of synthetic materials has altered the methods and benefits of mulching. The mulching not only conserves the moisture but also reduces the weed growth.

Ridge and furrow practices — Contour bunding — Broad bed furrows — Contour farming

Figure 1. Different types of *in situ* soil moisture conservation techniques in rainfed areas.

Deep tillage is the farming operation, performing tillage operations below the normal tillage depth to modify the physical or chemical properties of a soil. Deep tillage is mainly used for increasing the water holding capacity there by conserve the moisture for longer time. This practice applies to land having adverse soil conditions which inhibit plant growth, such as compacted layers formed by field operations, restrictive layers such as claypans, over wash or deposits from wind and water erosion or flooding, or contaminants in the root zone. The tractor operated tillage equipments such as chisels, subsoilers, bent-leg sub-soilers, or rippers are available to perform deep tillage. Broad bed furrows and ridge and furrows are used to conserve soil moisture and to control erosion to in the soil during rainy days. The broad bed and furrow system are laid within the field boundaries. The land levels taken, and it is laid using either animal drawn, or tractor drawn ridgers. Both acts as a drainage channel during heavy rainy days. Loss of biodiversity and ecosystem functioning due to agricultural intensification and deforestation is a major issue of threat in the SAT regions. Hence, alternate land use system has immense importance in the tropics especially in drylands areas. In the present study of agri-horti-silvi-pasture tree growth conserves soil moisture, improves soil fertility and protects crops against scorching and desiccating effects of wind. The trench cum bund also has helped to conserve soil and moisture during rainy period and led to 55% survival of plantings in spite of long drought spells.

Blue water management

In the rainfed areas, the generated run off can be directly collected in dug out or farm ponds and that water can be efficiently used as supplemental irrigation through drip or sprinkler/ rain-gun systems while in critical stages of crop period which will lead to high water productivity. Fig 2 illustrate the potential excess water available for harvesting as runoff in the states of the semi arid tropics in India.

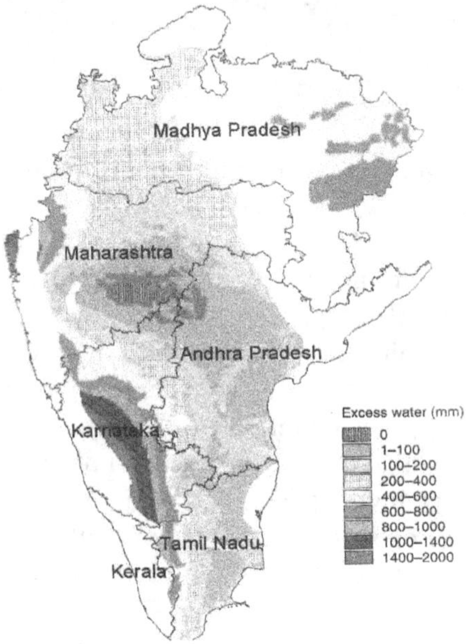

Figure 2. Excess water available for harvesting as runoff in the states of the semi-arid tropics, India (June-October).
(Sources: S. P. Wani, *et al.,* 2003. CAB International 2003. Water Productivity in Agriculture: Limits and Opportunities for Improvement)

The Nalla bunding, check dams, silt detention tanks, and percolation ponds are used to store the generated runoff. These structures will effectively reduce the soil erosion and are used for ground water recharge.

Figure 3. Farm pond hydrology and water application system in rainfed area.

The research conducted at ICAR-CRIDA research farm in alfisols indicated that there is ample scope for increasing water productivity in rainfed groundnut and Okra vegetable with supplemental irrigation integrated with soil amendments like tank silt application. The rain water harvesting gives high agriculture returns and it is a potential solution to problems of rural poverty and unemployment, resulting in an overall improvement in the nation's economy. Local people can be easily trained to implement such technology and construction methods. RWH is a convenient in the sense that it provides water at the point of consumption, which greatly reduces the operation and maintenance problem. It is sustainable due to decentralization and community participation.

Grey water management

All household wastewater, except toilet waste, is called grey water. This includes water from showering, bathing, and washing dishes and clothes. This grey water can be reused through available modern technologies. The treated grey water can be used for irrigation, ground water recharge. Grey water can be effectively used when it is used combination with surface water.

Technologies to Use Polluted Water in Agriculture

Saline water use in agriculture: Management practices for the safe use of saline water for irrigation primarily consist of selection of crops or crop varieties that will produce satisfactory yields under the existing or predicted conditions of salinity or sodicity, special planting procedures that minimize or compensate for salt accumulation in the vicinity of the seed, irrigation practices to achieve periodic leaching of the soil, land preparation techniques to increase the uniformity of water distribution and infiltration, leaching and removal of salinity, special treatments (such as tillage and additions of chemical amendments, organic matter and growing green manure crops) to maintain soil permeability and tilth.

Wastewater use in agriculture: The use of sewage sludge and effluents from wastewater treatment plants for irrigation of agricultural lands is on the rise particularly in peri-urban areas of developing countries. The reuse of nutrients and organic matter in treated wastewater and sewage sludge via land application is a desirable goal. Procedures for preliminary treatment of wastewater have been developed to make wastewater suitable for irrigation. Success in using treated wastewater for crop production will largely depend on adopting appropriate strategies aimed at optimizing crop yields and quality, maintaining soil productivity and safeguarding the environment. Several alternatives are available, and a combination of these alternatives will offer an optimum solution for a given set of conditions. The user should have prior information on effluent supply and its quality to ensure the formulation and adoption of an appropriate on-farm management strategy. In many parts of the world scientific information on components of an on-farm strategy in using treated wastewater which includes crop selection, selection of irrigation method, and adoption of appropriate management practices has been standardized. Furthermore, practices of blending conventional water with treated effluent, and using the two sources in rotation have also been standardized.

Production System

The sustainability of dryland area is endangered due to over exploitation of natural resources beyond their carrying capacity (Katyal and Das, 1992). Drylands have harsh environment (edaphic, biotic and abiotic) and suffer from a number of constraints such as low and uncertain rainfall, limited irrigation, low moisture, poor and degraded resource base, declining soil fertility, low income and low saving capacity, low productivity, and inadequate attention paid by policy makers and scientists, etc. An ideal system for SAT regions should have a judicious mix of crops, trees, and grasses only then the natural resources will be judiciously utilized and returns maximized without any detrimental effect to environment. Diversified farming system involving crops, fruit trees, agroforestry, animal husbandry and farm mechanization, is the only option for sustaining the livelihood and pave the ways for profitable agriculture in dryland. In this direction, agri-horti-silvi-pasture provides plenty opportunities of not only preventing further degradation of marginal lands and also obtain additional income (Pratap Narain, 2008). The promotion of dryland agri-horti-silvi-pasture trees has created great awareness among the farmers about the role of trees in improving soil and water conservation besides helping in improving micro climate and carbon sequestration in the long run. Further, the integration of fodder trees would provide the much-needed top feed for sustenance of livestock during lean period. In bio-industrial watershed production systems includes integrated management of agriculture, agri-horticulture, agri-forestry, agri-pasture, silvi-pasture and livestock. Instead of going for only agriculture production it is better to go for integrated management of available natural resources in sustainable manner. Figure 4 illustrates the Agri-horti systems in village watershed.

Figure 4. Combination of Agri-Horti-Forestry production systems in bio industrial watershed.

Indian agriculture is labour oriented and requires lot of man-power and energy but even after this hard work farmers are not able to earn their livelihood, especially small farmers because there is very little left after they pay for all inputs (seeds, livestock breeds, fertilizers, pesticides, energy, feed, labour, etc.). To fulfill basic needs of these farm families including food (cereal, pulses, oilseeds, milk, fruit, honey, meat etc.), feed fodder, fibre and fuel warrant an attention about integrated farming system (IFS). Livestock management such as dairy, poultry, piggery, fishery etc. would improve the prosperity of watershed people. The livestock management not only improves income benefit but also improves the nutritional availability to the rural people. Milk and milk products can be sold with valve addition it will increase the valve of the products. Figure 5 illustrate the different livestock in rural bio industrial watershed.

Figure 5. Livestock's in rural India.

The availability of nutrients from manure of different livestock species to be used in agriculture have been exhibited in Table 5. The nutrient content of the excreta varies with species. Apart from this, animal urine is very rich in nitrogen and potassium. The value in cattle urine has been reported to be 0.9-1.2% N, a trace level of P and 0.5-

10% K. Urine output per kg body weight is in the range 17-45 ml/kg body weight. The average body weight of cattle, if taken as 200-300 kg the average urine production per day per animal would be about six to nine liters. To use the animal manure effectively, the animals are to be stall-fed and housing should be designed in such a way that entire urine is collected (Pankaj *et al.*, 2020).

Table 5. Average N, P, and K level in the dung of different livestock

Source	N%	K%	P%	Ca%	Mg%
Poultry manure	2.12	1.16	1.40	2.32	2.45
FYM	0.65	0.50	0.18	3.82	3.96
Goat manure	1.16	1.09	0.40	3.20	3.65
Vermi-compost	2.15	0.95	0.55	4.15	3.60
Pig manure	0.50	0.28	0.35	2.15	2.85
LQM	0.56	0.18	0.45	0.10	0.07
Duck dropping	1.85	0.98	1.46	1.99	2.22

The production technologies include use of farm mechanization and use of high variety seeds and fertilizers. The farm mechanization in watershed level would reduce the human energy and it would improve the soil condition. In rural areas, the farm mechanization can be efficiently achieved through custom hiring basis. The produced product is to be processed for higher profit. The processing aspect is the installation of the processing industry, which should preferably be owned and managed privately by an experienced entrepreneur.

Rural Support Systems

Currently in most of the watershed programs in India, the community-based soil and water conservation are playing the key role in improving surface and groundwater availability and controlling soil erosion. To improve the prosperity of the village people in watershed level, the rural supporting systems are very important. The rural supporting systems include financial assistance from central and state governmental schemes, research and development institutions and social institutions. To meet the food demand for future generation we need to strengthen micro-credit system for establishing micro-enterprises in the watershed areas. The setting up of more *Gramin* Banks and Cooperative Societies in the rural sector is desired besides strengthening National Bank for Agriculture and Rural Development (NABARD).

Figure 6. Local market in watershed level.

The bigger financial institutions would seldom come to the help of small enterprises with little capacity to pay back the loans at higher interest rates. The smaller finance institutions, to the contrary, would extend small loans at low interest rates and flexible time limits. These supporting systems directly linking with both resources conservation technologies and also water resources management technologies. The schemes such as MNREGS, SHM, IWMP, APMIP are supplying inputs for crop production.

Processing

The processing of products in bio industrial watershed is based on need based and rural enterprises for valve addition. The financial support can be got from nationalized, Gramin banks, co-operatives, NGO's and Research and development institutions. The development of the cold storage/ cold chain to store and transport perishable produce to the markets at appropriate time to avoid distress sales should be part of the bio industrial development programme.

The processing or value addition of the produce at the watershed level would fetch more income to the farming community while generating employment for 15 million rural and tribal families. Likewise, the mushroom farming is big livelihood opportunities besides food and nutritional security to rural people.

Figure 7. Small scale agro industries in bio industrial watershed level

Markets

The generated produce can be sale in local markets and external markets. The marketing system is connecting with supporting systems, industries and production systems. In bio-industrial watershed, markets would provide backward linkages to develop crop production inputs like seeds, fertilizers, planting materials, manures, pesticides, drip / fertigation systems and technical know-how etc. and forward linkages in finding markets and remunerative prices for the produce. The bio-industrial watershed management is different from the contract farming in the sense that it does not only focus on production and buy back of the produce but takes care of natural resource conservation and protection also to have sustainable production.

Although implementation of bio-industrial watershed will enhance the income significantly, it will become futile unless the farmers' bargaining powers are also enhanced. Otherwise, they will remain predisposed to the unfair treatment from middlemen. Therefore, introduction of Community Based Organizations (CBO's) will be helpful for the farmers. The fresh or processed watershed produce can be sold directly to the consumers by these cooperatives, doing away with the middlemen. An industrialist or entrepreneur ready to invest for the bio-industry should also be welcomed provided they are ready for the benefit sharing with the cooperative. The bio-industry would take upon itself the role of running the affairs of bio- industrial watersheds. It would arrange for grants from the Government for soil and water conservation works and build-up of infrastructure and credit from financial institutions for production purposes.

In rural India, as livestock are essential components of farming systems, the bio industry based on poultry, milk, wool, meat and other animal products could be a great success. The water management in bio industrial watershed is very cruises as water is the major natural sources for small scale industry activities. In agricultural sector, improve water usage efficiency in the production of water-intensive crops such as rice, wheat and sugarcane; encourage adoption of techniques such as rain-water harvesting and watershed management in agriculture. Reduce subsidies on power and implement customized pricing models to counter groundwater exploitation through excessive withdrawal. In domestic sector, implementation of policies to make rain-water harvesting mandatory in cities with new construction projects and propagate efficient water usage practices through community-based education programs.

Beyond income improvement, livelihood diversification under BIWS also provides income security for the farmers. In the off season of certain crops, farmers can focus on another sub-sector. As farmers will be less dependent on a particular source of income, they will experience considerably greater income security. Simultaneously, the local village remains at the center of the farmers' economic dependence thereby reducing the seasonal migration also. The watershed farming community through cooperatives or corporate should wholly or substantially own these bio-industries. Even the landless people of the watersheds would be partners of the bio-industry. Sustainable productivity, production, marketed surplus, market arrivals, procurement, milling of rice, extended safe storage of stocks, distribution logistics, physical availability, access/affordability and farm portability provide security of food, social harmony and sovereignty.

Some successful stories of the bio-industrial watershed management system are the parlance of Amul cooperatives in Gujarat, Kuppam vegetable farming and processing cooperative in Andhra Pradesh, Kokam fruit concentrate production cooperatives in Maharashtra and Goa, sugar mills co-operatives in Maharashtra and Gujarat, Leh Berry fruit concentrate production cooperative in Leh (J&K) and many other ventures in different parts of the country.

Conclusions

The Government of India is implementing the Integrated Watershed

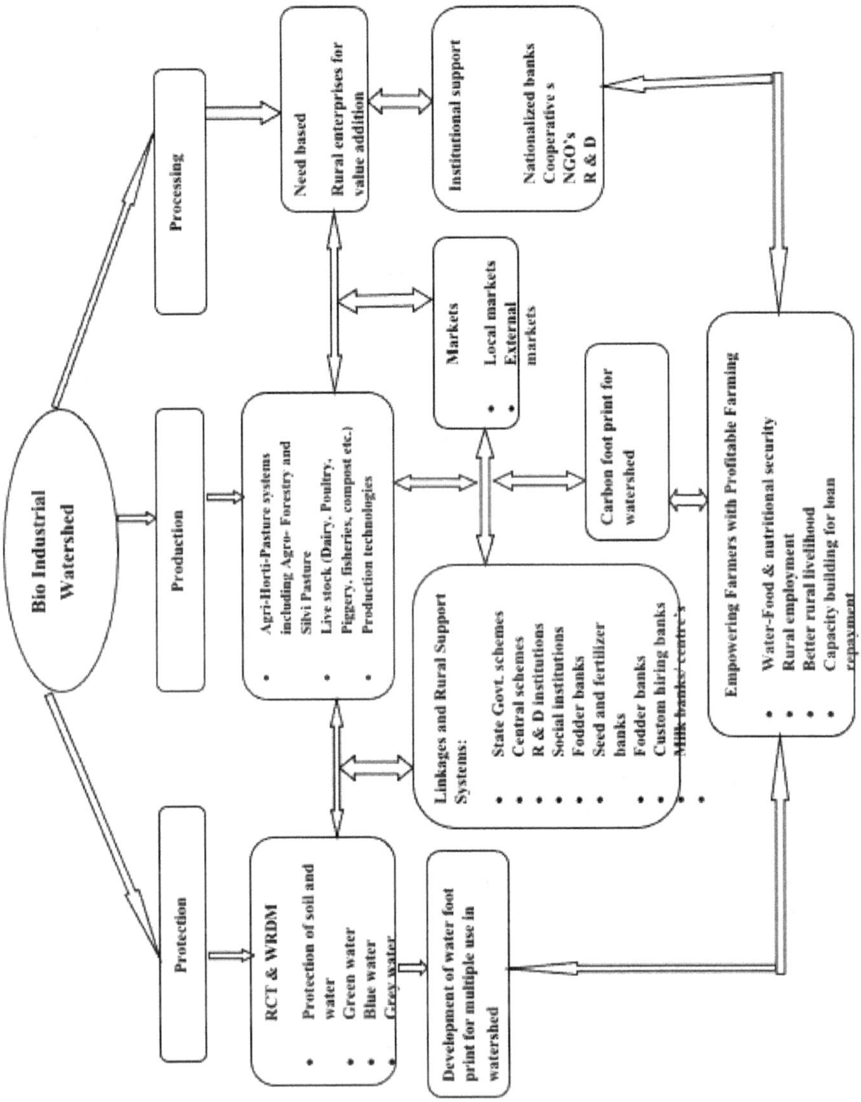

Figure 8. Bio industrial watershed and its components: A conceptual framework.

Management Program (IWMP) as per the new guidelines issued by NRAA. Though the programs suggested in bio industrial watershed are available in IWMP, the issues are treated independently without converging the other state Govt. schemes and central schemes, which would indeed bring the prosperity to the rural India. The bio industrial watershed concept is introduced to bring all the rural development schemes into convergence integrated with agri enterprises suiting to the local environment and markets, which otherwise country loses economic value of agri-horti products in the long transport and inadequate storage facilities, more so with malnutrition and inequity in the food distribution. Therefore, a holistic view of the IWMP system suggests that if we can generate employment using the concept of bio-industrial watershed in the rural areas, then soon every watershed in our country will become a bio-industrial watershed, to ensure work and income security to rural families with water, food and nutritional security at rural villages cluster. Bio-industrial watershed management may be adopted at the village level, integrating all the development programmes. The present chapter deals with all conceptual framework proposed for such new paradigm within the existing system calling for effective convergence and integration of the schemes for making the bio industrial watershed would become reality.

References

ACARA Challenge (2011). Introduction to food and water security in India. ACARA institute of Environment. Unit 3, pp. 1-42.

Addams, L., Boccaletti, G., Kerlin, M., Stuchtey, M. (2011). Charting our water future: economic frameworks to inform decision-making, 2030 Water Resources Group. McKinsey & Company, New York.

Bali, J.S. (2010). Bio-industrial Horticulture Watersheds. *Journal of soil and water conservation*, 9(4): 255-260.

Gandhi, V.P. and Bhamoriya, V. (2011). http://www.idfc.com/pdf/report/2011/Chp-8-Rainwater-Harvesting-for-Irrigation-in-India.pdf.

Gupta, R.K., Singh, N.T., Sethi, M. (1994). Ground water quality for irrigation in India (with 1: 1 million map). Technical Bulletin, 19, CSSRI Publication, CSSRI, Karnal.

John Rockström, Jennie Barron and Patrick Fox. CAB International (2003). Water Productivity in Agriculture: Limits and Opportunities for Improvement.

Katyal, J.C. and Das, S.K. (1992). Stability in dryland agriculture – Potential and Possibilities. *Indian Journal of Dryland Agricultural Research & Development*, 7(1):1-4.

Kumar, R., Singh, R..D. and Sharma, K.D. (2005). Water resource of India. *Curr. Sci.,* 85(5): 790-811.

Majumdar, D.K. (2002). Irrigation water management: principles and practices. Prentice-Hall of India Pvt. Ltd., New Delhi.

Menon, M.S. (2002). A case for inter-basin transfer of water. The Hindu. https://www.thehindu.com/thehindu/op/2002/11/19/stories/2002111900130200.htm.

Milesi, C., Samanta, A., Hashimoto, H., Kumar, K.K., Ganguly, S., Thenkabail, P.S., Srivastava, A.N., Nemani, R.R., Myneni, R.B. (2010). Decadal variations in NDVI and food production in India. *Remote Sens,* 2(3): 758-776.

Minhas, P.S., Gupta, R.K. (1992). Irrigation water quality: assessment and management. Information and Publication Division, Indian Council of Agricultural Research, New Delhi, pp.123.

Naraynamoorthy, A. (2002). Indian irrigation: five decades of development. *Water Resour. J.,* 212: 1-29.

NRAA, (2011). Challenges of Food Security and its Management. National Rainfed Area Authority. NASC Complex, DPS Marg, New Delhi, India.

Pankaj, P.K., Pourouchottamane, R. and Rai, B. (2020). Scientific pig production and management: A bio-treasure for rural society. Satish Serial Publishing House, New Delhi.

Pratap Narain (2008). Dry land management in arid agro-ecosystem. *J. Indian Soc. Soil Sci.*, 56(4): 337-347.

Samra, J.S. and Sharma, P.D. (2009). Food security - Indian scenario. Proceedings IPI-OUAT-IPNI International Symposium, pp. 15-135.

Shah, T., Singh, O., Mukherji, A. (2006). Some aspects of South Asia's groundwater irrigation economy: analyses from a survey in India, Pakistan, Nepal Terai and Bangladesh. *Hydrogeol J,* 14: 286-309.

Sharma, C.U. and Sharma, V. (2012). Sustainability of agroforestry systems on acid soils for livelihood security. *Indian Journal of agroforestry,* 14(2): 27-36.

Singh, M.J. and Khera, K.L. (2011). World association of soil and water conservation. Bulletin. 27(2): 15.

6

Agroforestry: A Potential Tool for Watershed Management

Saresh N.V. and Archana Verma

Introduction

The common perspective of agroforestry is nothing but "the deliberate cultivation of woody perennials with the agricultural crops on same unit of land in some form of spatial mixture or sequence". This is in broader and a distinct perspective for land use, but in real sense agroforestry is a key for the evolution of a more mature agro-ecosystem due to its interaction with the multiple component in an ecosystem in which it exist. So, agroforestry be considered as "dynamic, ecologically based natural resource management system that, through the integration of tree in farm and range land, diversities and sustains smallholders production for increased social, economic and environmental benefits" (Leakey, 1996). According to functionality there are two different types of agroforestry exist i.e. simultaneous system and sequential system (Sanchez, 1995) and further this may be classified according to two subcategories based on land management unit: field-based system at house hold level and landscape-based system at village or watershed level (Thomas, 1996). Field based systems received higher consideration than the landscaped system because these are closely related with the conventional farming practice. As early mentioned for watershed level we should go for landscape-based agroforestry system because as per the concept of watershed "a watershed is technically considered a geo-hydrological unit or an area that drains to a common point". The watershed, catchment and drainage basin are synonymous term indicating an area surrounded by a ridge line that is drained through a single outlet. More specifically, it supplies water by surface or sub-surface flow to a given drainage system or body of water, stream, river, wetland, lake or ocean. The watershed is the product of the interactions between land and water, particularly its underlying geology, rainfall patterns, slope, soils, vegetative cover and land use. "Watershed hydrology," as a term, encompasses these interactions. The availability of water and its flow is a critical determinant in the various production functions in a landscape, especially since it is open to interference by human agency.

Watershed is a terrestrial ecosystem consisting of interacting biotic and abiotic components such as land, water, plants, animals and minerals. It is not simply the hydrological unit but also social, political and ecological entity which plays crucial role in determining food, social and economic security in terms of life-supporting services to rural people. It is a major site for residential, commercial, industrial, agricultural, educational, experimental, environmental, and forest land uses which are often conflicting and competing with each other for the limited watershed land resource. Form above all there should be a proper watershed management plan to achieve diverse and conflicting goals. Watershed management is a process of creating and implementing plans, programs, and projects to sustain and enhance watershed functions that affect the plant, animal, and human communities within a watershed boundary". It involves management of land, water, energy and greenery integrating all the relevant approaches appropriate to socioeconomic background for a pragmatic development of a watershed. It also involves the greening of watershed through proper management of land water and energy resource. The main components of watershed management are resource conservation, crop production and alternate land-use systems (Tideman, 2007). As per the World Agroforestry Centre (ICRAF) who have identified six ways that agroforestry can contribute to achieving the Millennium Development Goals of combating hunger, poverty, disease, illiteracy, environmental degradation through conserving biodiversity and watershed services, and discrimination against women (Garrity, 2004).

Agroforestry Systems Across the Ecological Zones as a Tool for Watershed Management

Agroforestry has been explained as an age old land use system, which has the ability of yielding multiple products that sustain life at the same time it can conserve and rehabilitate the overall ecosystems. As such agroforestry is the practice which is consider to be the twin sister of agriculture itself and has been practiced by many countries from many centuries under subsistence farming scenario (Dhyani, 2018). Agroforestry systems combining tree/shrub, crop, grass and livestock have great scope and role in optimizing land productivity and environmental protection in the fragile ecosystems. Traditional agroforestry which are followed in many regions is a complete, ecologically sustainable livelihood system (Roy and Tewari, 2014). Besides providing livelihood securities, agroforestry is a tool for solving serious problems like reclamation of degraded soil and water resources.

Table 1. Agroforestry systems across the ecological zones (Dhyani, 2018)

Agro-ecological zone	Agroforestry system	Examples
Himalayan Region	Agri-horticulture	Intercropping of tuber crops, rhizome crops and cereals with orange, peach, guava, pecan nut, citrus crops, plum, in Meghalaya, Sikkim, Uttarakhand, and NE region.
	Agri-silviculture	Important trees for fodder in hill region of NW (Himachal Pradesh, Uttarakhand, Jammu-Kashmir) and NE (Sikkim and hill region of other states)
	Multi-storeyed and specialized system	In Nagaland and Sikkim cultivation of coffee, large cardamom with alder; in West Bengal and Assam tea with legume trees; in Uttarakhand, NE region including Meghalaya on black pepper and betel vine.
	Silvi-pasture	In the Himalayas from Celtis, Grewia, Quercus, Bauhinia and other species
	Horti-pastoral	Peach and other fruit trees with grasses of rainy and winter season
Indo-Gangetic Plains	Agri-silviculture	Poplar, *Tamarindus indica*, *Bombax ceiba*, *Eucalyptus*, *Dalbergia sissoo*, *Melia* spp. etc.

	Fodder banks and woodlots	Trees for fodder *Azadirachta indica, Albizia* spp., *Melia* spp., *Syzygium cuminii*, etc. Bamboo spp., *Casuarina, Dalbergia sissoo, Eucalyptus.*
	Silvi-pasture	On lands for grazing purpose *Bauhinia* spp., *Albizia* spp., *Dalbergia sissoo* and other important species
	Trees for soil reclamation (sodic/ saline / degraded soil)	For soils with problems *Acacia nilotica, Prosopis* spp., *Parkinsonia* spp. and other salt tolerant species
Arid and Semi-arid Region	Agri-silviculture	*Prosopis cineraria* (Khejri), *Acacia senegal, Ziziphus* spp. (Ber), *Acacia nilotica* (Kikar), *Ailanthus excels* (Ardu), *Acacia leucophloea*
	Fodder banks and woodlots	*Ziziphus nummularia, Prosopis* spp., *Acacia nilotica*
	Windbreaks and shelterbelts	*Salvadora oleoides, Salvadora persica Cassia* spp., *Acacia* spp. and other *Acacias, Prosopis chilensis, Azadirachta indica, Tamarix* spp.
	Silvi-pasture	*Ziziphus* spp., *Prosopis* spp., *Acacia senegal, Azadirachta indica, Acacia tortilis*
	Trees for reclamation ofdegraded soils	*Casuarina* spp., *Albizia lebbeck, Azadirachta indica, Acacia tortilis* and other tree species
Humid and Sub-Humid Region	Multi-tier system or plantation crop combination	Trees such as Coconut, Erythrina, Arecanut, and other trees with pineapple / papaya, coffee / banana, cacao /coffee black pepper / betel vine and large cardamom.

	Home gardens	A mixed plantation of trees, herbs, shrubs etc. in a form of home gardens in Kerala, West Bengal, Assam. Arecanut, Coconut, Gliricidia, Erythrina with cash crops such as black pepper, coffee, cacao, cassava etc., and others
	Multipurpose trees in agricultural fields	*Aegle marmelos (*Bel patra*), Acacia auriculiformis, Anogeissus latifolia, Albizia* spp., *Casuarina equisetifolia, Anthocephalus chinensis, Artocarpus, Diospyros melanoxylon* (Tendu leaves), etc.
Coastal and Island Region	Mangroves plantations as part of homesteads	*Azadirachta indica, Casuarina equisetifolia, Prosopis chilensis, Acacia Senegal, etc.*
	Trees on boundaries of Agricultural fields (agri-silviculture	*Aegle marmelos, Azadirachta indica*, Albizia spp., Bamboos, *Calliandra calothyrsus, Bombax malabaricum, Cassia* spp., etc.
	Plantation crop combination, Multi-storied	Composite fish culture in ponds and multipurpose trees in homesteads.
	Shelter belts and wind breaks	*Albizia spp., Aegle marmelos, Bamboos, Bombax malabaricum, Azadirachta indica, Cassia spp. Calliandra calothyrsus*
	Trees with aquaculture	*Acanthus ilicifolius, Avicennia officinalis, Carbera odollam, Rhizophora conjugata*, etc.

Agroforestry System for Maintaining or Improving Soil Fertility

Agroforestry systems are efficient techniques of restoring degraded soil system such as agro-horticulture, agro-pastoral, agri-silvipastoral systems and others are the prominent example (Pandey, 2007).

The leaf litter in an agroforestry systems falls on the floor which are decomposes and forms humus by the symbiotic effect of microorganisms ultimately enhances soil fertility, help in reducing losses of soil, water, organic matter and nutrients (Chakraborty *et al.,* 2015; ICAR, 2006; Jose, 2009). More over many tree species have ability to conserve moisture by adding leaf litter in and above ground and enhance soil fertility in agroforestry systems; legumes trees and crops has also have an effective role for promotion of soil fertility (Yadav *et al.,* 2008, Verma *et al.,* 2017). Systematic agroforestry enhances soil fertility, there are numerous evidence that agroforestry systems have potential for enhancing soil organic matter (SOM), crop productivity, water retention, carbon sequestration, biodiversity and soil sustainability (Lana *et al.,* 2018; Paul *et al.,* 2017). Trees improve soil fertility through a wide range of processes such as soil carbon pools by providing additional below- and above-ground biomass, reduction of losses and improvement of soil physical, chemical and biological conditions and nutrient cycling. Among mentioned processes soil fertility improvement is mainly augmented majorly due to the addition of tree litter fall, prunings and root residues.

Agroforestry for Erosion Control and Sand Dune Stabilization

Agroforestry components has always highlighted the protective role in controlling erosion or slope management. The perennial component in agroforestry system have extensive root morphology which binds and anchor the weaker upper horizon of the soil profile and provide a strong interlocking system to the soil binder which reduces slipping and sliding of the soils which is very much important for a watershed to reduce shallow loss of mass wasting, sedimentation and to perform stabilization function (Ziemer, 1981 and O' Loughlin, 1974).

Shifting of sand dunes are one of the important problems in desert areas throughout the world. Sand dunes are dominant features of western Rajasthan and occurs 62% area of the region. The height of these dunes varies from small (few meters) to more than 60 m. due to high mobility of sand dunes causes obstacle to agricultural operations, establishment of vegetation, roads and rail traffic, and other infrastructures. Thus the stabilization of sand dunes by establishing of plantation stand is of vital important for desert region for combating desertification, rehabilitating fragile dune ecosystem and to meet the growing demand for ever increasing human and livestock demand. The sand dune stabilization technique developed by ICAR-CAZRI consists of (a) fencing of the area, (b) establishment of micro-wind breaks on the windward side of the dune in 5 m chess board pattern or in 5 m parallel strips, and (c) sowing of grasses and transplanting of trees and shrubs with the onset of monsoon. For raising micro-wind breaks, locally available brushwood materials like *Leptadenia pyrotechnica* (Khimp), *Ziziphus nummularia* (Pala), *Crotalaria burhia* (Sania) and *Panicum turgidum* (Murath) were used. Suitable tree species for sand dune stabilization are: *Acacia tortilis, Prosopis* spp., *Acacia senegal, Parkinsonia articulata* and *Tamarix articulata.* Among grasses, *Lasiurus sindicus* and *Cenchrus ciliaris,* and among creepers *Citrullus colocynthis* were most suitable (Harsh and Tewari, 1997).

Agroforestry and Ground Water Recharge

The perennial component plays an important role in conserving soil through runoff control, increases soil infiltration, binding the soil by its root and so on. The greatest advantage of trees was in controlling soil erosion by checking the rate of soil runoff. The reduction in runoff was to a small degree, caused by canopy interception and direct transpiration but the greater part of it results from higher infiltration capacity under trees. It helps in improving recharge of ground water, besides perennial flow of stream and better moisture availability through micro-climatic interventions (Tyagi *et al.,* 2002).

Trees stand in any piece of land act as giant sponges, soaking up water during wet periods and gradually releasing it during dry periods, thereby maintains the dry season flows this theory is called as sponge theory (Malmer *et al.,* 2010; Hayward, 2005). In support to this sponge theory and related ideas, many governments and NGOs promote planting trees for reforest or afforest to restore springs, raise groundwater levels and increase dry season flows (Hayward, 2005; Calder et al., 2004; Kaimowitz, 2004). Similarly, forest protection and conservation is promoted to maintain water supplies (Galudra & Sirait, 2009; Hayward, 2005). In many cases, tree-based restoration, forestation and forest conservation measures aimed at enhancing or maintaining water supply are promoted through financing mechanisms such as Payment for Ecosystem Services schemes (Ogden *et al.,* 2013; Locatelli & Vignola, 2009). In turn, these measures are expected to improve the livelihoods of rural people through increased access to water (Ogden *et al.,* 2013; Calder *et al.,* 2004). Planting of trees in a degraded land could restore ground water level if the associated reduction in runoff, gains in infiltration opportunities exceed the extra evapotranspiration losses from the planted area. In Figure 1. The blue arrows illustrate different hydrological flows (ET: evapotranspiration, i.e., transpiration and interception; IFR: soil infiltration; RO: Runoff flow; GR: groundwater recharge). a) Fallow land a change in land use associated with soil disturbance can lead to diminished of ground water level due to heavy runoff of water. b) Agriculture land involving only a change in land cover, agriculture or mono cropping without soil degradation, more evapotranspiration will lead to enhanced ground water level compared to fallow land. c) Agroforestry system involving perennial and annual crop without soil degradation, more evapotranspiration will lead to enhanced ground water level compared to fallow and agricultural land.

a. Fallow Land b. Agricultural Land c. Agroforestry Land

Conclusion

Agroforestry is a useful tool in watershed management to promote sustainable use of various resources and to improve the economic well-being of the local people. Properly designed and managed agroforestry systems can provide various benefits and has potential to meet environmental and socioeconomic requirements. The agroforestry system aid in managing watershed by various ways through improve soil fertility, erosion control and sand dune stabilization, ground water recharging and so on. Overall agroforestry plays a major role in success of any established watershed area.

References

Calder, I.R., Amezaga, J., Bosch, J., Fuller, L., Gallop, K., Gosain, K., Hope, R., Jewitt, G., Miranda, M., Porras, I. & Wilson, V. (2004). Forest and water policies – The need to reconcile public and science perceptions. *GeologicaActa,* 2(2), pp. 157-166.

Chakraborty, M., M. Z. Haider & M. M. Rahaman. (2015). Socio-economic impact of cropland agroforestry: evidence from Jessore district of Bangladesh. *International Journal of Research in Agriculture and Forestry* 2: 11–20.

Dhyani, S. K. (2018). Agroforestry in Indian perspective. In: R. G. Rajeshwar, M. Prabhakar, G. Venkatesh, I. Srinivas & R. K. Sammi. (Eds.) *Agroforestry Opportunities for Enhancing Resilience to Climate Change in Rainfed Areas.* ICAR-Central Research Institute for Dryland Agriculture, Hyderabad, India.

Galudra, G. and Sirait, M. (2009). A Discourse on Dutch Colonial Forest Policy and Science in Indonesia at the Beginning of the 20th Century. *International Forestry Review,* 11(4): 524-533.

Garrity, D.P. (2004). Agroforestry and the achievement of the Millenniums Development Goals. *Agroforestry Systems* 61: p. 5-17.

Harsh L. N. and Tewari J.C. (1997). Desertification control in arid ecosystem of India for sustainable development. Ed. Surendra Singh & Amal Kar. Agro Botanical Publishers (India), Bikaner. 317-325.

Hayward, B. 2005. *From the mountain to the tap: how land use and water management can work for the rural poor.* Hayle, UK.: FRP-DFID. ICAR. (2006). Handbook of Agriculture, 5th edition. New Delhi: Indian Council of Agricultural Research.

Jose, S. (2009). Agroforestry for ecosystem services and environmental benefits: An overview. *Agroforestry Systems* 76: 1–10.

Kaimowitz, D. (2004). Forests and water: a policy perspective. *Journal of Forest Research,* 9(4), pp. 289-291.

Lana, A. M. Q., R. M. Q. Lana, E. M. Lemes, G. L. Reis & G. H. F. A. Moreira.

(2018). Influence of native or exotic trees on soil fertility in decades of silvopastoral system at the Brazilian savannah biome. *Agroforestry Systems* 92:415–424.

Leakey, R. R. B. (1996). Definition of agroforestry revisited, *Agroforestry Today* 8(1) 5-7.

Locatelli, B. & Vignola, R. (2009). Managing watershed services of tropical forests and plantations: Can meta-analyses help? *Forest Ecology and Management,* 258(9): 1864-1870.

Malmer, A., Murdiyarso, D., Bruijnzeel, L.A. & Ilstedt, U. (2010). Carbon sequestration in tropical forests and water: a critical look at the basis for commonly used generalizations. *Global Change Biology,* 16(2): 599-604.

O'Loughlin, C. L. (1974). The effect of timber removal on the stability of forest soils. Hydrology 13(2): 121-134.

Ogden, F.L., Crouch, T.D., Stallard, R.F. & Hall, J.S. (2013). Effect of land cover and use on dry season river runoff, runoff efficiency, and peak storm runoff in the seasonal tropics of Central Panama. *Water Resources Research,* 49(12): 8443-8462.

Pandey, D. N. (2007). Multifunctional agroforestry systems in India. *Current Science* 92: 455–463.

Paul, C., M. Weber and T. Knoke (2017). Agroforestry versus farm mosaic systems: comparing land-use efficiency, economic returns and risks under climate change effects. *Science of the Total Environment*587:22–35.

Sanchez, P.A. Science in agroforestry. *Agrofor. Syst.* (1995). 305–55doi: https://doi.org/10.1007/BF00708912
Thomas, S. (1996). Crafting better urban watershed protection plans. *Watershed Protection Techniques.* 2 (2): p. 329.

Tideman, E. M. (2007). Watershed management: guidelines for Indian conditions. Omega Scientific Publishers. New Delhi. 372p.

Tyagi, S., Mohan, K. and Kumar, S. (2002). Role of agroforestry in watershed management in the state of Jharkhand.. 12th ISCO Conference. Beijing. 6p

Verma A, Shiran K., J.C. Tewari, Kalia R. K, Saresh N. V., Gaur. M.K, and Kumar S. (2017). Agroforestry: A Sustainable, Multifunctional and Diversified Production System for Hot Arid Zone of India. In Climate Change and Agroforestry Editors: C.B. Pandey, Mahesh Kumar Gaur and R.K. Goyal, New India Publishing Agency, New Delhi, India, pp. 251-272

Yadav, R. S., B. L. Yadav & B. R. Chhipa. (2008). Litter dynamics and soil properties under different tree species in a semi-arid region of Rajasthan, India. *Agroforestry Systems* 73: 1–12.

Ziemer R. R. (1981). Roots and stability of forested slope. IAHS Publication 132. Christchurch, New Zealand.

7

Improving Productivity of Agroforestry System through Rainwater Harvesting in Arid Region of Rajasthan

Bilas Singh

Introduction

Water is the most important constituent of the ecosystem. The basis of life is water. The journey of life started in water. Human civilizations have progressed and flourished along the water courses and their development in interwoven with the development of water resources. This relationship has been manifest in arid regions in much more earthly terms. With decline in per capita availability of water, deterioration in water quality, increase in per capita water requirement and with increasing inter - sectorial competition for good water, water crisis is surfacing in several regions of the world, particularly in Africa and other under development and developing regions. It is projected that by the year 2030, one out of every three persons will suffer from water deficit. Presently 28 countries fall in the category of water deficit regions. By year 2030, other 18 countries will be added to this category, mostly African and gulf countries. Though, India is not in this category but water problem is certainly dominating the arid and semi arid regions. With increasing crop intensity, animal husbandry and dairying activities, urbanization, forestry/plantation activities, industrialization and changing life styles (more demanding on resources) inter - sectorial competition for water has increased. In 2019, about 70% of the global water uses was for agriculture, while 20% was used for industry. The remainder was for municipal water supply and other uses. Forestry is mainly dependent on rainfall.

The rainfall water that is stored as green water in the soil which is available to plants and organisms is the main source of water for natural ecosystems and for rainfed agriculture, which produces 60% of the world's food (WWC, 2019). Surface water runoff and groundwater recharge—is the main source for human withdrawals and the traditional focus of water resource management. Rainfall and runoffs occur in large amounts during very short periods during the monsoon periods in Asia, this runoff water is stored in reservoirs, or tanks and available to human use, which is a the traditional system in the Indian subcontinent. After the green revolution, developed technologies on irrigation increases agricultural productivity—particularly in Asia, which contains about 70% of the world's irrigated area. Depending on the technology, consumption can range from 30–40% for flood irrigation to 90% for drip irrigation.

Study Area

Rajasthan is the largest state in India covering an area of 10.5 percent of the country's geographical area, but sharing only 1.15 percent of its water resources. The state is predominantly agrarian as the livelihood of 70 per cent of its people depends on agriculture-based activities. The condition of water availability in arid region of the state is more critical where rainfall is very low, 100 mm in western sector Jaisalmer to 450 mm in eastern and southern part of arid zone (Rao and Singh, 1998). The annual estimated PET values range from 1600 mm in the eastern part to >1800 mm in the western part. Moreover the rainfall characteristic is such that it is confined to very short period of July – August. Out of 365 days in a year 350 days remain dry and only 10-15 days are the rainy days. Most of the rainfall is received in a few events of high intensity during the monsoon season resulting in considerable runoff losses. On average, 30-50% runoff losses have been reported in Indian arid zone. In some rainfall events runoff losses as high as 85 per cent have been observed from a bare land. According to the report Rajasthan Agriculture Department out of the 15.8 million acre feet rain water received in the state, half is lost as runoff. Rainwater harvesting provide an opportunity to stabilize agricultural landscapes in dry lands and to make them more productive and more resilient towards climate change (Narain and Khan, 2002). This is a formidable amount and if harvested, conserved and used judiciously, it will be invaluable in this region of acute water scarcity.

Traditional Rainwater Harvesting Practices

Traditionally people in arid zone were aware of need and importance to harvest and conserve rain water and they had developed very effective methods of harvesting the rain water. While the variety of water harvesting structures occurring in the arid regions of Rajasthan speaks volumes about the depth of traditional wisdom, the number of structures running into thousands in a single district of Jaisalmer alone demonstrates the dedication, devotion, hard work and social concerns of the people of this region. The remarkable feature of these structures is that they were largely made by individual and community efforts. Depending upon requirement, utility, topographic, physiographic features, soil and rock types and several local factors including climate, the nature, shape and size of water harvesting structure and systems varied. Thus efforts were made to harvest every drop of water and use it efficiently. In Jaisalmer district, which received only 160 mm average annual rainfall, there is thousands of different size of water harvesting structures. Their names varied according to their size/structure/location such as 'Talab', Tal, Talai Bandh, Bandha, Nadi, Johad, Jahadi, Sar, Jheel, Khadin, Bhe, Tanka, Dhari, Deibandh Jagah, Kua, Kui, Bavadi. 99.78 per cent villages of this hyper arid district Jaisalmer have ponds, wells and other water harvesting structures. What a miracle created by people that the rains received only in 10 days is managed to meet the water requirement of whole ecosystem round the year. These structures are 100-500 years old. Khadin system of water harvesting is 600-700 years old where bumper crop of wheat is being harvested for centuries. Jaisalmer town for centuries remained center for trade with Iran, Afghanistan and Russia. In addition to individual efforts there are exemplary efforts of water harvesting at the community/state level. Diversion of every drop of rain water to Kailana Lake and Takhat sagar from kailana hills around Jodhpur city of Indian Arid Zone was such a remarkable feat, but in our strides to develop the network of modern water engineering projects these traditional systems have been neglected and damaged.

Probably the mistake has been realized and now these old systems are getting revived in the form of watershed development programmes. Water harvesting structures such as anicuts, percolation tanks, check dams, gully plugging-ditches, contour trenches, etc., depending upon physiographic and topographic condition are being made to effectively use the water of a watershed unit (Goyal *et al.*, 2013). The state still has significant potential for harvesting and conserving water if an integrated water resources management approach is adopted, and proper policies and investment actions are implemented using recent technologies. Integrated water shed development Project of Rajasthan state has been engaged in erecting various types of structures in different water shades. These techniques have proved beneficial in improving forest growth, agricultural productivity, and animal husbandry and above all in improving quality of water resources in the water shed region (Narayan *et al.,* 2017).Water is single most important factor limiting growth of plantations because of paucity of water, large areas of arid land remain barren bereft of vegetation. Crop failures are frequent and tree growth owing to the poor availability of water. Rain water harvesting techniques have caused significant improvement in the yields of agriculture and horticulture crops world over.

Arid region of Rajasthan are facing scarcity of water and need resource diversification for improved land productivity. Alternate land use systems such as agroforestry are interface of agriculture and forestry. It could contribute to improve livelihood in arid region of Rajasthan, where people have practices and indigenous knowledge systems on tree-growing. The farmers of this region conserve and promote scattered trees and shrubs growing on their farmlands or grazing lands to sustain their livelihood. These multi-purpose woody perennials support livelihood in the region where harsh and dry climate creates agricultural instability and trees play as boon to the farmers particularly during drought when rainfed crops fail to provide any yield (Harsh *et al.,* 1992). These trees provide fodder, fruit, vegetable, fuelwood, timber, medicine and fiber for sustaining rural livelihood and improve soil fertility status and environmental services as well. Considering the privilege of agro-forestry systems are traditionally practiced in the dry lands of the Rajasthan, often with pearl millet, moong bean, moth bean, cluster bean, cotton, groundnut, soybean, maize, wheat, barley, mustard, cumin and ishabgol as major crops. In this cropping system with livestock as important component, naturally growing scatter trees are conserved with different tree density in rainfed dominant condition in the state (Bhati, 2017).

Agroforestry for Sustainability

In arid region of Rajasthan, poverty and food insecurity are interlinked and widespread and are strongly linked to the natural resource endowment (water, soils and vegetation). Land and water degradation, overgrazing, and low production of agricultural have led to significant environmental degradation and food shortages. Under erratic rainfall conditions in the arid zone of Rajasthan, a major contribution to improving crop production can be anticipated from improved and up-scaled soil water conservation and rain water harvesting devices. Runoff conservation in agroforestry has tremendous potential in arid and semi-arid region where irregular rainfall often results in heavy runoff losses from agriculture field. And if prevented, this excess rain water may cause temporary water logging in crop field and consequent crop damage. Laying water harvesting trenches along the tree rows of agroforestry field will solve this problem and at the same time harvested water could be beneficially utilized for tree growth.

Integration of tree on farmland facilitates ensure food, fodder, fuel wood, timber and generate income security for resource-poor rural households, and sustains their farming systems. Keeping in view of above, ecosystem services benefits from tree integration on agriculture lands, traditional and improved agroforestry models have evaluated by numerous workers from different institutions in Rajasthan state. The ecosystem services in multi-functional agroforestry systems were reviewed mostly in arid region of state as compared to semi arid region of the state because of 61% area covered under arid condition and most fragile ecosystem. The majority of the studies reported provisioning and supporting followed by regulating factors of ecosystem services (Reed *et al.,* 2017). Beneficial impacts of tree were dominating, particularly in arid region where these were associated with enhancing water and nutrient cycling. The traditional agroforestry systems in arid region of Rajasthan are based largely on indigenous knowledge and species selection, where farmers conserves and promote the woody and non-woody plant species of the cultural patterns of social and traditional significance. To meet the growing demand of population pressure for insecure food, sustainable soil nutrient and improved livelihood, the agroforestry system are way to stabilize productivity through soil moisture conservation, and soil nutrient buildup (Gupta, 1995; Singh *et al.,* 2013). Therefore, if planned appropriately, agroforestry has potential to address number of land use problems such as food security, soil fertility, soil and water conservation biodiversity, economic returns and environmental services (Tewari *et al.,*2014). Thus, agroforestry land use system has been found ecologically and economically viable in arid region of the state.

Water is a limiting factor for crop production and tree growth in hot arid zone of Rajasthan. While woody vegetation cover and density in agroforestry system reduces the velocity of runoff of soil surface, check soil and water erosion, evaporation from the soil surface, infiltration is enhanced and thereby the availability of water in the root zone increased (Vahland and Vary, 2009).In rainfed arid regions where soils are very poor in nutrients and loss of applied nutrients are more due to adverse climatic situations, the use of FYM, compost and vermin compost would improve soil productivity and moisture conservation (CAZRI, 2007).Thus, soil moisture conservation plays an important role to improve soil water availability which determines the sustainable nutrient fertility to enhance land productivity of an agro-ecosystem (Bayala *et al.,* 2012). The sustainable productivity of a soil mainly depends upon its ability to supply essential nutrients to the growing plants. Integration of rain water conservation in agroforestry system is one of the best management options to sustainable land production in rainfed agriculture (Rathore *et al.,* 2018).

Rainwater Harvesting for Agroforestry

Rainwater harvesting practices not only trap surface water but also reduce runoff to maintain and sustain their ecosystem functionality. It has been observed that rainwater harvesting practices have most significant improvement on biomass and grain yield of agriculture crops (Kayombo *et al.,* 2004).To improve the early establishment and growth of plantation various techniques of *in situ* water harvesting and conservation have been studied by the Arid Forest Research Institute, Jodhpur. It involves rainwater harvesting through intermittent bunding and planting of trees species along the bunds. Among three tree species tested, *Albizzia lebbeck* performance was best followed by *Azadirachta indica* and *Acacia nilotica* (Anonymous, 1998). Just after planting, bunds were prepared along the row of the plants at 6 m interval and down slope to harvest excess rainwater from the cropped area into the trench upside of the bund. Bunding at proper interval facilitated runoff collection and help conserved soil moisture. This is useful to enhance agriculture production as well as cultivation of tree for fodder and firewood along the bunds. Agriculture intercropped *Cymopsis tetragonoloba* produced grain yield 86 kg per ha with *A. indica* plot compared to 72 kg per ha with *A. nilotica* plots and 60 kg per ha in *Albizia lebbeck* plot in 1996. The crop yield was highest with *A. indica* plant in initial years. The highest pruned biomass was for *A. indica* (3.2 kg plant^{-1}), followed by *A. lebbeck* (0.95 kg plant^{-1}) and *A. nilotica* (0.81 kg plant^{-1}). Improved water use efficiency, improved nutrient use efficiency (4-7 times), thus give a good start to crop and young plantation. The results of demonstration trials are in conformity with experimental results at Jodhpur and Rohat (Pali) where tree height attained by *Albizia lebback, Dalbergia sissoo* and *Azadirachta indica* was 18-20 feet within 18 months of planting on runoff catchments area (Anonymous, 1998). The water harvesting techniques prevent runoff losses, maintain higher soil moisture regime and facilitate better tree establishment and growth owing to better development of root system (Gupta *et al.,*1995).

Water conservation practices have been developed to enhance water availability for fruit yield of horticulture plants in arid region. Runoff collection was obtained higher in 5% slopes of Circular catchments (Sharma *et al.,*1983). Circular catchments around each tree (1.5 m radius) with 5% slope towards the tree trunk and covering the catchment with black polyethylene sheet have been found effective to conserve moisture. Presently collection of ex-situ water in ponds is very useful in the arid region where this water is use as supplementary irrigation through drip irrigation system for commercial fruit production in *Z. mauritiana, Punica granatum* and *Emblica officinelis* species (CAZRI, 2007). Saroj *et al.,* (2003) has reported that *C. tetragonoloba - B. juncea* and Indian aloe as the ground storey component in ber to optimize productivity and profitability in the arid ecosystem. Integration of cluster bean, moong bean and sesamum with *Z. mauritiana* (cv. Seb) increased fruit yield three-fold compared to a sole orchard (Singh 1997). Intercropping of mung bean with Zizyphus gave higher grain yield (by 5-20%) (Singh *et al.,* 2003) and intercropping is promising particularly during the juvenile period of fruit plantation. Initial first three -four year of the plantation of these fruit species, intercropped moong bean and cluster bean provides substantial economic benefit to the farmers in the region (Anon., 2019).

In-situ run-off collection are very effective in boosting early establishment and growth of trees and soil moisture availability found beneficial to enhancing production of grasses and survival and growth of tree species in silvi-pastoral system in dry areas of Rajasthan (Singh and Singh, 2010). Such differences in clump diameter might be due to variations in soil resource like soil water and nutrients mobilized through water conservation. Resources diversification between trees and grasses can be optimized through collection of *in-situ* runoff to enhance biomass production as well as improves people livelihood. Various in-situ rainwater harvesting practices helped in conserving higher rainwater, reduced soil and nutrients loss that in turn resulted in higher growth of tree species besides higher yield of intercrops over control (Kumar *et al.,* 2014; Narayan *et al.,* 2017).

Figure 1. Growth of tree seedlings planted along water conservation structure in silvipastoral programme at Goramgarh in Bhilwara District.

The focus has to be on how to harness every drop of water and transform it into green gold. Let us promote water conservation ethos using improves techniques of harvesting and conserving water and frame all our production activities within this philosophy of save water resource and improve land productivity and livelihood of the people.

References

Anonymous (1998). Project progress report, Arid Forest research Institute, Jodhpur.

Anonymous (2019). Annual report. Indian Council of Forestry Research and Education, Dehradun. Uttara Khand, India.

Bayala, J., Balesdent J., Marol C., Zapata F., Teklehaimanot Z. and Ouedraogo S.J. (2006). Relative contribution of trees and crops to soil carbon content in a parkland system in Burkina Faso using variations in natural 13C abundance. *Nutrient Cycling and Agroforestry Systems,* 76:193–201.

Bayala, J., Sileshi G.W., Coe R., Kalinganire A., Tchoundjeu Z., Sinclair F. and Garrity D. (2012). Cereal yield response to conservation agriculture practices in drylands of West Africa: A quantitative synthesis. *Journal of Arid Environment*, 78: 13-25.

CAZRI (2007). CAZRI Perspective Plan: Vision 2025. Central Arid Zone Research Institute, Jodhpur, 94 pp.

Goyal, R.K., Khan, M.A., Bhati, T.K., Pandey C.B. and Roy, M.M. (2013). Watershed Management for Development of Hot Arid Zone of India. Central Arid Zone Research Institute, Jodhpur, India, 46 p.

Gupta, G.N. (1995). Rain water management for tree planting in Indian desert. *Journal of Arid Environment*, 31:219-235.

Gupta, G.N., Limba N.K. and Gupta P.K. (1995). Micro-catchment water harvesting for raising neem in arid region. *Indian Forester*, 121: 1022-1032.

Harsh, L.N., J.C. Tewari, U. Burman and S.K. Sharma (1992). Agroforestry in Arid Regions. *Indian Farming* (Special issue on Environmentally sound Biotechnology) 42(5):32-37

Kayombo, B., Hatibu N. and Mahoo H.F. (2004). Effect of micro-catchment in rainwater harvesting on yield of maize in a semi-arid area. In: 13th International Soil Conservation Organisation Conference. Conserving Soil and Water for Society, Sharing Solutions, Brisbane, July 2004. http://www.tucson.ars.ag.gov/isco/ isco13/PAPERS%20F-L/KAYOMBO.pdf (online).

Kumar, D., Ahmed N. and Hussan A. (2014). Impact of rainwater harvesting and mulching on growth and yield of apple (*Malus domestica*) under rainfed condition. *Indian Journal of Soil Conservation*, 42(1): 74-79.

Narain, P. and Khan, M.A. (2002). Water for food security in arid zone of India. *Indian Farming*, 52 (7): 35-39.

Narain, P., Khan M.A. and Singh G. (2005). Potential for water conservation and harvesting against drought in Rajasthan, India. Working Paper 104 (Drought Series: Paper 7). Colombo, Sri Lanka: International Water Management Institute (IWMI).

Narayan, D., Kumar P., Kumar M and Yadav R.S. (2017). Rainwater harvesting practices and resource conservation in agroforestry system in red soils of Bundelkhand. *Indian J. of Agroforestry*, 19(2): 34-37.

Rao, A.S. and R.S. Singh (1998). Climatic features and crop production. In: Fifty Years of Arid Zone Research in India (Eds. A.S. Faroda and Manjit Singh). Central Arid Zone Research Institute, Jodhpur. pp.17-38.

Rathore, V.S., Goyal, R.K., Yadava, N.D., Saxena A. and Soni, M.L. (2018) Enhancing productivity through water management in hot arid Rajasthan. *Indian Farming*, 68(09): 21–25.

Reed, J., van Vianen J., Foli S., Clendenning J., Yang K, MacDonald M., Petrokofsky G., Padoch C. and Sunderland T. (2017). Trees for life: The ecosystem service contribution of trees to food production and livelihoods in the tropics. *Forest Policy and Economics*, 84: 62–71.

Saroj, P.L., Dhandhar G., Sharma B.D., Bhargava R. and Purohit C.K. (2003). Ber (*Ziziphus mauritiana* L.) based agri-horti system: A sustainable land use for arid ecosystem. *Indian Journal of Agroforestry*, 5:30-35.

Sharma, K.D., Verma C.M. and Shankarnarayan K.A. (1983). Soil moisture storage under contour furrows. *Journal of Indian Society of Soil Science* 31(2): 110-114.

Singh, R.S., Gupta J.P., Rao A.S. and Sharma A.K. (2003). Micro-climatic quantification and drought impacts on productivity of green gram under different cropping systems of arid zones. Pages 74-80 in Human Impact on Desert Environment (Narain, P., Kathju, S., Kar, A, Singh, M.P., and Praveen Kumar, eds.). Scientific Publisher, Jodhpur (India).

Singh, G. and Singh, B. (2010). Assessment of growth and biomass production of *Cenchrus setigerus* based silvipastoral system in community pasture land in Bhilwara District of Rajasthan. *Indian Forester*, 136(7): 898-909.

Singh, G, Rathod, T.R., Komara, S.S. and Limba, N.K. (2013). Rainwater harvest influences habitat heterogeneity, nutrient build up and herbage biomass production in Aravalli hills of Rajasthan, India. *Tropical Ecology,* 54(1): 73-88.

Tewari, J.C., Ram, M., Roy, M.M. and Dagar J.C. (2014). Livelihood improvements and climate change adaptations through agroforestry in hot arid environments. J. C. Dagar et al. (eds.), Agroforestry Systems in India: Livelihood Security & Ecosystem Services, Advances in Agroforestry 10, DOI: 10.1007/978-81-322-1662-9_6, _ Springer India 2014.

WWC (2019). The Use of Water Today. https://www.worldwatercouncil.org/fileadmin/wwc/ Library/WWVision/Chapter2.pdf.

8

Agroforestry for Income Enhancement in Arid Region

Keerthika A., A.K. Shukla, Dipak Kumar Gupta, M.B. Noor Mohamed and R.S. Mehta

Introduction

Arid region in India is spread over 38.7 million hectare, out of which 32 million hectare is occupied by hot arid zone and remaining 7 million hectare lies in cold arid region. It forms a continuous stretch in the north-western states of Rajasthan (61%), Gujarat (20%), Punjab (5%), Haryana (4%) and scattered landmasses in the peninsular states namely Maharashtra, Karnataka and Andhra Pradesh. In the recent decades, semi- arid region is expanded by 10% and accounts for 34% of the total land area. Reduction in rainfall and increased evapotranspiration has led to increased aridity (IMD, 2018). Indian arid region is the most populated desert of the world with 27.5 million human (Census, 2011). The human population density of this part of the country is quite high (127 person/km^2) as compared to global average of 6–8 persons per km^2 for arid zones. About 40% of the total populations of Rajasthan lives in Thar Desert (Gupta, 2008). The state stands 2nd in livestock population (56.9 millions), 1st in goat (20.84 millions) and camel (0.213 millions) population. The arid ecosystem is most fragile ecosystem which is characterized by frequent drought and extremes of aridity due to low to erratic rainfall (100-420 mm/year), extremes of temperature (-5 to 50°C), long sun shine duration (6.6-10 hours), low relative humidity (30-80%), high wind velocity (9-13 kmph) and high evapotranspiration (1500-2000 mm). Under such adverse situations, traditional agroforestry plays a paramount role in providing food, fodder, fruits, fuel etc. and also provides security to local people and animals/livestock during extreme drought conditions. The traditional agroforestry system is a permanent feature of landscape, where farmers without disturbing the natural tree component, integrate crops, grasses and animal. Especially, there is clear cut demarcation in distribution of agroforestry systems based on rainfall gradient in arid and semiarid region of India. Areas receiving <250 mm rainfall have predominance of grasses and shrubs; hence range/pasture development with livestock rearing becomes the major proposition. In 250–350 mm rainfall zone, besides grasses and shrubs, multi-purpose tree species dominate the landscape. Mixed farming encompassing agroforestry system, mixed cropping, livestock and pasture management are main livelihood options. In the place, where rainfall is more than 300 mm, crops and cropping system diversification, agroforestry and livestock rearing are the major systems of sustenance of arid zone farmers (Bhati and Joshi, 2007). Over all, agriculture and animal husbandry is the main occupation of people of arid region.

Climate and Soils of Arid Region

Arid region has erratic rainfall distribution with the annual average of 200-400 mm mostly occurring during the July to September which is the main cropping season. During summer, mean maximum daily temperature is around 40°C (maximum 52°C) and during winter it is 22-28°C. The minimum daily temperature varies from 4-10°C during winter and 24-26°C during summer. The mean evaporation rate during summer exceeds 10 mm/day (Gupta, 2000). The soils of the north-western arid region described as 'desert soils', and 'grey brown soils'. It falls under soil orders of ardisols, alfisols and entisols. Most of the arid areas (about 64.6%) are duny, where the soils often contain only about 3.2-4.0% clay and 1.4·-1.8% silt. The soils are poor in organic matter having 0.03% OC in bare sand dunes to 0.1 % in the stabilized dunes. Soils are generally rich in total potassium and boron, but are low in nitrogen, phosphorus and micro-nutrients such as copper, zinc and iron. The soils often have high salinity. The ground water is not only limited owing to poor surface and subsurface drainage, but is also saline in quality (Sharma *et al.,* 2013). Soil texture is sandy with average sand content of 89%. In the extreme arid part of Jaisalmer, Bikaner, and Barmer, sand content is > 90 for most of the places. Soil color is mainly yellowish brown. Despite varying factors, agroforestry is playing a paramount role for sustaining productivity and improving income over a period of time.

Agroforestry Area of Rajasthan

Rajasthan is the geographically largest state located in the north-western part of the India, constituting 10.41% of the total geographic area. The area under agroforestry in Rajasthan was estimated to be

1.49 million ha. The figure arised for Rajasthan was on higher side than Forest Survey of India (2013) reports, which has mentioned 0.84 million ha. This is due to significant variation in the methodology adopted by the institutes (Dhyani, 2014; Chavan *et al.,* 2015). Agroforestry area for state level in the selected districts was estimated using the methodology followed by Rizvi *et al.,* 2016 as mentioned in Table 1. Based on the survey made in selected districts, the data was extrapolated for the state (20% of total districts of state).

Table 1. Area of agroforestry in Rajasthan

S.N.	Districts	Agroforestry area (ha)	Agroforestry area (%)	Tree density (tree ha^{-1})
1.	Bikaner	79,752.21	2.64	1.40
2.	Dausa	22430.59	6.54	12.87
3.	Jhunjhunu	37,100.97	6.27	6.95
4.	Pali	84,149.97	6.71	14.90
5.	Sikar	38,792.56	4.99	12.42
	Rajasthan	**1497,593.01**	**4.38**	**9.71**

(Chavan *et al.,* 2020)

Traditional agroforestry of Arid Rajasthan

Traditional agroforestry is a typical parkland system, where growing of annual crops / livestock in association with scattered trees is common practice. People of this region have recognized the importance of tree on farm from ancient time to reduce the adverse effect of climate and soil. Their century old experiences of growing tree on farm, under diverse socio-economic and cultural conditions have been passed from generation to generation and people have established well-contained systems of production (sivlipastoral/agri-silvicultural/agri-silvipastoral/livestock husbandry) that accrue maximum benefits from their woody components. In accordance with distribution of rainfall pattern, the agroforestry system exists in these areas are presented in Table 2.

Table 2. Distribution of traditional agroforestry in arid region of Rajasthan

District	Rain-fall (mm)	Soil	Tree/shrub species	Crops	Prominent grass species
Jaisalmer	100 - 200	Sandy plain, dune	*Ziziphus nummularia, Prosopis cineraria, Acacia senegal, Capparis decidua*	Mung bean, pearl millet and cluster bean	*Lasiurus sindicus*
Sri-ganganagar	200-300	Indus alluvial	*Prosopis cineraria, Acacia nilotica Acacia tortilis*	Pearl millet, mung bean and cluster bean (rainfed), Wheat, cotton, rice and mung bean (irrigated)	*Lasiurus sindicus*
Bikaner	100-400	Sandy plain	*Prosopis cineraria, Ziziphus nummularia, Calligonum polygonoides, Acacia jacquemontii*	Mung bean, moth bean, cluster bean and pearl millet	*Lasiurus sindicus*
Barmer	200-400	Sandy plain, dune	*Prosopis cineraria, Tecomella undulata, Ziziphus nummularia, Capparis decidua*	Pearl millet, mung bean and cluster bean	*L. sindicus, Cenchrus ciliaris*
Jodhpur	200 - 400	Sandy plain, brown light	*Prosopis cineraria, Ziziphus nummularia,*	Pearl millet, mung bean	*Cenchrus ciliaris*

		loam, grey brown loam	*Capparis decidua, Acacia senegal*	and cluster bean (rainfed)	
Churu, Jhunjhunu and Sikar	300-600	Sandy plain, dune	*Prosopis cineraria, Gymnosporia montana, Ziziphus nummularia*	Pearl millet, mung bean and cluster bean	*L. sindicus, C. ciliaris*
Jalore	400-500	Grey brown loam, sandy plain	*Prosopis cineraria, Salvadora oleoides, Acacia nilotica, Punica granatum* (fruit tree)	Pearl millet, mung bean, isabgol, sorghum and cumin	*C. ciliaris*
Pali	400-600	Sandy plain, brown light loam, grey brown loam	*Acacia nilotica.* var. *cupressiformis, Acacia leucophloea, Acacia catechu, Salvadora oleoides*	Sorghum, pearl millet, mung bean and cluster bean	*Cenchrus ciliaris, Cenchrus setigerus*

(Harsh *et al.,* 1992)

Prosopis cineraria (Khejri) based agroforestry practices is the most popular and found to be the most dominant tree species in farmers' fields at Bikaner, Jhunjhunu, Sikar and Pali districts, comprising of 45.17–64.78% total tree species (Chavan *et al.,* 2020). *P. cineraria* belongs to the family Leguminosae and sub family Mimosaceae and well suited to arid climate with several environmental and socio-economic benefits. *P cineraria* based agroforestry system is one of the most popular remunerative systems in arid region associated traditionally with people and also meets livelihood opportunities. Experimental evidences showed its complementary association with cow pea, cluster bean, moth bean, mung bean, mustard, fenugreek and pearl millet as intercrops. Through trial and error and by experience of local people and scientific community, it is proven that the integration of khejri with crops does not reduce yield of agricultural crops (CAZRI and AFRI experiments) rather has positive effect. The common practice of 100% canopy pruning just before winter due to its heavy pruning bearing ability ensures the availability of sunlight to the understorey crops, which makes this as the best agroforestry tree (Ratha Krishnan and Jindal, 2015). The yield of the annual crops can be optimized in combination with Khejri at optimum tree densities of 278 trees ha^{-1} at 6 and 7 years, 208 trees ha^{-1} at 10 year and <208 trees ha^{-1} at 11 years of age (Singh *et al.,* 2007). Besides achieving good yield of dryland crops, bonus yield of dry leaves (25–30 kg/tree) and fuel wood (40–60 kg/tree) could be obtained from these farm trees through annual lopping (Bhati *et al.,* 2008). The leaves are palatable and a nutritional source to small ruminants like sheep and goats.

The economic analysis of khejri based traditional agroforestry system shown that, bajra sole crop alone provides gross income of Rs. 1600 per ha whereas bajra as sole crop intercropped with khejri (30 trees ha^{-1}) is capable of providing an income of Rs. 4600 per ha. Thus, it is revealed that khejri + bajra based agroforestry system is much more remunerative (Harsh and Tewari, 2007). Another study conducted at AFRI, Jodhpur indicated that, at 12th year (208 tree ha^{-1}) *Prosopis cineraria* provides utilizable biomass of 19.96 tonnes ha^{-1} including leaf fodder of 0.85 tonnes ha^{-1} as the additional output. In addition to the crop harvested, farmers may obtain an annual return of Rs. 3000-3500 per hectare as per the present rate of leaf forage and firewood (Singh, 2009).The Khejri fruits are used as vegetable by the local people which costs around Rs. 300-400 per Kilogram in the markets. The sale of the fruits has brought about an improvement in economy of this region. Green fruits are also used as vegetable. Central Institute of Arid Horticulture (CIAH), Bikaner has developed high yielding variety called 'Thar Shoba'. It has been recommended for uniform tender pod especially for vegetable use. This yields a harvest of about 4.25 kg tender pods and 6 kg dry fodder per year (CIAH, 2017). Over all, it is a multi-utility tree, wherein each and every part makes a purpose in day to day activities of local people.

Salvadora oleoides is a multipurpose tree growing on dry, saline and sand dune areas in Indian desert, commonly called as jhal, pilu, peel, meethajhal etc., belonging to the family Salvadoraceae. The tree is known to tolerate very dry environment with mean rainfall of less than 200 mm which is mainly distributed in Barmer, Jalore, Jodhpur and Pali districts of Rajasthan (Bhandari, 1978). It has also very high soil salinity tolerance. Fruits are sweet, edible and have reportedly saved the lives of inhabitants in famine or crop failure. *Salvodora ole-*

oides (density 10 trees ha^{-1}) are intercropped with bajra has reported to produce an average 9.5 q ha^{-1} (Harsh and Tewari, 2003). The average fruit production is 30 kg and which is sold at the rate of Rs. 10/kg. Farmers can earn Rs. 13000 ha^{-1} yr^{-1} from *Salvodora* sp. based agroforestry system if the fruits are utilized properly for making squash whereas the income from sole agricultural crop will be Rs.10000 ha^{-1} yr^{-1} (Harsh and Tewari, 2007). Apart from this, leaves are also good source of feed for goat and camels. The tender leaves are highly preferred by grazing animals.

Tecomella undulata (Rohida or Marwar teak) belonging to the family Bignoniaceaeis mainly found growing in community land and forestland along with *Prosopis cineraria* (L.) Druce (Singh, 2009). It is also state tree of the Rajasthan. Under tree crop interaction studies of *P. cineraria* and *T. undulata* and yields of *Vigna radiata* (L.) R. Wilczek in agroforestry systems indicated that the yield of *V. radiata* increased with increase in distance from the trees. During establishment phase, *T. undulata* neither competed with the associated crops nor was its growth affected by the intercrops. The density of 278 stems/ha was most favorable for crop production at the age of 6-7 years and 208 stems/ha later, at 10-11 years (Tewari and Singh, 2006).The leaves are used as fodder for cattle and goats (Rokde, 2001). The flowers and pods are relished by camel, goats and sheep (Singh, 2010).

Acacia nilotica commonly known as 'babul' belongs to the family Mimosaceae, occurs in most part of the arid region. *Acacia nilotica var. indica and Acacia nilotica var. cupressiformis* (Khajur babul) is distributed in Pali Marwar region. Khajur babul have erect cypress shaped crown, straight bole and less shade casting are the characteristics that naturally fits in to agroforestry system. It is associated with rain fed crops such as pearl millet, maize, sorghum, groundnut, black gram (*Vigna mungo*), green gram (*Vigna radiata*), etc. during *kharif*(summer) or with irrigated crops like wheat, barley, mustard, gram, etc. during *rabi* (winter), returns vary from model to model. The tree is lopped for fodder as s feed for livestock.

Shelterbelt and windbreaks: Trees also act as shelter belt and windbreaks in the high wind velocity area providing protection to the crops and also reducing speed of wind regimes. Prasad et al. (2013) has reported morphological characteristics of different shelterbelts

(*Acacia tortilis, Eucalyptus camaldulensis, Dalbergia sissoo, Tecomella undulata*) that affect wind regimes, air temperature and soil properties in arid western Rajasthan. They also found that all the shelterbelts had caused a maximum reduction (21.5 to 36.0 %) in wind speed on the downwind side at a distance of 2H (H is the average height of shelterbelt). An increase of 305.6% in net returns has been observed by shifting from non-shelter belt to shelterbelt in farms of Jaisalmer district (Gajja *et al.,* 2008).

Agroforestry Models for Arid and Semi- Arid Regions

Agri-horticulture

Farmers intercrop *Ziziphus nummularia* and *Ziziphus rotundifolia* with Bajra or legume crops. In the month of October - November, after harvest of bajra crops, *Ziziphus* spp. are cut from the ground level and allowed to dry. This dry leaf called 'pala' is stored for lean period to feed livestock. This agroforestry system can give gross return of Rs. 7000 – 8000 ha^{-1} (Harsh and Tewari, 2003). CAZRI, Jodhpur has made some efforts to improve the existing *Ziziphus* spp. based agroforestry system for a higher return. Some of the Ziziphus plants were budded with improved cultivars of 'Gola' and 'Seb'. These budded Ziziphus plants produce fruits in the first year itself and from 2nd year onward their fruit yield increases to 30 kg tree^{-1} yr^{-1}. A farmer can earn gross returns of 12,000-15,000 ha^{-1} if farmers maintain 50 trees ha^{-1} besides natural growing Ziziphus (Faroda and Harsh, 1999). The economics of traditional agroforestry systems have been depicted in Table 3.

Table 3. Economics of traditional agroforestry system of Arid Region

System	Component	Cost of cultivation (10^3 /ha)	Net returns (10^3 /ha)	Benefit –cost ratio		
				2012-13	2013-14	2014-15
Agro-forestry	*P.cineraria* + crops	19.4	24.5	2.02	2.34	2.41
Agri-silviculture	*Hardwickia binata* + crops	19.5	19.0	1.52	1.82	2.31

Agri-horti-culture	*Zizyphus mauritiana* + crops	47.4	66.1	2.42	2.39	2.36
Horti-pasture	*Zizyphus mauritiana* + grass	29.2	29.8	1.56	2.54	2.12
Sil-vipas-ture	*Colophosper-mum mo-pane* + grass	16.4	39.6	2.94	3.75	3.48
Agri-pasture	Grass alone	16.3	32.2	2.9	3.29	2.65

(Tanwar *et al.*, 2016)

Silvipastoral systems

In Rajasthan, the majority of the agricultural farmers and livestock farmers have been predominantly following traditional methods of livelihood and most of them are economically vulnerable. Therefore, to maintain ecological balance between human and livestock populations, it is necessary to use at least 20 to 25% of the cropland for silvipastoral activity.

Similar to trees, grasses are also classified based on rainfall gradient (Table 1). Depending on region wise, combination of fodder trees, shrubs and grasses can be included. The main species to be utilized are *Cenchrus ciliaris* (*dahman*) in Jodhpur and Nagaur, and *Lasiurus sindicus* (*sewan*) in Jaisalmer, Barmer, and Bikaner, while *Cenchrus setigerus* in Pali (Tewari and Singh, 2006). Fodder shrubs like *Grewia tenax* can be incorporated in village land along with fodder trees (Khejri, Ardu and Babool) which will solve the fodder scarcity in arid regions. Apart from this, fodder banks are also viable option to meet the fodder scarcity and tree based fodder from different trees of arid region are given in Table 4.

Recently, *Ailanthus excelsa* commonly known as 'ardu' is emerging as promising fodder source in arid and semi- arid region. It grows successfully in the region where minimum average annual rainfall is <500 mm. It is well distributed in the central part of the Rajasthan and along the Aravali mountain range. High density of ardu is found in Sikar, Jaipur, Ajmer, Tonk and Dausa region of Rajasthan. Different forage crops, viz. *Cenchrus ciliaris, Cenchrus setigerus* and *Panicum*

Table 4. Tree based fodder in arid region

S.N.	Tree/shrub	Spacing (m)	Yield (Green fodder in kg)	Harvesting age (Years)
1.	*Prosopis cineraria*	10 x 10	100 kg (app.) between 8-10 years	8-10
2.	*Colophospermum mopane*	8 x 8	5-7 (3- 4 years)	3-4
3.	*Zizyphus nummularia*	5 x 5	3-5 (3-4 years)	3-4
4.	*Salvadora oleoides*	12 x 12	80-100 (8 -10 years)	8-10

(Sharma, 2013)

antidotale can be grown in association with *Ailanthus excelsa* (Jat *et al.*, 2011). Studies at CSWRI, Avikanagar, Rajasthan showed that three tier silvi pastoral system (Cenchrus + *Ailanthus excelsa* + *Dichrostachys cinerea*) provided maximum average forage production (2.78 t ha^{-1} dry forage from pasture + 0.95 green tree leaves) over other systems (Sharma *et al.,* 2001).

Silvipasture systems responds positively to fertilizer application during rainy season and 40 kg N/ha was reported optimum for enhanced biomass production and water-use efficiency in *Hardwickia binata + Cenchrus ciliaris* system (Patidar and Mathur, 2017). In an established silvi-pasture (eleven years), leaf fodder of *Colophospermum mopane* and *Hardwickia binata* trees contributed 15 and 9%, respectively to the total system productivity, but the animals preferred *Hardwickia binata* leaves over *Colophospermum mopane* (Patidar and Mathur, 2017; Patil *et al.,* 2011). Carrying capacity of silvi-pasture system was observed almost double over that of pure pasture (Tewari and Harsh, 1998). *Ziziphus nummularia* + grass (strips in 1:2 ratio) gave higher economic returns from grazing goats due to weight gain of the animals and higher wool production (Bhati, 1997a). The silvipasture system with *Zizyphus rotundifolia* and *Cenchrus ciliaris* sustained 554 cattle days/ha (Tharparkar breed) with 60% pasture utilization (Bhati, 1997b).

Horticultural-pastoral systems

Ziziphus nummularia (Jharberi) is an important top feed species of arid region and almost every part of it has a use. The leaves make excellent fodder having high crude protein (13–17%) and fibre (15%). It generally dominates the common village grazing lands of Bikaner, Barmer and Jaisalmer districts with a relative dominance of 75–90% (Tanwar *et al.,* 2018). *Zizyphus nummularia* when grown with *Cenchrus ciliaris* between trees rows of six year old ber spaced at 6 x 6 m yielded 1.55 tonnes per hectare per year of grass fodder was obtained. *Cenchrus ciliaris* yielded 1.55 tonnes per hectare per year of grass fodder when grown between trees rows of six year old *Zizyphus nummularia* with 6 x 6 m spacing. Higher returns are expected from third year of establishment of horti-pasture system (Sharma and Diwakar, 1989; Gajja *et al.,* 1999).

Role of Agroforestry in Soil Conservation and Carbon Sequestration

Under arid regions of Rajasthan, the common trees in traditional agroforestry are mostly leguminous (*Prosopis cineraria, Tecomella undulata* and *Acacia nilotica),* thereby enhancing soil productivity through biological nitrogen fixation, nutrient cycling and capture of nutrients and water from soils through deep tap root system. Tree based system also increase infiltration rate, bind the soil, reduce runoff and erosion thereby reduce moisture stress. They act as buffer against income risks associated with climatic variability by providing food (eg. Pods from khejri), and fodder (khejri, babul and ardu). It also act as shelter belt and windbreaks in the high wind velocity area providing protection to the crops and also reducing speed of wind regimes and air temperature. Trees in arid agro-ecosystem acts like insurance against climate vagaries to buffer crop-yield losses, and enhance resilience and diversify the system (Chavan *et al.,* 2015)

In arid zone of Rajasthan, *Prosopis cineraria, Acacia nilotica, Tecomella undulata, Azadirachta indica, Ziziphus mauritiana* and *Ailanthus excelsa* are scattered on *farmers* field in varying density and geometry (Roy *et al.,* 2011; Keerthika *et al.,* 2015; Tanwar *et al.,* 2019). The agroforestry of Rajasthan has huge potential to offset global warming i.e. GHG emission at a rate of 1.42 million tonnes of CO_2-eq year^{-1}. *Hardwickia binata* Roxb. based silvipasture system has also reported to sequester carbon at the rate of 2.24–3.44 Mg C ha^{-1}yr^{-1} (Dhyani *et al.,* (2016). This system can sequester about 116.1 ±46.2 Mg CO_2 ha^{-1} in biomass under semi- arid region (Gupta *et al.,*

2019). CSP of existing agroforestry system has been assessed to be as 0.36 Mg C ha^{-1} $year^{-1}$ in Pali, 0.29 Mg C $ha^{-1}year^{-1}$ in Dausa, 0.28 Mg C ha^{-1} $year^{-1}$ in Sikar, 0.22 Mg C ha^{-1} $year^{-1}$ in Jhunjhunu and 0.14 Mg C ha^{-1} $year^{-1}$ in Bikaner. The estimated value of CSP for the entire Rajasthan state was 0.26 Mg C ha^{-1} $year^{-1}$ at average tree density of 9.71 ha^{-1}(Chavan *et al.,* 2020).

Constraints and Way Forward

In the past, the traditional agroforestry systems of this region did function satisfactorily in time and space by providing fuel, fodder, fruit and other products but, due to increasing biotic and abiotic pressure and associated demand, these systems have started deteriorating to a considerable extent along with decrease in number of tree per unit area due to improper tillage and lopping operations as well as lack of awareness on the importance of tree. Moreover, there are very meagre organized models or any kind of agroforestry plantations developing in this region. In spite of multiple benefits from agroforestry practices, majority of farmers have been hesitant to adopt these systems on large scale primarily because of certain apprehensions about the tree component such as long rotation, reduction in gross area, poor income and complicated legal procedures involved in tree farming trade and market fluctuations. Farmers are protecting and conserving trees in farmlands but are very reluctant to go for any new plantation in terms of trees since they have to wait for years together to get economic yield. Apart from yield, these traditional agroforestry are not much concern in carbon policies but having huge potential to contribute in offsetting GHG emission.

Conclusion

During past four decades agroforestry has come of age and attracted scientific community for sustaining production and solving problems (salinity, contamination of water and also excessive usage of fertilizers) in long term. The main component to increase production are proper tillage operation, appropriate lopping, appropriate grazing, appropriate tree density, application of bio-fertilizers, genetic improvement of native tree species for better yield. Apart from this, the National Agroforestry policy and Sub- Mission of Agroforestry has widened scope and also encouraged farmers to open up new ventures like nursery and plantation establishment. Integrated farming system model developed at CAZRI is also new eye opener under the climate change scenario for increasing production and income in a sustained way.

References

Bhati, T.K. and Joshi, N.L. (2007). Farming systems for the sustainable agriculture in Indian Arid zone. (In) Dryland Ecosystem: Indian Perspective, pp. 35–52. Vittal, K.P.R., Srivastava, R.L., Joshi, N.L., Kar, A., Tewari, V.P. and Kathju S. (Eds.). Central Arid Zone Research Institute and Arid Forest Research Institute, Jodhpur, India.

Gupta, M. L. (2008). *Rajasthan Gyan Kosh*. 3rd Edition. Jojo Granthagar, Jodhpur. ISBN 81-86103-05-8

Census (2011). Office of the Registrar General & Census Commissioner, India, Ministry of Home Affairs, Government of India, http://censusindia.gov.in/

Sharma S.K, R.S. Singh and B. Bhargawa (2013). Arid Horticulture- An overview. Annals of arid zone 52 (3 & 4): 251- 264.

Rathakrishnanan and Jindal (2005). Khejri, the king of Indian Thar desert is under pheno phase change. *Current Science*, Vol.108 (11):1987-1990.

Bhati, T.K., Tewari, J.C. and Rathore, S.S. (2008). Productivity and dynamics of integrated farming systems in Western Rajasthan. (In) Diversification of Arid Farming System, pp. 23–30. Pratap Narain, Singh, M.P., Kar, A., Kathju, S. and Praveen Kumar (Eds.), Arid Zone Research Association of India and Scientific Publishers, Jodhpur, India.

Singh, G., S, Mutha and N, Bala (2007). Effect of tree density on productivity of a *Prosopis cineraria* agroforestry system in North Western India, *Journal of Arid Environments*, 70 (1): 152-163.

Harsh, L.N. and J.C.Tewari (2007). Agroforestry systems in arid regions of India. 2007. In: Agroforestry: Systems and Practices. (Eds.) Puri, S and P. Panwar. pp. 175-189.

Rizvi, R. H., Newaj, R., Karmakar, P. S., Saxena, A., & Dhyani, S. K. (2016a). Remote sensing analysis of agroforestry in Bathinda and Patiala districts of Punjab using sub-pixel method and medium resolution data. *Journal of Indian Society of Remote Sensing*, 44, 657–664.

Chavan S.B, Ram Newaj, R. H. Rizvi *et al.*,(2020). Reduction of global warming potential vis-à-vis greenhouse gases through traditional agroforestry systems in Rajasthan, India. Environment, Development and Sustainability.https://doi.org/10.1007/s10668-020-00788-w.

Chavan, S. B., Keerthika, A., Dhyani, S. K., Handa, A. K., Newaj, R., & Rajarajan, K. (2015). National Agroforestry Policy in India: A low hanging fruit. *Current Science*, 108(10), 1826–1834.

Keerthika, A., Chavan, S. B., & Shukla, M. (2015). Khejri agroforestry for addressing issues of soil health. Life science leaflet, 64, 102–108

Gupta, D.K., R. K. Bhatt, Keerthika A, M. B. Noor Mohamed, A. K. Shuklaand B. L. Jangid (2019). Carbon sequestration potential of *Hardwickia binate* Roxb. based agroforestry in hot semi-arid environment of India: an assessment of tree density impact. *Current Science*, Vol 116 (1): 112-116

Dhyani, S. K., Ram, A. and Dev, I. (2016). Potential of agroforestry systems in carbon sequestration in India. *Indian J. Agric. Sci.*, 2016, 86, 1103–1112.

Tanwar, S. P. S., Kumar, P., Verma, A., Bhatt, R. K., Singh, A., Kanhaiya Lal, M., *et al.*, (2019). Carbon sequestration potential of agroforestry systems in the Indian arid zone. *Current Science*, 117, 2014–2022

Singh, G. (2009). Comparative productivity of *Prosopis cineraria* and *Tecomella undulate* based agroforestry systems in degraded lands of Indian Desert. *J Forest Res.* 20(2): 144-150

CIAH, (2017). Varieties Released by CIAH, Biknaer. ICAR-Central Institute of Arid Horticulture, Bikaner, Rajasthan-334006.

Bhandari, M.M. (1978). Flora of Indian Desert, Scientific Publisher, Jodhpur, pp 471.

Harsh, L.N. and J.C.Tewari (2003). Agroforestry systems in hot arid regions with special reference to value addition. In: Agroforestry: Potentials and Opportunities. (Eds.) Pathak P.S and Ram Newaj. 159-165

Singh, G. (2009). Comparative productivity of *Prosopis cineraria* and *Tecomella undulata* based agroforestry systems in degraded lands of Indian Desert. *J For Res* 20:144–150.

Tewari, V.P and Singh, Mahenda (2006). Tree-crop interaction in the Thar Desert of Rajasthan (India). Secheresse. 17(1-2):326-32

Rokde, S. N. (2001). Grazing behavior of Marwari goats in Marwar tract of semi-arid Rajasthan. *Indian J Small Rumin* 7:72–74.

Singh, A. K. (2010). Probable agricultural biodiversity heritage sites in India: VII. Arid West *Reg Asian Agric Hist* 14:337–359.

Prasad, R., Mertia, R.S., Gajja, B.L., Singh, D. and C. Prakash (2013). Morphological characteristics of tree shelterbelts and their impact on wind regimes, air temperature and soil quality. *Indian J. Agroforestry,* 15(1):74-81.

Gajja, B.L., Rajendra Prasad, Mertia, R.S., Khem Chand and Samra, J.S. (2008). Impact of shelterbelt on net returns from agricultural production in arid Western Rajasthan. *Agricultural Economics Research Review* 21(1): 118–122.

Tanwar, S.P.S., Singh, A., Patidar, M., Mathur B.K. and Lal, K. (2016). Integrated farming system with alternate land uses for achieving economic resilience in arid zone farming. (In) Extended Summaries of Fourth International Agronomy Congress 22–26 November 2016, IARI, New Delhi, pp. 246–247.

Harsh, L.N. and J.C.Tewari (2003). Agroforestry systems in hot arid regions with special reference to value addition. In: Agroforestry: Potentials and Opportunities. (Eds.) Pathak P.S. and Ram Newaj. 159-165.

Faroda, A.S. and L.N. Harsh (1999). From empty bowl to sell-reliance: Success of farmers of Desert Villages. Environmental News. June-July pp.14.

Tewari, V. P. and M. Singh (2006).Tree-crop interaction in the Thar Desert of Rajasthan (India). *Article scientifique, Sécheresse* 2006; 17 (1-2): 326-32.

Jat, H.S., Ranjay, K. Singh and J. S. Mann (2011). Ardu (*Ailanthus sp*) in arid ecosystem: A compatible species for combating with drought and securing livelihood security of resource poor people. *Indian Journal of Traditional Knowledge*.Vol.10 (1), pp.102-113.

Patidar, M. and Mathur, B.K. (2017). Enhancing forage production through a silvi-pastoral system in an arid environment. *Agroforestry Systems* 91: 713–727.

Patil, N.V., Mathur, B.K., Patel, A.K. and Bohra, R.C. (2011). Nutrition evaluation of Colophospermum mopane as a fodder. *Indian Veterinary Journal* 1: 87–88.

Tewari, J.C. and Harsh, L.N. (1998). Forestry research in arid tract of India. (In) Agroforestry Systems in India, pp. 307–322. Faroda A.S. and Manjit S.(Eds.), Central Arid Zone Research Institute, Jodhpur.

Bhati, T.K. (1997a). Integrated farming systems for sustainable agriculture on drylands. (In) Sustainable Dryland Agriculture, Central Arid Zone Research Institute, Jodhpur, pp.102–105.

Bhati, T.K. (1997b). Management of dryland crops in Indian arid ecosystem. (In) Desertification Control in the Arid Ecosystem of India for Sustainable Development, pp. 298–307. Surendra Singh and Amal Kar (Eds.), Agro Botanical Publishers (India), Bikaner.

Sharma, N.K. (2013). Fodder Strategy for Sustainable Animal Production in Arid Rajasthan. Ann. Arid Zone 52(2): 101-108.

Sharma, S, C., Mann, J. S. and R. S. Mehta (2001). *Ardh-sushk Kshetro mein Tri-stariye Van Charagahon ka Vikas*, In: *Prakratik Sansadhano ke Parbhandhan mein Krishi-vaniki ka Yogdan*, edited by Dadhwal KS & Solanki KR, (Indian Grassland and Forage Research Institute, Jhansi), 115-116.

Tanwar, S.P.S., T.K. Bhati, Akath Singh, M. Patidar, B.K. Mathur, Praveen Kumar, Ando. P. Yadav (2018). Rainfed integrated farming systems in arid zone of India: Resilience unmatched. *Indian Journal of Agronomy* 63 (4): 403-414.

Sharma, S.K. and Diwakar, G.D. (1989). Economic evaluation of horti-pastoral system in arid region of Western Rajasthan. *Indian Journal of Range Management* 10(2): 119–122

Gajja, B.L., Bhati, T.K., Harsh, L.N. and Khan, M.S. (1999). Comparative economics of silvipasture, hortipasture and annual crops on marginal agricultural lands of arid zone of Rajasthan. *Annals of Arid Zone* 38(2): 33–37.

Harsh, L.N., Tewari, J.C., Burman, U. and S. Sharma (1992). Agro-forestry in arid regions. *Indian Farming* 1992; 45: 32-7.

Hardwickia binata based silvipastoral system in semi-arid region of Rajasthan.

9

Livestock-based Diversification in Watershed for Sustainability in Drylands

Prabhat Kumar Pankaj, K. Sreenivas Reddy, V. Maruthi and G. Nirmala

Introduction

The Dry Climates are characterised by less precipitation than evaporation. Dry climate zones cover about 26 percent of the world's land area. They have abundant sunshine (summer temperatures are high; winters are cooler and longer than Tropical Moist climates) and rainfall is irregular (several years of drought are often followed by a single year of abundant rainfall). Almost 2.5 billion people live in dry areas and about one-third depend on dryland agricultural production systems for their food security and livelihoods. Dryland production systems cover about three billion hectares, or 41% of the Earth's land area, and employ a highly diverse mixture of crops used for food, feed and fibre; rangeland and pasture species; trees used for a multitude of purposes; and fish and livestock. The rainfed agro-ecosystem has a distinct place in Indian agriculture, occupying 67% of the cultivated area, contributing 44% of the food grains and supporting 40% of the human and 65% of the livestock population (Misra, 2006).

Farmers in dryland (arid and semi-arid) regions of India suffer from scientific integration of natural resources leading to deterioration in soil quality and low resource use efficiency. Most of them keep one or another type of livestock (Cattle/buffalo/sheep/goat/pig/poultry) with few acres of cropping system. The farming systems in rainfed areas are quite diverse with a variety of crops and cropping systems, agroforestry and livestock production. Traditionally the farmers are growing crops such as sorghum, castor, pigeon pea, finger millet, groundnut and paddy depending on the soil type, rainfall and availability of ground water. The changes in the quantity of rainfall received and its distribution pattern at the project location are leading to intermittent droughts during the crop period resulting in crop failures leading to debts and migration. Livestock plays an important role in the economy of India in general and sustainable livelihood of poor people of dryland region in particular. Dairy animals enable poor and landless farmers to earn income using common-property resources (World Bank, 1999), and provide a constant flow of income and reduce the vulnerability of agricultural production (Holmann *et al.,* 2005). The expanding market for livestock products offers an opportunity for augmenting their income, even for those who do not have access to land and capital resources (FAO, 2000). In India, milk production contributes on an average 27 percent of the household income; its contribution varies from about 19 per cent in the case of large farmers to about 53 percent in the landless category in India (Shukla and Brahmankar, 1999). The residues from both crop as well as livestock are treated as waste under field conditions and contribute majorly to the harmful gas emissions by burning in the open fields.

The agriculture sector in India witnessed a skewed development since early 70's with much of the development-taking place in the irrigated regions at the cost of rainfed areas. For example, the green revolution was confined to the irrigated and better-endowed regions of the country. To make up for this lacuna and also because the dry lands account for more than 60% of the cropped area in the country, several programs have been initiated for the development of dryland agriculture, like for instance, the introduction of the Integrated Wasteland Development Program (IWDP) of 1989-90 and the National Watershed Development Program for Rain-fed Areas (NWDPRA) of 1990-91. Improving agricultural production and restoring ecological balance are the twin objectives of these programs. Watershed approach allows for a more holistic development of the agricultural sector i.e., crop and allied sectors like, horticulture, livestock, fisheries, etc., with focus on integrated farming systems and management of common property resources to augment family income and improve nutritional levels of communities participating in watershed programs. The watershed program focuses on soil and water conservation and is expected to improve crop yields and green fodder availability. This, in turn, is likely to have an impact on milk production.

Recent projections indicate that the world population could be increased to 12.3 billion people by 2100 (Gerland *et al.,* 2014). To satisfy the food requirement of this population, research has suggested that agriculture must double yields (Tilman *et al.,* 2002, 2011; Tilman and Clark, 2015). The increased agricultural yields may be an achievable task through the intensification and expansion of farming accompanied by sufficient water supply to ensure improved crop growth and livestock survival (McNeill *et al.,* 2017). This inextricable link between food production and water resources is vital to the subsistence of humankind. However, agricultural production drives unacceptable environmental impacts that include soil degradation, greenhouse gas emissions, and water quality impairment (Swain *et al.,* 2018). Therefore, production systems that can increase agricultural yields while reducing adverse impacts on the environment, especially preserving water quantity and quality, are the major issue in the research. In agricultural systems which is gaining popularity among researchers, producers, and agricultural land managers because it can improve both soil productivity and environmental quality (Rakkar and Blanco-Canqui, 2018; Sulc and Franzluebbers, 2014).

Watersheds are the logical study units for understanding land-water connections and managing land use impacts. Watershed development refers to the conservation; regeneration and the judicious use of all the natural resources particularly land, water, vegetation and animals and human development within the watershed. All land-based productive activities are dependent on the topography of the area, soil type, available biomass and water and call for adopting an integrated management approach, which can be better evolved within a natural domain such as the watershed. This is because a watershed is an intricate, dynamic and natural functional unit established by physical relationships and social communication and actions. Thus, watershed as a unit enables planners and implementing agency to consider all inputs, processes and expected outputs systematically which are essential for a holistic development approach. In a dry climate situation, where the majority of the population is engaged in livelihoods largely dependent on the natural resource such as agriculture, factors like quality, availability and access to natural resources occupy a crucial role in influencing incomes earned by households. In a situation like this, watershed development and management practices play a dual role of natural resource conservation as well as livelihood enhancement. For instance, improved soil quality and water availability facilitate higher crop productivity and more fodder for livestock, which in turn augments income earned by the people engaged in such occupation.

All regions of the world are experiencing, and will continue to experience, the effects of climate change with varying magnitude and consequences. Climate change is already affecting the four dimensions of food security (i.e. the physical availability of food, its economic and physical accessibility, its use, and the stability of these three dimensions over time), and its implications extend across all determinants of malnutrition. Climate change affects food security in complex ways. It impacts crops, livestock, forestry, fisheries and aquaculture, and can cause grave social and economic consequences in the form of reduced incomes, eroded livelihoods, trade disruption and adverse health impacts. However, it is important to note that the net impact of climate change depends not only on the extent of the climatic shock but also on the underlying vulnerabilities. With changing climate, increasing human population and changes in land use/land cover, a comprehensive quantification of livestock feed is needed at different level in order for countries to develop policies for maintaining or increasing livestock production.

Earlier, each component of agricultural production has been described in isolation from each other. A holistic approach of interaction among land use-watershed-livestock will unfurl many new areas of study in the dry areas to achieve food and nutritional security. The inter-relationship and inter-dependencies among water, land, vegetation and animal resources determine the nature and kind of livelihood support systems particularly in rural areas. Land use changes, human-livestock population dynamics, climate change variability and lack of livelihoods diversification aggravate watershed degradation through inappropriate land use methods resulting to water scarcity, land and water pollution, and food and nutritional security. Soil erosion and siltation has led to land denudation, habitat loss and farm lands losing their soil fertility and compromising food security. Food security, malnutrition, climate change can no longer be addressed independently of one another, it is only possible through scientific land use/land cover, watershed management and livestock interaction. The present chapter will be reviewed works of connecting watershed with livestock production system in dry climate for tackling food and nutritional security.

Livestock Production System and its Linkage with Watershed in Dryland

Livestock sector plays an important role in the rural economy of India with a high contribution to the gross domestic product (GDP) and a high absorption of female labour. The sector accounts for 5.59% to the GDP and 27.7% of the income from agriculture in India in 2001-02. In absolute terms, the sector has contributed 84.6 million tons of milk, 50.7 million tons of meat and 34 billion eggs and significant amount of organic manure.

Most of the farmers in watershed keep more than one species of domestic animals and consider them as an indicator of wealth. Farmers also perceive that livestock is a mean of risk aversion in case of natural disaster in watersheds. Manure from cattle, sheep and goats is used as farmyard fertilizer, while cattle dung is also used as fuel source for cooking food, and as material for decorating house. Watershed development while improving the crop sector also improve the feed and fodder situation and thus facilitate dairy development. Ban on grazing in forest areas, and the restrictions to grazing in natural resource conservation and management programmes as a policy matter may cause reduction in rearing indigenous cattle and non-dairy livestock. In watershed areas, poverty alleviation programmes and animal husbandry schemes/programmes that promote cross-bred cattle for improving economic returns through increased milk production have accelerated a shift towards rearing crossbred cows.

In dryland, a success story was reported by Wani *et al.,* (2012) about crop-livestock linkages in watershed of six villages (Malleboinpally, Mentapally, Nandavaram, Devanakonda, Thirumalapuram and Kacharam), Medak district, Telangana in a study period of 5 years (2000-2005). They reported that firstly, bovine activity is higher in the watershed village indicating that improvement in soil and moisture conditions leads to development of the livestock sector. This is because of the improvement in the availability of green fodder after implementing the soil and moisture conservation measures undertaken as a part of the program. The proportion of households maintaining bovines increased from 60% in the control village to 68.3% in the watershed village. Secondly, there was a shift from small ruminants to bovine activity. Studies showed that small ruminant activity is confined to resource-poor areas (Hanumantha Rao, 2000). The shift from small ruminant to bovine activity in the watershed village indicates improvement in the resource base of the village due to the watershed program. The proportion of households maintaining small ruminants declined from 30.9% to 26.3% and this shift came because of the shift of small farmers from small ruminants to bovine sector. It is to be noted that though small ruminant production is more in resource-poor areas, it is not high among resource-poor farmers. The proportion of milch holdings increased from 22.2% to 39.0% and the proportion of work holdings declined from 47.2% to 22.0%. The share of mixed holdings also increased from 30.6% to 39.0%. As a result of these shifts, the proportion of bovine holdings producing milk increased steeply from 52.8% to 78.0%. Fourthly, the improvement in the bovine sector comes through productivity improvement and not through increase in the size of the herd. The improvement in the green fodder availability in the watershed village improved milk production and this improvement came through spread of the activity and improvement in milk yield. There was no increase in the number of milch animals per household.

Livestock Feed Resources in Watershed

The cattle and buffaloes are normally fed on fodder available from cultivated areas, supplemented to a small extent by harvested grasses and top feeds. The three major sources of fodder supply are crop residues, cultivated fodder and fodder from common property resources like forests, permanent pastures and grazing lands. At present, the country faces roughly a net deficit of 63 % green fodder, 23.5 % dry crop residues and 64 % feeds.

Mondal *et al.,* (2014) analysed the impact of watershed (2010-11 as compared to 2000-2001) in Bundelkhand region of India which is an extensive dryland area and reported that Labour force participation rate among farmers in watershed villages was relatively higher implying that the higher cropping intensity and crop production generated new employment opportunities. The herd size per family and per ha of cultivated area in watershed villages (3.76 and 2.27 unit) were comparatively bigger than that of control villages (3.53 and 1.99 unit) which was attributed to improvement of grass/ fodder land and development of non-arable areas with grasses that might have encouraged the households to maintain more number of livestock in watershed villages. The large farmers maintained the larger herds compared to other categories in the watershed areas. Although the watershed development programmes look indifferent to the livestock but the results indicate that the farmers in the watershed villages are likely to maintain larger herd sizes compared to control areas due to better vegetation and fodder availability.

Crop residues, an important component in the livestock feed, are available from food grain crops and pulses. The yield of crop residues is expected to increase with increase in the crop yield and shift in cropping pattern. Cropping pattern is more favourable to livestock feed in the watershed village as compared to the control village. The share of food grains is higher in the watershed village than in the control village due to a larger extent of area under crops. Thus, the availability of crop residues per ha of cultivated land as well as per adult bovine unit in the watershed village is reported to be more than that of the control village both due to shift in cropping pattern towards food grains and higher crop yields. The level of feeding dry matter is higher by 75% in the watershed village than in the control village. The improvement in the feeding of dry fodder is only 35%. Thus, feeding levels improved through mostly green fodder and a little bit of dry fodder. These two types of feeds more than compensated the decline in the feeding of concentrates. It should be noted that farmers always try to manage with home-grown feeds rather than purchased feeds. The watershed program is expected to reduce the demand for concentrates because of the higher availability of green fodder. As the quality of animals improves, demand for concentrates will again increase. There is a close linkage between the crop and livestock sector in the selected watershed villages. Livestock sector makes a significant contribution to the income in villages with well-developed dairy sector. The success of the dairy sector depends on several aspects but feed availability is one of the critical factors influencing dairy development. Impact study for a completed watershed village indicates that due to implementation of watershed program the availability of feeds and fodder increases and, in this case, particularly green fodder that stimulated the growth of dairy sector. At the same time the feeding of concentrates has declined indicating farmers' preference for home grown feeds/fodder. Clearly the watershed program is beneficial to poor and small-scale livestock keepers.

Watershed activities improves the water availability in the dependent area which can be utilized by allotting some of the area for forage production and management. Watershed programme is an important intervention in dryland areas to improve livestock productivity through increased feed and fodder supply. Availability of higher quantity of forages from diversified land-use, improved use of common property resources and improved livestock management system enhance the livestock productivity.

Feed and fodder cost constitute about 60-70% of cost of milk production, thus cultivated fodder has an important role in meeting requirement of various nutrients and roughage in our country to produce milk most economically as compared to concentrates. Feeding not only meets nutrient requirement but also fills the rumen to satisfy the animals. Feed has to meet requirement of cattle maintenance, production and requirement of microbes to promote digestion. Fodder production and its utilization depend on the cropping pattern, climate, social-economic conditions and type of livestock. The study revealed that there was a drastic change in the area under cultivation and the type of crop with the implementation of watershed project. The crops which required less amount of water were cultivated more prior to project implementation in Bunga watershed, these were maize, jowar fodder, bajra and groundnut in the monsoon and pulses like gram and pigeon pea in the post-monsoon season. With the availability of water after the implementation of the project, the area under these crops reduced drastically and was replaced by water demanding crops like sugarcane and paddy, besides maize in the monsoon season. In post-monsoon season, about 59 per cent of the area was under pulses in 1982-83, which reduced to 1 per cent in 2006-07. The area under wheat increased from 37 to 73 per cent and berseem from nil to 6 per cent. During the survey conducted in 2007-08, it was reported that more than half of the storage capacity of water harvesting reservoir had been lost due to siltation from the catchment area. Consequently, water availability to the crops also reduced. Significant changes in livestock sector as a whole can be achieved only when there has been a substantial increase in irrigation intensity. It means that availability of irrigation is central to increased cropping intensity, crop and fodder productivity and forage resources availability from common forest lands. The study by Chopra and Gulati (2001), Shah (2001) and Deshingkar and Start (2003) have also highlighted that irrigation intensity and regeneration of forests have direct impact on strengthening the crop-livestock linkages. There is also a need to explore a large part of the available water as common pool resources and thereby expand its net of beneficiaries. At the same time, emphasis should be on efficient use of resources on private lands (Birthal *et al.,* 2005). Successful watershed village has lowered gap in the availability of fodder and demand for livestock which is a major source of livelihood for small and marginal farmers. Long-term study in Bunga has shown effectiveness of watershed development programme in reducing the gap in demand and supply of livestock fodder. The gap, however, widened after reaching the saturation point of watershed interventions. The Bunga watershed has reached its saturation level after a period of 20 years. The reservoir needs to be de-silted and renovated to revive its rainwater storage capacity. The history of ponds and 'johars' in India shows that desiltation was a regular activity undertaken by the local communities. Unfortunately, such a tradition has slowly died down.

Sustainable Watershed Management Using Livestock in Climate Change Scenario

Livestock products are an important agricultural commodity for global food security because they provide 17% of global kilocalorie consumption and 33% of global protein consumption. The livestock sector contributes 14.5% of global Green House Gas (GHG) emissions (Gerber *et al.,* 2013), which is the major reason of global warming and climate change (IPCC, 2013). The primary livestock GHG emissions are CO_2, CH_4, and N_2O, where CH_4 contributes the most to anthropogenic GHG emissions (44%), followed by N_2O (29%) and CO_2 (27%) (Gerber *et al.,* 2013). Higher concentrations of these gases, can be explained by lower efficiency and productivity of livestock system due to excess loss of nutrients, energy, and organic matter (Gerber *et al.,* 2013). Thus, there is need to explain the system which will be able to enhance the productivity ultimately improving its mitigation capacity.

The livestock sector contributes directly and indirectly to GHG emissions, including through animal physiology, animal housing, manure storage, manure treatments, land application, and chemical fertilizers (Monteny *et al.,* 2001). Direct emissions from animal sources include enteric fermentation, respiration, and excretions. Indirect emissions refers to emissions derived from feed crops, manure application, farm operations, livestock products processing, transportation, and the land use allocation for livestock production (e.g. deforestation, desertification, carbon released from cultivated soils) (IPCC, 1997; Mosier *et al.,* 1998). In the livestock sector, indirect emissions play a greater role in the release of carbon to the atmosphere than direct emissions (Steinfeld *et al.,* 2006).

Adaptation measures involve production and management system modifications (IFAD, 2010). Research is needed on assessments for implementing these adaptation measures and tailoring them based on location and livestock system. Similarly, mitigation means implementing policies to reduce GHG emissions and enhance sinks'. There is potential to reduce livestock sector GHG emissions through the implementation of different technologies and practices (Gerber *et al.,* 2013). Adaptation and mitigation are complementary to each other. For example, if mitigation measures are undertaken effectively, lesser will be the impacts to which we will need to adapt. Similarly, if adaptation measures (or the degree of preparedness) are strong, lesser might be the impacts associated with any given degree of climate change. Still limited research have been carried out on livestock production systems and its simple adaptive and mitigative means to be suggestive to farmers, this paper aims to observe management system of sheep as an adaptation strategy and to quantify methane emission for identifying the possible mitigation strategies in dryland regions of India.

Sustainability of Participatory Watershed Management Using Livestock

It is being recognised more and more that to increase the equitable impact of watershed development, more attention needs to be paid to livestock production, which is a major livelihood activity in most rain fed areas. In addition to providing income through animal and dairy sales, livestock plays an important role in crop production as a source of manure and draught power. Especially in the ecologically fragile areas of semi-arid watersheds, livestock makes an important contribution to the survival of the economically weaker sections. It is even ironic that the watershed guidelines and policies developed by both the central and the state Governments, which for over 30 years have been considered the main catalysts for the development and growth of drought prone areas, have paid only lip service to livestock development.

Social Empowerment Using Livestock in Watersheds

Though the importance of livestock in watershed management is generally recognized, there is a conspicuous absence of data and analysis on watershed management — livestock interaction at strategic, institutional as well as operational levels. As a matter of fact, livestock does not seem to be considered a priority in the relevant strategy papers. There are a number of studies that have looked at the impact of watershed programmes on rural livelihoods with a focus on crops and related activities, but few have considered the importance of livestock, its implications to rural livelihood and especially, feed and fodder availability and requirement (Birhtal and Taneja, 2006; Subrahmanyan *et al.,* 2006). Livestock production is an integral part of the farming system in the rainfed areas, especially in the ecologically fragile semi-arid watersheds (Puskur and Thorpe, 2005). Livestock plays a vital role in the economy of Shivalik foothill villages in Haryana state and is the primary source of livelihood contributing 50-65 per cent of the total income (Arya *et al.,* 1994; Arya and Samra, 2001; 2006). It is increasingly being recognised that to increase the equity of watershed development, more attention has to be paid to livestock production, which is a major livelihood activity in most rainfed areas. In addition to providing income through animal and dairy sales, it has an important function for crop production by providing manure and draught power. Especially in the ecologically fragile areas of semi-arid watersheds, livestock makes an important contribution to the survival of the economically weaker sections (India Task Force 1987), small ruminants in particular playing an important role in ensuring rural livelihoods against drought (Pasha, 2000).

It is well established that as the aridity increases, the choice of rearing small ruminant gets prioritized due to its high drought tolerant capacity, better feed utilization ability, sustainability on low water supply due to low water excretion and low requirement and low feed requirement. In dryland region of India, sheep are the people choice due to its tender meat for local preparations and ease in management. Mostly farmers rear small ruminants to supplement their income and regular monetary supply as and when required. Thus, in drylands, a systematic study on sheep production system is required to suggest suitable adaptive production system.

Watershed-Crop Intensification and Diversification-Integrated Nutrient Management

Sustaining intensification of smallholder livestock system should take into account both social and environmental welfare and be targeted to sectors and areas of most positive social welfare returns and where natural resource conditions allow for intensification. However, in fragile environments, de-intensification is required to avoid irreversible damage to the ecosystems. Smallholder livestock keepers dominate crop-livestock systems, with livestock playing an essential role in highly diversified livelihood strategies that typically combine crops and livestock with off-farm activities. Combination of unsustainable land use, over-exploitation of resources and climate change can potentially desertify millions of hectares. It is hard to deny the capital benefits of an increase in global trade and expansion of the livestock production and consumption. Although various regulations have been implemented around the world there is still more that can be done to mitigate environmental damage and keep the growth of industry. Intensification may be the answer for increasing supply to meet demand while lessening impact to our ecosystem. Confinement at high stocking density is one part of a systematic effort to produce the highest output at the lowest cost by relying on economies of scale, modern machinery, biotechnology, and global trade. The biotechnological method of intensification is genetically modifying livestock to be more cost effective. Resulting in a single cattle or chicken producing a larger quantity of product than before. These intensifications can produce a larger quantity of milk with fewer emissions due to an increase in productivity per cow. There is a continuing debate over the benefits, risks and ethical questions of factory farming. The issues include the efficiency of food production; animal welfare; whether it is essential for feeding the growing global population; and health risks. However, with world's trend in recent technological advancement these costs to society can be diminished. Intensification is very efficient in providing for increase in demand. It also benefits the market as secondary industries will continue to grow. Organic food markets will benefit from the increase of intensification as some consumers may feel more confident in purchasing organic products. They will pay a premium to cover for the higher cost that organic livestock production yields, but essentially it brings more options to the consumers and opens market entry of new businesses. To ensure intensification is beneficial in developing countries, as there will be farmers pushed out of the market. Government authorities should provide the secondary producers that will eventually leave the market support transition into labour jobs in the factors of production or provide a short-term subsidy to these individuals to provide a smooth economic conversion process. International collaboration should be a focus between the major players in producing livestock products. There should be an international body like the IMF sort of speak, that would set standards of production and international policies to ensure upgraded monitoring of the factors of production. Having one central larger facility where the meat or poultry gets converted in product would also help to pin point potential problems and will make it easier to implement tight restrictions on production. Due to the rapid expansion of global trade having central facilities may make it easier for authorities to shut down production if need be.

The overall impact of watershed activities have resulted into crop diversification (cropping system, vegetables and fodder with small ruminants) combined with social engineering through capacity building measures and participatory institutional mechanism resulted into not only doubling the farmers' net income through technological package, but also improved the overall water productivity of crops by 77%, improved nutrient availability in terms of protein (pulses and goat meat) and micro-nutrients (vegetables and goat meat), reduction in forest migration, reduction in mortality and morbidity of livestock, improved biomass for animals and small ruminants, etc. in the Petrallachenu village, Nagar Kurnool district, Telangana, India (Reddy *et al.*, 2018).

Livestock based Integrated Farming System in Watersheds

Under a project on Agri-consortium Research Platform on water, ICAR-CRIDA, Hyderabad, India has established a dynamic integrated farming system model at Tribal village of Nagar Kurnool district of Telangana, India. The selection of major crop in the integrated farming system was done by a local village institution named Integrated Rain Water Management Association (IRWMA). During rainy season of 2016-2017, pulse-based farming system which includes black gram (1 hectare), vegetables (0.2 hectare), fodder (0.05 hectare), and small ruminants (10 animals) was taken up in the farmers' field (Fig-1). Due to the long dry spell in the onset of monsoon period, the farmer sow this system late in the field and two irrigation periods are considered, one in the vegetative stage and the other in the pod development stage. The vegetables and fodder were given regular deficit irrigation of 20 mm at 10 days' interval. During the rainy season of 2017-2018, the IFS module of cotton-based system which include cotton (0.2 hectare), vegetables (0.2 hectare), fodder (0.05 hectare), and small ruminants (10 animals) was taken up by the farmer. During vegetative stage and bowl formation stage, critical irrigations were given from the OFR with irrigation depth of 30 mm. Similarly, vegetable and fodder were maintained with the 10 days irrigation interval with 30 mm irrigation depth due to more collection of runoffs into the OFR. According to Kumar et al. (2018), Integrated Farming System (IFS) ensures the highest standard of food production with minimum environmental impact even under highly vulnerable climatic condition.

Figure 1. Pulse-based integrated farming system.

The overall impact of this intervention (crop diversification (cropping system, vegetables and fodder with small ruminants) combined with social engineering through capacity building measures and participatory institutional mechanism resulted into not only doubling the farmers' net income through technological package, but also improved the overall water productivity of crops by 77%, improved nutrient availability in terms of protein (pulses and goat meat) and micro-nutrients (vegetables and goat meat), reduction in forest migration, reduction in mortality and morbidity of livestock, improved biomass for animals and small ruminants, etc.

Figure 2. Farming system experimental site in ICAR-CRIDA.

A long term (2004-2011) study was carried out to document farm pond (650 m^3 capacity) based integrated farming system module for nutrient management on an Alfisol sandy loam textural class (sand: 79.1%, silt: 7.6% and clay: 13.3%) covering an area of 1.65 ha of Hayathnagar Research Farm, ICAR-Central Research Institute for Dryland Agriculture, Hyderabad (17º20'N latitude and 78º35"E longitude). The crop and vegetable components were grown on three pediments consisting of upper soil depth of 8.0±4.3 cm (D_1) with low water requirements of grasses (19 % of area), middle basin of depth 15±4.4 cm (D_2) with arable crops and cropping systems as sorghum + pigeon pea (2:1) and castor + cluster bean (1:1) (48% of area) and lower basin of depth 30±12 cm (D_3) with agro-forestry systems (25 % of area) and vegetables (8% of area), respectively (Fig-2). The rainfall was categorized as deficit (<20% of normal as in 2004, 2009 and 2011), normal (750 mm long term rainfall as in 2006 and 2007) and excess (>20% of normal as in 2005, 2008 and 2010) and evaluation of farming system in different categories of rainfall years were carried out based on net income (Rs.), B:C ratio, nutrient recycling, etc.

The net income (Rs.) using one dairy animal (17471-25002) or two sheep (18681-24334) were the highest among all the combination over different rainfall years. These benefits over different year were unaffected by the rainfall of corresponding year due to the fact that biomass obtained from the cropping system was sufficient to cater the need of these animals and the production was nearly unaffected. Among the cropping systems, sorghum + pigeon pea (3:1) were the most performing combination with a net income of Rs. 10941-17918.

B:C ratio of dairy animals (2.15-2.22) and sheep (2.63-2.88) were the highest among all the combination of farming systems in normal and excess rainfall years, but the ratio was highest in sorghum+pigeon pea (3:1) combination during deficient rainfall years (Figure 3). This change in B:C ratio was observed due to increased cost of production of keeping livestock. B:C ratio in the minor components were excellent (1.73-3.27) during normal and excess rainfall years due to water dependency nature of these components. In the deficient rainfall years also farm pond was able to provide sufficient water to the system, but profitability was severely compromised if livestock component was not taken (B:C ration of 1.56).

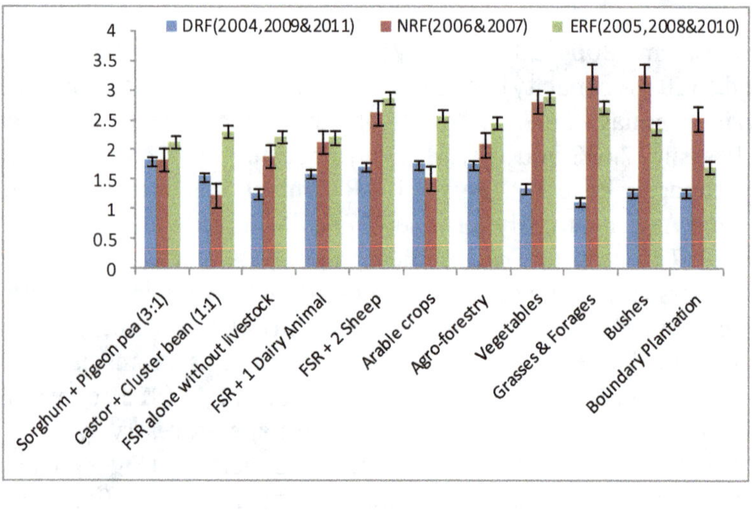

Figure 3. B:C ratio of farming system model with cropping system and minor components in variable rainfall years.

Number of days that can be supported for one dairy animal was only 307 days during deficient rainfall years, however, it can sustain upto 391 days during normal rainfall years and even more during excess rainfall years. The urea fertilizer can be saved to the tune of 21.43 Kg during deficient rainfall years and 26.84-27.33 Kg during normal/excess rainfall years (Figure 4). Similarly, SSP fertilizer can be saved to the tune of 25.2 Kg during normal and 31.56-32.14 Kg during normal/excess rainfall years. Even muriate of potash fertilizer can be saved upto 11.9 Kg during deficit and 14.87-15.15 Kg during normal/excess rainfall years. Thus, the meta-analysis can suggest that using this type of farming module could save huge national money on fertilizer even using unproductive cattle.

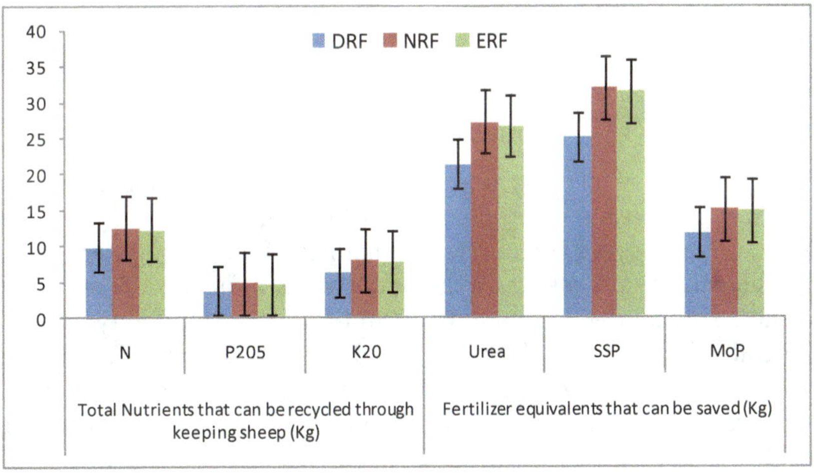

Figure 4. Nutrient recycling though dairy in rainfed watershed-based farming system during various year.

Sheep are the poor man's cow in arid and semi-arid regions of drylands. This module will be able to save huge money spent every year on soil nutrients. The suggested module will be able to save the urea fertilizer to the tune of 54.9 Kg during deficient rainfall years and 68.8-70.1 Kg during normal/excess rainfall years (Figure 5). Similarly, SSP fertilizer can be saved to the tune of 84 Kg during normal and 105.1-107.1 Kg during normal/excess rainfall years. Even muriate of potash fertilizer can be saved upto 34.1 Kg during deficit and 42.7-43.5 Kg during normal/excess rainfall years.

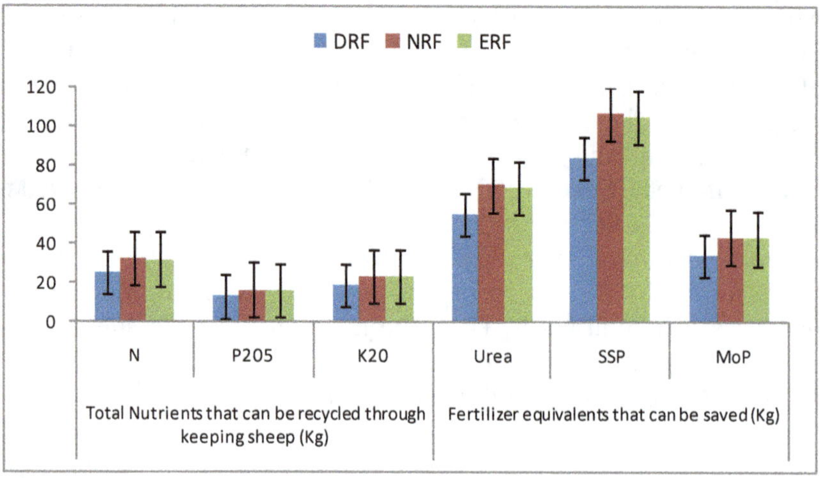

Figure 5. Nutrient recycling though sheep rearing in rainfed watershed-based farming system during various years.

Thus, this type of farming module could save huge national money on fertilizer even using unproductive cattle. Overall, this study was undertaken to observe the nutrient dynamics of farming system under rainfed conditions to sustain the nutrient need for the soil of a small farmer. This model was supportive in terms of profitability and less dependency on external inputs, etc. during excess rainfall years as compared to normal/deficit rainfall years. Scaling-up of such models is necessary to address the livelihood security of small farmers in rainfed areas.

Impact of Livestock Production System to the Environment

One major environmental concern is how the livestock industry is affecting water quality. When phosphorus and nitrogen in manure are applied to land in greater amounts than can actually be used by growing plants, excess nutrients can leach into ground water or run off into surface water with heavy rainfall, floods and melting snow. Excessive levels of phosphorus and nitrogen in surface and ground water fuel the production of algae and aquatic plants that can change aquatic life habitat, reduce essential levels of oxygen, clog fishing nets, interfere with drinking water treatment facilities and cause taste and odour problems in drinking water (Foster and Reimer, 2007). These effects on water quality from the livestock industry is at a cost to other markets. There will be a cost associated with purifying and filtering the water so that it may be safe for consumer use and or even to maintain a healthy environment in an ecosystem. The pollutants from the run off that enter a nearby river will affect the fish in that river, most likely poisoning the fish, thus, decreasing the demand for fish and negatively impacting the fishing market and economies around the world.

Apart from this, livestock industry has on easily spread pollutions such as gas emissions and pathogens. Natural, accidently or deliberate deaths occur all the time in the livestock industry, with regulation the bodies or excess can be disposed of properly. However, cases have occurred were diseases and infections have spread from animals to human population. As the industry grows further in the developing world it is hard for regulators to meditate and attain efficiency in the market. Animal pathogens are a natural occurrence and tend to stay within livestock populations. Most pathogens carried by animals are harmless to people but without the regulations or even sometimes with regulation the pathogens being transmitted to humanity does present itself as lethal, leaving a cost on the society and economy to bare.

Perhaps overlooked by many is the contribution that the livestock industry has made on air pollution. Air pollution can be generated by buildings, manure handling and storage systems and during land application in the form of odours and gases (in particular ammonia) generated from anaerobic manure decomposition. (Paton, 2003)Dust can also be a concern arising from feed systems and the animals. Two related aspects of air pollution are odour and ammonia. Most problems arise with intensive pig and poultry units, and when animal manures are stored for lengthy periods: anaerobic (oxygen deprived) storage of manure, common practice in intensive livestock production, results in the production of malodorous compounds, which are intensified under severe winter conditions when biological activity is minimized.

Watershed-Land Use-Livestock Connection

Watershed activity factors influenced by livestock activities which ultimately influences the land use are frequency (no. of repeated activities), intensity/ magnitude of use (animal unit months), seasonality (biomass availability varies with season), motivation (socio-cultural/ recreational activities). Livestock grazing is the most extensive use of majority of watersheds under public lands, if poorly managed, can modify the sediment content across different slopes affecting compactness and infiltration properties of soil. Physically, the grazing also leads to surface disruption, trampling of vegetation leading to plant species composition on upland areas. Watersheds, when properly functioning, provide habitats with multiple niches, but poorly managed watersheds do not support species diversity nor do they provide quality natural resources. Concentrated Animal Feeding Operations (CAFOs) are agricultural operations where large numbers of livestock or poultry are kept and raised in buildings or confined feedlots that occupy a relatively small land area. Manure and wastewater from CAFOs can contribute pollutants such as nitrogen and phosphorus, organic matter, sediment, pathogens, heavy metals, hormones, antibiotics, and ammonia to the environment.

Grazing patterns, livestock numbers and distribution, and manure management all have implications for soils, water quality. Pastoralist systems that fulfilled the requirements of grazing without over-grazing have been badly misunderstood by the society. We need to understand the relationships where livestock contribute a major part of the solution to environmental health that are critical to the health of the land in arid and semi-arid parts of the world. Part of this health means more plant growth and thus more carbon capture. In developing countries, grassland and rangeland vegetation comprise a large component of the feed that is potentially available for livestock use. Worldwide, rangelands occupy almost 50 percent of the terrestrial land cover, and provide almost 75 percent of the forage used by domestic livestock.

Land management practice that can have a surprisingly large impact on the relationship between soil and water on grasslands or common property resources is grazing. It is established that switching from conventional grazing to "multi-paddock" grazing, wherein land was grazed quickly then rested for long periods (a process that mimics the natural movement of grazing animals), led to an almost 40% reduction in quick surface runoff, 5% increase in infiltration into the soil, and a reduction in immediate stream flow of roughly 30%. An increase in infiltration rate has been reported (from 0.5 inches per hour to more than eight inches per hour) as a result of similar changes in grazing practices. Thus, in an extensive area of pasture, grassland and common pool resources, judicious land management with livestock grazing can rebuild soil health and may be an untapped tool for re-shaping the living watersheds in the drylands. Compared with cropland, forage systems protect the soil surface from erosion, and, if fertilizer and animal waste inputs are low to moderate, both surface and ground water quality under grazed areas may be better than that under cropped areas. Apart from this, an update on water hoof print will improve the value of the subject. Prevailing version of land use planning with respect to livestock sustainability in dry climate will be discussed and future desirable modifications in those areas will also be suggested.

Watershed development is crucial to build resilience in agriculture. Watershed is defined as the bio-hydro-geological unit consisting of land, water, vegetation, livestock, climate, people and its socio-economic environment (Mangurkar and Kumar, 2001). Livestock (all farm animals - large and small ruminants and poultry) management is one of the major components of integrated watershed management. Livestock contributes not only to the income generation, livelihoods security thereby reducing poverty but also to food and nutritional security. Watersheds are the integral part of farm land in drylands which are primarily meant for soil and water conservation. The integration of livestock in watershed was considered as an essential pre-requisite to the most economic use of fertilizer, family labour and to increase the farm income (Singh *et al.*, 1993). In dry areas, crop production system is always at a higher risk and under such scenarios, livestock rearing plays an important role in livelihood security of poor farmers.

The dying link in the harsh and unproductive environments of the arid and semi-arid rain-fed systems, livestock rearing goes beyond the purpose of mere milk and meat production and supporting rural livelihoods (Ashley *et al.,* 1999): it plays a critical role in providing draught power for agriculture and transportation, biodiversity regeneration and maintenance, and most importantly, manure for agriculture. This is a low-carbon system. It is increasingly being recognised that to increase the equity of watershed development, more attention has to be paid to livestock production, which is a major livelihood activity in most rainfed areas. In addition to providing income through animal and dairy sales, it has an important function for crop production by providing manure and draught power. Especially in the ecologically fragile areas of semi-arid watersheds, livestock makes an important contribution to the survival of the economically weaker sections (India Task Force, 1987), small ruminants in particular playing an important role in ensuring rural livelihoods against drought (Pasha, 2000).

Watershed and Livestock Policy

While large ruminants, particularly milch buffaloes, remain dependent on irrigated fodder or crop residues, the small ruminants and to a lesser extent non-lactating or non-draft cattle remain heavily dependent on biomass produced on the common lands. This raises an important policy and management challenge vis-a-vis the relative investment (financial, government, NGO, or community management) in WDPs on private vs common or public lands. When livestock owners do not take part in the planning and implementation of watershed development, the direct benefits of increased biomass availability and access to common lands are not recognised as such.

Watershed treatments generally increase crop output and therefore output of crop residues. With watershed projects a shift in cropping pattern has also taken place, from low to high value crops or from mono to mixed and inter-cropping. An increase in the extent of irrigated area and the well command area, and also an increase in ground water table, ranging from 3 to 38 m in different watersheds have been reported. Beneficial impact of watershed programme on moisture availability has been observed (Ninan, 1997). In the study watersheds, 100 per cent of the sample villages reported an increase in productivity of both irrigated and rainfed crops.

It is felt that there has been a general decline in the quality and quantity of crop residues for fodder, especially so during drought years. While 35 per cent of the villages report improvement in the availability of crop residues for some households, 27 per cent report a decline in availability for most households in the village. As a result, the households have responded by decreasing their livestock holdings, but would like to acquire livestock again when the situation becomes comfortable. There is a large import of fodder from outside the watershed during drought periods. The water harvesting structures though not planned explicitly with a focus on livestock, have served the livestock drinking water purposes. In several places they eased the constraint of water on livestock production. While 31 per cent of the villages report shortages of livestock drinking water and conflicts with crop production during drought years, 27 per cent report the same being true even in normal years. It has been observed in many areas that due to lack of water, and in some areas due to consumption of dirty water by livestock, there is high incidence of livestock diseases. It has also been pointed out that in many places farmers consider construction of farm ponds as failure since they may not be able to use such water for various reasons (Ramanna, 1991) and for livestock consumption.

In most villages where watershed development takes place, attention is paid to soil erosion caused by overgrazing on common lands. The grazing restrictions that are usually imposed to reduce the pressure on the land do not necessarily have to affect marginal, small ruminant holders negatively: in the study sites innovative ways of equitable common and pasture land management were found. In Kalyanpur watershed, fodder from common pasture lands is collectively harvested and equally shared. In fact in all sites, even small ruminant holders seemed to prefer controlled grazing to open access, since the availability of fodder is increased. However, only in 16 per cent of the villages does controlled grazing takes place. Establishing arrangements for controlled grazing to improve fodder availability on common lands could be an important instrument to increase the productivity of livestock in semi-arid watersheds.

Rangeland degradation is globally widespread in recent decades primarily due to climate change and overgrazing, especially in arid and alpine environments (Geerken and Ilaiwi, 2004; O'Reagain *et al.,* 2014). Rangeland conservation and forage shortage become a dilemma faced by the development of animal husbandry. The key to this problem is to foster forage production from other multiple sources. Sown pasture for forage production is a major supply of livestock rations in grazing systems due to its intensive production and high nutrient. But the establishment of a sown pasture needs to occupy a certain area of agricultural land. Incorporating forage production during the fallow periods in croplands is a possible option to solve the conflicts of feed deficits and limitation of land resources.

Indian agriculture currently faces a host of diverse challenges due to the ever growing population, increasing food and fodder needs, natural resources degradation, higher cost of inputs and climate change. Supplemental irrigation and water harvesting are the most important and proven technologies for improving crop productivity and the efficient of use of water in the dryland areas of semi-arid tropics. These technologies integrated with crop diversification can increase farmers' income in dryland agriculture systems with improved nutrient availability and food security among the farmers. Results from experiments in tribal farmer field during 2016-2018 with two integrated farming system models (pulse and cotton) in Telangana region highlighted how the maximum net benefits were obtained in the cotton-based IFS model under the OFR from Rs. 51,000/- to 69,000/- as compared to pulsed based IFS model from Rs. 17,000/- to 35,000/- under different investment options with and without government support. There was an increase in the water productivity of cotton by 77% over the rainfed. The traditional system before the project followed by the farmer with mono cropping of sorghum and cotton are not sustainable with negative net returns with cotton and Rs. 12,000/- in case of sorghum. This research highlighted how farmers' income can be doubled with IFS model with technological support of On Farm Resources (OFR) for runoff storage and utilization in the farmers' field. Moreover, such structures would restore the soil loss along with nutrients into the farmers' fields. From this study, it can be concluded that the adoption of the OFR and IFS models not only ensured economic returns but provides regular employment even in less than one hectare of land, which is usually non-sustainable if mono-cropping is being practiced. The approaches used in this work might be useful for guiding producer's decision in optimizing resources and selecting crops and cropping systems based on economic and water use efficiency. Small and marginal farmers can take suitable crops along with horticulture, animals and other components that would minimize water and soil risks, and provide additional income and employment from same piece of land. In addition, future technicalized crop production trends may also adopt the IFS where potentials of environmental degradation and global warming effects are substantially reduced along with reduction/minimization of agricultural operational cost.

References

Arya, Swarn Lata and Samra, J. S. (2001). Re-visiting Watershed Management Institutions in Haryana Shiwaliks, India. Central Soil and Water Conservation Research and Training Institute, Research Centre, Chandigarh. 360 p.

Arya, Swarn Lata and Samra, J. S. (2006). Impact analysis of watershed development projects in Shiwalik foothill villages in Haryana state. Agricultural Situation in India, 62(2):711-722.

Arya, Swarn Lata, Agnihotri, Y. and Samra, J.S. (1994). Watershed management induced changes in animal population structure, income and cattle migration in Shiwaliks. Ambio, Journal of Royal Swedish Academy of Science, 23(7):446-450.

Ashley S, Holden S and Bazeley P (1999). Livestock in Poverty Focused Development. Livestock in Development. Crewkerne. United Kingdom.

Birthal, P.S., Mishra, S.N, Dixit, A.K. and Tripathi, G. (2005). India's Livestock Feed Balance and its Environmental Implications. Project Report. National Centre for Agricultural Economics and Policy Research, New Delhi.

Chopra, Kanchan and Gulati, S. C. (2001). Migration, CPRs and Environmental Degradation: Inter-linkages in India's Arid and Semi-Arid Regions. Sage Publications, New Delhi.

Deshingkar, P. and Start, D. (2003). Seasonal Migration for Livelihoods, Coping, Accumulation and Exclusion. Working Paper No. 220. Overseas Development Institute, London.

F.A.O. (2000). Agriculture Towards 2015/30. Technical Interim Report, April 2000. Economic and Social Department, FAO, Rome.

Foster, C. and Reimer, E. (2007). Rare Plant Surveys by the Manitoba Conservation Data Centre, 2006. MS Report 07-01. Manitoba Conservation Data Centre, Winnipeg, Manitoba. 53 pp.

Geerken, R. and Ilaiwi, Md. (2004). Assessment of rangeland degradation and development of a strategy for rehabilitation. Remote Sensing of Environment 90(4):490-504.

Gerber, P.J., Hristov, A.N., Henderson, B., Mak-kar, H., Oh, J., Lee, C., Meinen, R., Montes, F., Ott, T., Firkins, J., Rotz, A., Dell, C., Adesogan, A.T., Yang, W.Z., Tricarico, J.M., Kebreab, E., Waghorn, G., Dijkstra, J. & Oosting, S. (2013). Technical options for the mitigation of direct methane and nitrous oxide emissions from livestock: a review. Animal, 7 (2): 220–234.

Gerland, P., Raftery, A.E., Ševčíková, H., Nan Li, Danan Gu, Thomas, S., Leontine, A., Bailey K.F., Jennifer, C., Nevena, L., Guiomar, B., Thomas, B., Gerhard, K.H. and John Wilmoth. (2014). World population stabilization unlikely this century Science. 346(6206): 234-237.

Hanumantha Rao, C.H. (2000). 'Watershed Development in India: Recent Experiences and Emerging Issues', Economic and Political Weekly, November 4, pp.3943-47.

Holmann, F., Rivas, L, Urbina, N., Rivera, B., Giraldo, L.A., Guzman, S., Martinez, M., Medina, A. and Ramirez, G. (2005). The role of livestock in poverty alleviation: An analysis of Colombia. Livestock Research for Rural Development. 17 (1):1-7.

India Task Force (1987). Report of the Task Force to Evaluate Impact of Sheep and Goat Rearing in Ecologically Fragile Zone, GOUI, Ministry of Agriculture, Department of Agriculture and Cooperation, New Delhi.

International Fund for Agricultural Development (IFAD). (2010). IFAD's livestock position paper: Livestock planning, challenges and strategies for livestock development in IFAD. Rome: IFAD.

IPCC (1997). IPCC Working Group I Stabilization of Atmospheric Greenhouse Gases: Physical, Biological and Socio-economic Implications (IPCC Technical paper III, 1997).

IPCC (2013). Climate Change 2013: The Physical Science Basis. Contribution of Working Group I to the Fifth Assessment Report of the Intergovernmental Panel on Climate Change. Cambridge University Press, Cambridge, United Kingdom and New York, NY, USA, pp. 1535.

Kadam, M.M., Ganvir, B.N. and Lamtule, J.A. (2013). Impact of watershed project on farmers' economy in Nagpur district. Internat. Res. J. agric. Eco. & Stat., 4 (2): 176-179.

Mangurkar, B R and C Ravi Kumar (2001). 'Livestock-Environment Interactions in Watersheds in India', CALPI/SDC, New Delhi.

McNeill, S.H., Belk, K.E., Campbell, W.W. and Gifford, C.L. (2017). Coming to terms: meat's role in a healthful diet. Animal Frontiers, 7(4): 34-42.

Mishra, S.N. and Sharma, R. (1990). Livestock Development in India - An Appraisal. Vikas Publishing House Pvt Ltd, New Delhi. 53 p.

Misra, A.K. (2006). Integration of livestock in land use diversification. In: K.D. Sharma and B. Soni (Eds) Land Use Diversification for sustainable rainfed agriculture. Atlantic Publishers and Distributor's, New Delhi. pp. 277-299.

Mondal, B., Singh, A., Singh, D.R. and Sekar, I. (2014). Watershed-livestock linkages: A study in Bundelkhand region of Madhya Pradesh. Indian J. Anim. Res., 48 (3): 262-269.

Monteny, G.J., Groenestein, C.M. and Hilhorst, M.A. (2001). Interactions and coupling between emissions of methane and nitrous oxide from animal husbandry. Nutrient Cycling in Agro ecosystems. 60(1-3):123-132.

Mosier, A., Kroeze, C., Nevison, C., Oenema, O., Seitzinger, S. and Oswald van Cleemput. (1998). Closing the global N_2O budget: nitrous oxide emissions through the agricultural nitrogen cycle. Nutrient Cycling in Agro ecosystems, 52(2-3):225-248.

Ninan, K. N. (1997). 'Economic and Environmental Benefits of Watershed- Base Dry Land Development Programmes in India' in Anil Agarwal (ed), The Challenge of the Balance, Centre for Science and Environment, New Delhi.

O'Reagain, P., Scanlan, J., Hunt, L., Cowley, R., Walsh, D., (2014). Sustainable grazing management for temporal and spatial variability in north Australian rangelands – a synthesis of the latest evidence and recommendations. Rangel. J. 36, 223–232.

Pasha, S.M. (2000). Economic and Ecological Dimensions of Livestock Economy, Commonwealth Publishers, New Delhi.

Paton, D. (2003). Disaster preparedness: a social-cognitive perspective. Disaster Prevention and Management, 12(3): 210-216.

Planning Commission, Government of India (2001). Draft Report on the Working Group on Animal Husbandry and Dairying for Five Year Plan (2002-2007), New Delhi.

Puskur, Ranjitha and Thorpe, William (2005). Crop and noncrop productivity gains: Livestock in water scarce watersheds In: Watershed Management Challenges, Improving Productivity, Resources and Livelihoods, Eds: Bharat R. Sharma, J.S. Samra, C.A. Scott and Suhas P. Wani. Department of Agriculture & Cooperation, Ministry of Agriculture, Government of India, New Delhi.

Rakkar, M.K. and Blanco-Canqui, H. (2018). Grazing of crop residues: Impacts on soils and crop production. Agriculture, Ecosystems & Environment. 258: 71-90.

Ramanna, R (1991). 'Watershed Approach to Dryland Agricultural Development', Indian Journal of Agricultural Economics, No 3, July-September.

Reddy, K.S., Maruthi, V., Pankaj, P.K., Kumar, M., Pushpanjali, Nagasree, K., Sammi Reddy, K. and Sai Krishna, T. (2018). Doubling farmers' income through On Farm Reservoirs (OFR) based IFS models for small landholders of SAT Regions: A Success Story in Tribal Region of Telangana, India, ICAR-Central Research Institute for Dryland Agriculture, Hyderabad, India. pp-12.

Shah, Amita (2001). Water scarcity induced migration: How far watershed projects can help. Economic and Political Weekly, 36(35): 3405-3410.

Shukla R.K. and Brahmankar, S.D. (1999). Impact evaluation of operation flood on rural dairy sector. National Council of Applied Economic Research, New Delhi pp.58-60.

Singh, B. and Narang, M.P. (1993). Indigestible cell wall fractions in relation to lignin content of various forages. Indian J. Anim. Sci., 63 (2): 196-200.

Singh, K., Sandhu, H.S., Singh, N. and Kumar, B. (1993). Kandi Watershed Development Project: A Critical Evaluation. Economic and Political Weekly, 28(52): A122-A128.

Steinfeld, H., Gerber, P., Wassenaar, T., Castel, V., Rosales, M. and de Haan, C. (2006). Livestock's long shadow, FAO, Rome.

Subrahmanyam, S., Bezkorowajnyj, Peter, Shiferaw Bekele, Wani, S.P., Parthasarathy Rao, P. and Nageswara Rao, G.D. (2006). Crop-Livestock Linkages in Watersheds of Andhra Pradesh. Global Theme on Agro ecosystems Report No. 29. International Crops Research Institute for the Semi-Arid Tropics, Patancheru, Andhra Pradesh, India. 52 p.

Sulc, R.M., Franzluebbers, A.J., (2014). Exploring integrated crop–livestock systems in different eco regions of the United States. Eur. J. Agron. 57, 21–30.

Swain, M., Blomqvist, L., McNamara, J., Ripple, W.J. (2018). Reducing the environmental

Tilman, D., Balzer, C., Hill, J., Befort, B.L., 2011. Global food demand and the sustainable intensification of agriculture. Proc. Natl. Acad. Sci. Unit. States Am. 108: 20260-20264.

Tilman, D., Cassman, K.G., Matson, P.A., Naylor, R., Polasky, S. (2002). Agricultural sustainability and intensive production practices. Nature 418, 671–677.

Tilman, D., Clark, M. (2015). Food, agriculture & the environment: can we feed the world & save the earth? Daedalus 144, 8–23.

Wani, S.P., Garg, K.K., Singh, A.K., Rockstrom, J. (2012). Sustainable management of scarce water resource in tropical rainfed agriculture. In Soil Water and Agronomic Productivity, ed. R. Lal and B.A. Stewart, 347–408. Advances in Soil Science. CRC Press (Taylor and Francis): United Kingdom.

World Bank (1999). India: Livestock Sector review: Enhancing growth and development. The World Bank and Allied Publishers: New Delhi.

10

Improving Production System of Small Ruminants under Watershed in Dry Areas

A.K. Patel

Introduction

Small ruminants production systems plays a vital role in sustenance of livelihood of rural poor of dry lands agro-ecosystem and particularly important for landless, small and marginal farmers, where crop production is a risk prone enterprise due to scanty rainfall and frequent droughts. Livestock contributed 16% to the income of small farm households that indicates importance of livestock sector as a whole. Agriculture sector grew at low rate of 4.7% during 2013-14 and contributed 13.9% to National GDP (Economic Survey 2013-14). The livestock sector contributed over 4.1 per cent of the total GDP in 2012-13. According to estimates of the Central Statistics Office (CSO) the value of output livestock sector at current prices was about Rs. 591691 Crores during 2015-16 which is about 28.5 % of the value of output from agricultural and allied sector.

Sheep and goats are an important livestock species for a large number of rural poor, especially socially backward, marginal and landless labourers in India. Sheep are mostly reared for wool and meat purpose. India ranks third in sheep population and account for nearly 6% of world sheep population, currently India has 65.06 million sheep as per Livestock census 2012. The wool production is almost stagnant in the country for last one decade, however, the slight increase was achieved in wool production to about 48 m kg in 2013-14 from last year 46.1 m kg (BAHS, 2014). Despite of low wool production in country the handmade carpet exports have crossed Rs. 8559.01 crores during 2014-15 fiscal year. Woolen industry provides employment to about 2.6 million people. Out of total production of raw wool in the country about 70% is carpet grade, 20% coarse grade and 10% apparel grade (Karim and Shakyawar, 2011). Besides this sheep also significantly contributes to mutton production which increased to 485.53 m kg in 2015-16. The goat population over the past few decades has increased 186.4% (from 47.2 m in 1951-52 to 135.2 m in 2012). According to Compound Annual Growth Rate (CAGR) of sheep and goat during (1992-2003) in the country shows that their population has been on increase at a fast rate in some of the districts (Kumar *et al.,* 2010). This is a main factor that makes goat as most desired species of animal for meat production in the country. The meat production from goats in India as estimated is 942.91 m kg (BAH & FS 2016). The goats and their products contribute Rs. 17661.7 crores annually to the national economy.

The dry areas of arid and semi arid region of the country are more favourable for livestock and pasture production system. The harsh climatic conditions prevailing in these regions, e.g. low and erratic rainfall, high evapo-transpiration, high solar radiations, frequent droughts etc make it unsuitable for crop farming. Majority of the lands is covered under arid zone (31.8 lakh square km) in the country. The 62% of hot arid region lies in the state of Rajasthan and 1/3rd area of arid zone of Rajasthan are wastelands, of which 50% are grazing lands and 45% are sandy wastes (Balak Ram *et al.,* 2005). The production from available grazing lands is hardly 300-400 kg/ha biomass. This low productivity is due to deterioration and over grazing. Further, the occurrence of droughts are common in arid zone which affects agricultural and forage production. This coupled with degraded grazing lands caused feed and fodder deficit for large population of livestock in arid zone. The fodder deficit of western Rajasthan is estimated to be as high as 60% of the demand and ranges from 55% in western districts to 69% in central districts and 72% in eastern districts (Pratap Narain and Kar, 2005). This situation is further aggravated during drought years. The deficiency of dry and green fodder was estimated to be 35.9% and 79.9% during drought year (Patidar and Saxena, 2007).

The dry areas of arid and semi arid nature require a planned effort to create sufficient feed base in terms of feed and fodder to sustain the productivity of livestock in the region and it is more essential to rejuvenate the natural rangelands by practicing the methods of reseeding, protection and grazing management. In order to combat the problems of overall nutritional deficiency, there is need to develop and popularize agroforestry system in arid and semiarid lands for enhancing forage biomass even in low and erratic rainfall years.

Climate Change and Agriculture, Forestry & Livestock Productivity

Climate change has become a serious environmental, social and economic threat particularly to natural ecosystems, biodiversity, and livelihood of agricultural and forest fringe communities. Climate change is becoming a key driver in agriculture and forest management, but its complexity and magnitude threaten the sustainability. Agriculture along with forestry and animal husbandry can play an important role in climate change adaptation, through diversified land-use practices, livelihoods and sources of income. Sustainable agroforestry has a tremendous potential to serve as a tool in adapting climate change, protecting ecosystems and livelihoods, and creating a foundation for economic and social development. On the contrary another issues of climate change is expected to alter temperature, precipitation, atmosphere carbon dioxide levels and water availability which will affect the productivity of crops and livestock systems (Hatfield *et al.,* 2008). Climate change could increase thermal stress for animals affecting animal health, reproduction; production and overall productivity with reduced profitability. Climate changes, could impact the economic viability of livestock production systems worldwide. Heat stress negatively affects reproductive function. Normal oestrus activity and fertility are disrupted in livestock during summer months. The livelihood of the rural poor in developing countries depends critically on local natural resource-based activities such as crop and animal production. Due to negative weather impact on livestock rearing, the poor farmers whose principal livelihood security depends on these animal performance is directly on stake. Housing and other management technologies are available through which climate impacts on livestock can be minimized. Under the changing climatic scenario, the concept of multiple stresses emerges as a potential threat to livestock production and its survival. Hence, there is a need to prioritize the research to tackle the multiple stress including nutrition stress and sustain the productivity of animals in change climate scenario.

Forage Scenario of Arid Rajasthan

The fodder production vary widely depends on the cropping pattern, socio-economic conditions, and type of livestock. The cattle and buffaloes are normally fed on the fodder available from cultivated areas, supplemented to a small extent by harvest-harvested grasses and top feeds. The three sources of fodder supply are crop residues, cultivated fodder and fodder from common properly resources, viz. forests, permanent pasture and grazing lands. The deficiency in field and fodder is identified as one of the major constraints in achieving desired level of livestock productivity. The requirement and availability of fodder during different situations is described in Table 1. The acute fodder shortage in central and eastern districts is due to extensive cropping, fragmentation of land holding, intensive cultivation and shirking of pasture & grazing lands.

Table 1. Requirement and availability of fodder in arid zone of Rajasthan

Fodder	Requirement	Availability (mt)		Deficit (%)	
		During normal year	During drought year	During normal year	During drought year
Dry fodder	22	14.00	7.68	36	65
Green fodder	41	12.78	10.71	69	74
Concentrate	13	2.66	0.67	80	94

The gap between the demand and supply will continue to widen if appropriate strategies are not initiated to enhance biomass and forage production substantially.

The research outcome of last five decades of research institutions in country for pasture establishment and utilization techniques for various sub-agro-climatic regions showed that pasture grasses like *Cenchrus ciliaris, Cenchrus setigerus, Lasiurus sindicus, Dichanthium annulatum* and *Panicum antidotale* are promising for higher forage production (Daulay and Bhati, 1989).

Silvipasture Systems in Arid Zone

Traditionally silvipasture is used to grass farming in forest areas, which accommodate millions of livestock for grazing on the herbaceous growth. Extending this principle, silvipasture is the conscious practice of cultivation of forage annuals along with fodder trees and shrubs somewhat similar to agroforestry. Farmers have encouraged and protected the Khejri (*P. cineraria*) and Bordi (*Z. nummularia*) in their crop fields to lop them in winter for top feed and fuel wood. Recently Nair (1993) defined the silvipastoral system as the land use system in which trees and/or shrubs are combined with livestock and pasture for forage and fuel wood production on the same unit of land. Silvipastoral system has special significance in arid and semi-arid regions, where there is lower vegetation and high rate of deforestation and land degradation due to low and erratic rainfall.

The productivity level of ruminants is by and large extremely low, because of poor quality and seasonal nature of forage supply in the arid region. The problem becomes more acute in dry seasons. In such situations, forage based silvipastoral system could be a promise. Besides, trees in silvipastoral system also provide shade to the grazing animals and serve as fences, hedge, etc.

Suitable silvipastoral models for different land forms and rainfall situations

The selection of suitable species of trees/shrubs, grasses and legume for establishment of silvipastoral system depend on agro-climatic condition of the regions. The following trees/shrubs and pasture species foe different land forms and rainfall zones are identified.

(A) Suitable species for silivipastoral system under different landform			
Type of land	**Suitable species**		
	Grasses	**Legumes**	**Trees/shrubs**
Desert land and sand dunes	*Lasiurus sindicus, Cenchrus ciliaris, Cenchrus setigerus*	*Lab lab purpureus, Clitoria ternatea, Atylosia scarabaeoides*	*Acacia tortilis, Acacia nilotica, Acacia senegal, Prosopis cineraria, Prosopis juliflora, Azadirachta indica, Zizyphus nummularia, Colophospermum mopane, Dichrostachys nutan, Colligonum polygonoides*
Ravine lands	*Cenchrus ciliaris, Cenchrus setigerus,*	*Stylosanthes* spp, *Stizolobium*	*Acacia tortilis, Acacia nilotica, Acacia cate-*

	Dichanthium annulatum, Saccharam spontaneum, Chrysopogon fuluus	*decrigenum, Macroptilium atropurpureum, Atylosia* spp	*chu, Albizia lebbek, Albizia amara, Dalbergia sissoo Zizyphus spp., Ficus spp., Emblica officinalis, Eugenia jambolana*
Cultivable waste lands	*Cenchrus ciliaris, Cenchrus setigerus, Pennisetum polystachyon, Panicum antidotale, Sehima neruosum, Chrysopogon fuluus, Dischanthium annulatum*	*Macroptilium atropurpureum, Glycine jaranica, Clitoria ternatea*	*Albizia spp, Hardwickia binnata, Leucaena leucocephala, Acacia* spp. *Sesbania* spp., *Dichrostachys nutans, Prosopis cineraria, Ziziphus mauritiana*
Salt affected area	*Cynodon dactylon, Paspalium notatum, Chloris gayana, Lasiurus sindicus, Brachiaria mutica, Sporolobus marginatus, Urochloa*	*Glycine jauanica, Macroptillum* spp. *Phaseolus junatea, Stylosanthes*	*Acacia tortilis,* A. *nilotica, Prosopis juliflora, Saluadora spp., Ziziphus spp. Sesbania spp., Albizia amara, Atriplex spp*
(B) Suitable species of grasses and trees on the basis of rainfall			
Rainfall Zone	**Grasses**	**Legumes**	**Trees/shrubs**
150 to 300	*Lasiurus sindicus, Cenchrus ciliaris*	*Lab lab purpureus, Clitoria ternatea*	*Prosopis juliflora, Acacia tortilis, Acacia senegal, Ziziphus nummularia. Colligonum polygonoides, Dichrostachys nutans*
300 to 500 mm	*Lasiurus sindicus, Cenchrus cilliaris, Cenchrus setigerus, Panicum antidotale*	*Lab lab purpureus, Clitoria ternatea*	*Acacia. tortilis, A. nilotica,* *P. cineraria, Tecomella undulata, Colophospermum mopone, Ziziphus spp., Colligonum polygonoides, Azadirachta indica, H. binnata, Ailanthus exelsa,*

			Albizia lebbek
Above 500 mm	*C. ciliaris, Dichanthium annutatum, Chrysopogon fulvus, Sehima nervosum, Panicum antidotale, Heteropgon contortus*	*Stylosanthes*	*Acacia nilotica, Acacia catechu, Dalbergia sissoo, Leucaena leucocephala, Albizia spp., Alianthus excelsa, Hardwickia binata*

Forage Production of various Silvipastoral Models

In traditional silvipastoral system, trees/shrubs are permitted to grow at random in fields on the boundary of the fields or community grazing land with grasses of low productivity and nutritive value. In arid zone of Rajasthan, Ahuja (1980) recorded highest grass yield (1.1 to 1.5 t ha^{-1}) under Khejri and lowest (0.6 to 0.7 t ha^{-1}) under Kumat plantation of 14 to 18 year old with substantial contribution of *Cenchrus* and *Eleusine* species as under storey perennial grasses. The productivity and carrying capacity of grazing land can be increased through improved silvipastoral system. At Jhansi, Pathak *et al.,* (1996) reported that the degraded lands producing hardly up to 1 t ha^{-1} $year^{-1}$ biomass, which can be improved up to 10 t/ha through silvipastoral system. A study conducted at NRCAF, Jhansi revealed that an average total biomass yield of 12.62 t ha^{-1} $year^{-1}$ recorded under silvipastoral system consisting of *A. amara* and *L. leucocephala* as trees, *D. cinerea* as shrub and *C. fulvus* + *S. hamata* as pasture was about 4 time higher than yield (3.16 t ha^{-1} $year^{-1}$) obtained from natural grassland (Rai *et al.,* 1999). Silvipastoral studies undertaken at CAZRI, Pali involving four tree species, (*Acacia tortilis, Azadirachta indica, Albizia lebbek and Holoptelia integrifolia*) and four grasses (*C. ciliaris, C. setigerus. D. annlatum and Panicum antidotale*) revealed non significant difference in the dry fodder yield under different tree species, however, the mean dry forage yield was maximum in *Dichanthium annulatum* (2.8 t ha^{-1}) followed by 2.51 t ha^{-1} in *Cenchrus ciliaris* and 2.2 t ha^{-1} in *Panicum antidotale* (Muthana and Shankarnarayan,1978). Silvipastoral studies at Jodhpur indicated that sowing of *Cenchrus ciliaris* under the cover *H. integrifolia* did not interfere with normal growth with an additional dry forage yield of 0.7 to 1.2 t/ha. In another study Harsh *et al.,* (1992) reported that a mixture of *C. fulvus* and *Sehim nervosum* grass produced 3.25 t/ha forage under the canopy of five year old *Hardwickia binata*. Similarly Patidar (2006) recorded maximum herbage yield with *L. sindicus*+ cowpea (9.0 and 3.1 t ha^{-1}) green and dry forage yield) followed by sole (8.9 and 3.08 t ha^{-1}) green and dry fodder yield and *C. ciliaris* + cowpea system (8.46 and 2.86 t ha^{-1} green and dry forage yield grown in association with *C. mopane* and *H. binnata*. The forage yield was not different due to tree species however, application of 40 kg N/ha enhanced fodder yield by 13-14% over control. Yadav and Poonia (2003) recorded maximum biomass yield of grasses and tree from silvipastoral system with *Dichrostachys nutans* followed by *Prosopis cineraria* at existing stock trees at farmers' field in arid climatic condition of Rajasthan. In a horti-pastoral study on sandy rain fed of Rajasthan, Sharma and Diwakar (1989) reported that *Cenchrus ciliaris* in association with *Ziziphus mauritiana* system produced 1.2 t ha^{-1} forage yield and did not affect the fruit yield.

A lot of work on silvi-pasture and agroforestry systems was carried out to get maximum forage yield in semi-arid zone at CSWRI, Avikanagar. The highest dry fodder production (59.06 q ha^{-1}) was recorded with mixture of *Cenchrus and Dolichos*. Whereas lowest yield (28.82 q ha^{-1}) from *Cenchrus setigerus* alone was recorded. From mixture of *Cenchrus* and *Clitoria*, 45.2 q ha^{-1} dry fodder was obtained. The intercropping of cowpea and moth found to increase 5 and 2 times higher dry fodder yield of *Cenchrus* (Sharma and Chand, 2012). Maximum dry fodder (41.0 q ha^{-1}) from single cutting was obtained when carpet legume was sown at spacing of 30 x 30 cm. In low rainfall year (375 mm) the highest production was recorded from mixture of pearl millet and carpet legume in the ratio of 1:2, whereas, in high rainfall year (1000 mm) the maximum fodder was obtained from the mixture of maize and carpet legume. Harvesting of *Clitoria* forage at 40 days interval produced maximum dry matter yield. Maximum grain yield was produced with pure pearl millet followed by pearl millet + *Clitoria* and maximum dry fodder yield was obtained when pearl millet inter-cropped with *Cenchrus*. Introduction of cowpea in rainfed cereal gave maximum crude protein yield followed *Dolichos* and cluster bean. Paired cropping was superior to normal cropping. Among *Cenchrus* spp., *Cenchrus ciliaris* 358 produced higher yield. *Cenchrus ciliaris* and *Clitoria ternate* produced maximum DM during December.

Leaf Fodder Production - An Additional Source of Feed in Dry Areas

Studies on lopping intensity (Bhimya *et al.,* 1964) revealed that heavy intensity of lopping adversely affected the growth of *Khejri.* Further recurrent lopping reduced that leaf fodder yield irrespective of lopping perhaps due to successive reduction in the overall surface of new shoots. Need for a period of rest between two loppings of *Khejri* is, therefore, prima facie obvious for sustained yield of *loong* (leaf fodder of *Khejri).* It has been suggested that in arid area lopping of *Acacia* spp., particularly *A. nilotica* (ex. *Aarabica)* should be regulated on a cycle of 4 years with restriction on lopping of thicker (diameter of 1.9 cm and above) branches. In Madhya Pradesh and Tamil Nadu lopping of *Hardwickia* spp was permissible only in times of scarcity. Sharma and Gupta (1981) studied the effect of seasonal lopping on the top feed production and growth of *P. cineraria.* They observed that winter was the ideal season for lopping as it improved plant height and bole diameter of the tree as compared to trees lopped in rest three seasons i.e. spring, summer and monsoon.

Natural, single, double and multi-tier systems were studied for fodder production in Avikanagar (Sharma and Chand, 2012). *Cenchrus* yield from all the systems were higher over natural pasture. Growth of ardu was better in two-tier system. The production for 5 years in natural land, single tier, two tier and three tier silvi-pasture was to the tune of 16.59, 19.20 and 23.74 q ha^{-1}, respectively. The three and two tier systems provided 9.5 and 6.25 q ha^{-1} tree fodder. The average net income over four years was highest in ardu pasture then neem, siris and babool pasture systems. The highest green fodder yield (154.69 q ha^{-1}) was in multi-tier system (Ardu + pearl millet + cowpea) followed by two-tier, however in multi-tier, maximum dry fodder yield was with sole pearl millet followed by *Cenchrus* + cowpea. The highest dry fodder yield from crop and grasses and their combinations was in single-tier system followed by two-tier system. Growth parameters and yield of *Cenchrus* were comparatively higher in association of ardu tree in comparison to other fodder trees.

Alternate Feed Resources for Arid Livestock

Arid regions have many feed resources in the form of the agricultural produce, weeds, byproducts from agro-processing methods etc. Tumba (*Citrullus colocynthis*) seed cake a byproduct of the oil extraction industry is nutritionally rich as it contains 16-22 per cent crude protein. Inclusion of 25 per cent tumba seed cake (TSC) in concentrate feed lowers the cost of animal feeding by 18-35 per cent without any adverse effect on the production, general health and reproductive performance (Mathur *et al.*, 1989, Mathur *et al.*, 2000). Salty shrub Lani (*Salsola baryosma*) at vegetative stage can be a good source of fodder to animals. The acceptability and palatability of fresh cut shoots showed that Lani is palatable (Mathur *et al.*, 2007). Lana (*Haloxylon salicornicum*) seeds having 18.60 per cent crude protein could replace about 25 per cent of the conventional sesamum cake in the concentrate for lactating cows (Anonymous, 2007). The powder of *Prosopis juliflora* pods can be used up to 35 per cent in the concentrate of lactating goats with maintenance of milk production without having any detrimental effect on health (Mathur *et al.*, 2003). Similarly *P. juliflora* pod husk can be used up to 50 percent level in the concentrate along with tumba (*Citrullus colocynthis*) seed cake as low cost ration in the sheep without any adverse effect on animal health (Mathur *et al.*, 2002).

Fresh leaves of *Hardwickia binata* are palatable to goats and sole feeding supports the body weight growth of growing kids (Patil *et al.*, 2009). Feeding fresh leaves of *Colophospermum mopane* is possible in goats but it can be fed at 35-40 per cent of whole ration (Patil *et al.*, 2007). The palatability of dry *C. mopane* leaves was low and decreased with progress of feeding from 15 to 5 per cent from 1st to 5th week and the traditional local *P. cineraria* leaves were better source of supplementation even in the dry form supporting the milk yield of goats. (Mathur *et al.*, 2006).

Thornless cactus *Opuntia ficus indica* introduced in Indian arid region has fodder value as maintenance feed and was observed to reduce the water requirement if fed along with the dry roughages in goats, sheep and growing cattle. In addition, its high mineral content may reduce the mineral requirement, as arid animals often suffer from mineral imbalance (Mathur *et al.*, 2009 and Meghwal *et al.*, 2010).

The weeds in the rocky and sandy habitat of arid region have a distinct relative preference index for grazing sheep and goats. Patil *et al.* (2005) studied the preference of these animals to different weeds and found the same to be in the order: Kanti (*Tribulus terristris*), Kagio (*Tetrapogon tenellus*), Santo (*Trianthema protulacastrum*), Lolaru (*Digeria muricata*), Bekario (*Indigofera cordifolia*) and Gangan (*Grewia tanax*).

Animal Performance on Desert Top Feeds

By and large, the consumption of *Z. nummularia* leaves is more than that of *P. cineraria* leaves, in both sheep and goats, the latter consuming considerably higher than the sheep (Ghosh and Bohra, 1984). The leaves of *A. excelsa, Z. nummularia* and *P. cineraria* are high in TDN and total digestible nutrients (64%, 50%, and 41.0% respectively), while the digestible crude protein (DCP) value of *A. excelsa* leaves is higher (16%) than the more palatable *P.cineraria* in sheep (3.1%) and goats (5.5%) (Bohra, 1980). The DCP of *Z. nummularia* in sheep and goats were 4.1 and 3.6 per cent, respectively (Bohra, *et al.*, 1999). From the point of nutritive value, *A. excelsa* leaves can be considered the best feed among all desert tree leaves. However, this tree does not grow as extensively as *P. cineraria* or *Z. nummularia* in this region. On the basis of per unit of feed material in general, top feeds provide more energy than the grasses, as the digestibility of cellulose from desert grass *C. ciliaris* (Bohra and Ghosh, 1977) is higher than that of the tree leaves presumably because of the high lignin content of the top feeds.

The nutritive value in terms of digestible crude protein and total digestible nutrients of different feeds differ with animal species. *Z.nummularia* has high nutritive value for camel followed by the sheep and goats, while *P.cineraria* has the highest value for goats, followed by camel and the sheep. The browsing rams and grazing bucks reared under a silvipastoral system having *C. ciliaris* as a dominant grass and supplemented with loppings of *Ziziphus rotundifolia* exhibited substantial body weight gains (Patil *et al.*, 2009). In the arid zone a substantial area is salt affected, where nothing can come up, except a few halophytic plants like *Salicornia bigelovii,* which can be cultivated under flood irrigation of brackish well water (Bohra *et al.*, 2009).

The desert top feeds, especially, *P. cineraria* and *Z. nummularia* leaves although have appreciable quantities of crude protein but their DCP value is very low because of tannins and lignin in these feeds. (Bohra *et al.*, 1999), which not only reduces the palatability, but also hinders utilization of dietary proteins by the animals as indicated by *in situ* degradability study of *P. cineraria* leaves in sheep rumen (Mathur *et al.,* 1998). Attempts have been made to improve the nutritive value *P. cineraria* leaves by treatment with ferric chloride. For this purpose, overnight soaking the leaves with 0.5N aqueous sodium carbonate solution, followed by washing with water, proved to be the best method for detanning *P. cineraria* leaves removing about 94 per cent tannins of these leaves (Bohra and Goyal, 1986). Other desert plants also yield appreciable quantity of palatable pods. *P. juliflora* leaves are not relished by the animals, but its pods are highly palatable and relished by almost all the species of the livestock. Feeding of pods after suitable processing like crushing and grinding is possible. *A. tortilis* pods contain 5.7 per cent DCP and 62 per cent TDN and the *P. juliflora* pods, 7 per cent DCP and 75 per cent TDN (Mathur and Bohra, 1993). These pods contain appreciable quantity of micro-minerals, too.

Supplementary Feeding in Grazing Animals

Supplementary feed blocks

As the livestock in arid regions are mostly range managed except during monsoon, when dry grasses in the ranges and pastureland, and crop residues in the fallow lands are available, the animals suffer from deficiency of essential nutrients including fermentable energy, protein, minerals as well as carotene. There exist different means of supplementation of essential nutrients to the livestock. Appropriate formulations of multi-nutrient feed blocks (MNB) using locally available feed resources wheat bran, guar korma, *Bajra* husk and *Ardu* leaves, *Prosopis juliflora* ground pods, sugar cane molasses, urea, vitaminized mineral mixture, dolomite, common salt, deoiled soya bean meal have been developed (Bohra et al., 2012) and as a binder locally available organic product-guar gum powder is being used as a binder. The supplementary feed blocks meant for the grazing ruminants have been tested in cattle, sheep and goats and were found to be economical source of nutrient supplementation to support the production functions of milk yield, body weight gain and wool yield. During 16 weeks feeding trial the MNB supplemented sheep group recorded 3.6 per cent gain over the sheep maintained on roughage diet. The supplementary feed blocks were also found to improve the nutrient utilization in cattle. In a digestibility trial study on Rathi cows, the digestibility coefficient for DM and crude protein were found better (Mondal and Bohra, 2001). The supplementary feed blocks in the arid areas were found to improve the production performance of lactating animals especially cows, buffaloes and goats. The farmers also reported improvement in the conditions of Pica in the animals suffering from mineral deficiency along with increase in feed and water intake. In case of cattle and buffaloes, a 2 kg block offered as a lick lasted for about 7 and 5 days, respectively (Patel *et al.*, 2003). Supplementation of MNB enhanced 11.5% in daily milk yield of buffaloes under field in Nagaur district of Rajasthan (Patel *et al.*, 2012-14).

Supplementary multi-nutrient mixture

For the small grazing ruminants that do not lick the above blocks, supplementary multi-nutrient mixture formulations were developed using the same ingredients. Feeding these mixtures to goats and sheep after grazing hours was found to improve body weights and milk yield appreciably (Rohilla *et al.*, 2009 and Patel *et al.*, 2006)

Complete feeds and feed/fodder banks for arid livestock

A low cost balanced diet has been formulated for ruminants, which does not require these animals to be sent for grazing in the rangelands. The alternate feeding diet for ruminants is based on crop residues and locally available agro-industrial products like straws, fallen tree leaves, by-products from other crops like cotton seeds, maize, groundnut, subabul, forest products like dry grass, pods, which can be blended to enable balanced supply of nutrients to the animals. The complete feed rations have been formulated area wise as per availability of various agricultural residues.

To improve the livestock productivity total mixed ration (TMR) or complete feed blocks (CFB) have been developed. In an evaluation trial, the lactating Tharparkar cows fed TMR prepared out of local fodder *Cenchrus ciliaris* and concentrates available in the area were found to produce milk more economically compared to cows fed on local grass and supplemented with pelleted cattle feed (Mathur *et al.*, 2006). The results of trials conducted at Avikanagar under intensive feeding on 50:50 roughage and concentrate ratios of complete feed indicated that the feed efficiency was higher than 20 per cent because of higher carcass fats. In this study, the crossbred weaner lambs maintained on these diets had higher ADG of 180 g compared to 127 ADG in native lambs. In another study the lambs fed intensively on 50:50 RC based complete feed attained 33.5 kg finishing weight with 11 *per cent* feed efficiency at 6 months of age (Shinde *et al.*, 1995). Higher growth 20% was also observed in Magra male lambs due feeding of complete fodder block under field condition (Sawal et al., 2015). In a study conducted on Marwari kids, Patil *et al.,* (2006) observed that the kids weaned at 2 months age had comparatively better growth rate than the kids weaned at 3 and 4 months age when these were fed on complete feed diets having roughage to concentrate ratio 30:70 utilizing Khejri leaves and Masoor straw in equal proportion as a roughage source and local material in the concentrate mixture. The cost of feeding was economical for kids reared on CFB after weaning at 2 months age (Rs. 65.37 kg^{-1}) as compared to those weaned at 3 months (Rs. 70.40 kg^{-1}) and at 4 months (Rs.71.27 kg^{-1}). Similarly in the intensive management system the Marwari kids fed on CFB along with goat paneer whey gained higher body weight than the kids fed on CFB only. The results of the above studies indicate that goat and sheep can be reared intensively without sending the animals for grazing and it can achieve the targeted body weights during minimum possible time. It also helps to achieve the expanded use of crop residues and locally available agro-industrial by products by formulating complete rations.

The technology of complete feeds is appropriate for the arid region livestock encountering frequent droughts. In such situations, advanced planning could be of great help to have enough stores of complete feeds as feed banks, which can be mobilized to places of deficiency.

References

Balak, Ram, Narain, P., Sharma, J.R., Nagaraja, R. and Jayanti, S. (2005). Wasteland management of arid western Rajasthan- A joint publication of CAZRI, Jodhpur and NRSC, Hyderabad.

Bhimya, C.P. and Kaul, R.N. and Ganguli, B.N. (1964). Studies on lopping intencities of Procsopis cineraria. Indian Forester 90 (1) :19-23.

Bhagmal, P.S. Pathak, V.S. Upadhyaya, J.N. Gupta, and G. Suresh (2006). Forage crops and Grasses. In Hand book of Agriculture ICAR, New Delhi, pp 1128-1161.

Bohra, H.C. (1980). Nutrient utilization of *Prosopis cineraria* (Khejri) leaves by desert sheep and goats. *Annals of Arid Zone* 19, 73-81.

Bohra, H.C. and Ghosh, P.K. (1977). Effect of restricted water intake during summer on the digestibility of cell wall constituents, nitrogen retention and water excretion in Marwari sheep. *Journal of Agricultural Science* (Camb.) 89, 605-608.

Bohra, H.C. and Goyal S.P. (1986). Chemical detaining of *Prosopis cineraria* leaves. Presented at the 5th Animal Nutrition Workers Conference, Udaipur. Karnal: Animal Nutrition Society of India, July 14-17, 1989.

Bohra, H.C., Ghosh P. K., and Goyal, S. P. (1999). Nutrient availability from *Ziziphus nummularis* leaves in desert sheep and goats. *Journal of Eco-physiology*, 2:129-134.

Bohra, H.C., Patel, A.K. and Kaushish, S. K. (2009). Palatability, digestibility of various constitutes and nitrogen retention in Marwari sheep offered *Salicornia bigelovii* biomass and *Cenchrus ciliaris* straw mixed diet. *Indian journal of Animal Sciences.* 79 (10): 1050-1053.

Bohra, H.C., Patel, A.K., Rohilla, P.P., Mathur, B.K., Patil, N.V. and Misra, A.K. (2012). Feed production technologies for sustainable livestock production in arid areas. CAZRI, Publication, Jodhpur.

Daulay, H.S. and Bhati, T.K. (1989). Crops and pasture production potential of sandy soils. In the book on *Review of Research on Sandy Soils,* CAZRI, Jodhpur, pp. 245-262.

Deb Roy, R. and Pathak, P.S. (1974). Silvipastoral management. *Indian Farming* 24: 41-42.

Ghosh, P.K. and Bohra H.C. (1984). Palatability, digestibility and nutritive value of some important top feeds of arid and semi-arid regions of India. In: Agroforestry in Arid and Semi-arid Zones (Ed. K.A. Shankarnarayan). Jodhpur: Central Arid Zone Research Institute.

Harsh, L.N. Tewari, J.C., Burman, U and Sharma, S.K. (1992). Agroforestry in arid regions. *Indian Farming* 40: 32-37.

Hatfield, J., Booke, K., Fay, P., Hahn, L.C., Izaurralde, B.A., Kimball, T., mader, J., Morgan, D., Ort, W., Polley, A., Thomson and Wolfe, D. (2008). Agriculture. In: the effects of climate change on agriculture, land resources, water resources and biodiversity in the United States. A Report by the US Climate Change Science Program and the Subcommittee on Global Change research. Washington, DC, USA, 362.

Karim, S.A. and Shakyawar, D.B. (2011). National Seminar cum Workshop on Recent R& D Initiatives and Development Schemes of Wool and Woollens 28th May 2011, Mumbai.

Kumar, S., Kareemulla, k. and Venkateswarlu, B. (2010). Small Ruminant Production and Sustainable Rural Livelihood. Climate Change & Stress Management: Sheep & Goat Production. Pp 687-702.

Mathur, B. K. (2003). Livestock: human need for sustainability in arid environment. In: *Human Impact on Desert Environment.* (Eds. Pratap Narain, S. Kathju, Amal Kar, M. P. Singh and Praveen- Kumar). *Arid Zone Research Association of India* and Scientific Publishers (India), Jodhpur, India. pp. 506-514.

Mathur, B. K. and Bohra H.C. (1993). Nutritive value of Prosopisjuliflora leaves and pods. Presented at the workshop on potential of *Prosopis* spp. for Arid and Semi Arid Regions of India, Jodhpur. Jodhpur: Central Arid Zone Research Institute. November 22-23, 1993.

Mathur, B.K., Patel, A.K., Kaushish, S.K. (2000). Utilization of non-conventional resource: Tumba (*Citrullus colocynthis*) seed cake of Desert for Goat Production. *Indian journal of Animal Sciences.* 70(4): 431-433.

Mathur, B.K, Bhati, T.K. and Tewari J.C. (2006). Comparative palatability of *Prosopis cineraria* v/s *Colophospermum mopane* leaves in lactating Marwari goat in arid region. *In:* National Symposium on *"Livelihood Security and Diversified Farming Systems in Arid Region"*. January 14-16, 2006 organized by Arid Zone Research Association of India, CAZRI, Jodhpur. No.3-2 pp. 51-52.

Mathur, B.K., Bohra H.C., Patel A.K. and Kaushish, S.K. (1998). In situ Rumen degradability of Khejri (*Prosopis cineraria*) leaves in Marwari sheep. Presented at the Golden Jubilee Seminar on. "Sheep, Goat and Rabbit Production and Utilization for Maximizing Production Efficiency, pp., 27. "Indian Society for Sheep and Goat Production and Utilization. Jaipur: CSWRI, April 1998.

Mathur, B.K., Mittal J.P. and Prasad, S. (1989). Effect of tumba cake (*Citrullus colocynthis*) feeding on cattle production in arid region. *Indian Journal of Animal Sciences* 59: 1464-1465.

Mathur, B.K., Mittal J.P., Shiv Prasad and Mathur, A.C. (1991). Need of conserving Tharparkar breed of cattle in Indian desert. *Indian Farming* 41 (3): 4-6.

Mathur, B.K., Patil N.V., Mathur A.C., Bohra H.C., Bohra R.C. and Sharma K.L. (2009). Comparative Mineral Status of Jodhpur District Villagers' Cattle v/s Institute Farm Managed Tharparkar Cattle in Hot Arid Zone. Proceedings of Animal Nutrition World Conference, 14-17 Feb., New Delhi, India pp.61.

Mathur, B.K., Patil, N.V., Mathur A.C., Bohra H.C., Patel A.K and Bohra R.C. (2006). Effect of total mixed ration prepared from local feed resources on feed intake, milk production efficiency and blood profile of Tharparkar cattle in arid region. *In: VI Biennial Conference of Animal Nutrition Association on "Strengthening Animal Nutrition Research for Food Security, Environment Protection and Poverty Alleviation"* September15-17, 2006. Sher-e-Kashmir University of Agricultural Sciences & Technology, Jammu, J & K pp.77.

Mathur, B.K., Shiv Prasad and Mittal, J.P. (1989). Effect of tumba cake (*Citrullus colocynllus*) feeding on cattle production in arid region. *Indian Journal of Animal Science* 59:1464-65.

Mathur, B.K., Singh J.P., Beniwal R.K. and Singh, N.P. (2007). Utilization of salty shrub-Lani (*Salsola baryosma*) of arid region as drought feed for goats.FR-55.In: International Tropical Animal Nutrition Conference, October 4-7, 2007 NDRI, Karnal pp.33.

Mathur, B.K., Siyak S.R. and Bohra H.C. (2003). Feeding of Vilayati babool (*Prosopis juliflora*) pods powder to Marwari Goats in Arid Region

Current Agriculture. 27(1-2):57-59.

Mathur, B.K., Siyak S.R. and Paharia, Shailesh (2002). Incorporation of *Prosipis juliflora* Pod husk as Ingredient of low cost ration for sheep in arid region. *Current Agriculture* 26:1-2, 95-98.

Meghwal, P.R, Patel A.K., Patidar, M, Roy, M.M. (2010). Introduction, utilization and further potential of cactus pear in Indian arid zone. *In the VIIth International Congress on Cactus Pear & Cochineal,* Agardir, Morocco from October 17-22, 2010. PP: 175-176

Mondal, B. C. and Bohra, H. C. (2001). Effect of Multi-nutrient feed block supplementation on feed intake and nutrient utilization in Rathi heifers. Proceedings of X Animal Nutrition Conference on "Emerging Nutritional Technologies for Sustainable Animal Production and Environment Protection", abst. No. 153, p. 95. N.D.R.I., Karnal, November 9-11, 2001 285 + vii p.

Muthana, K.D. and Shankaranarayan, K.A. (1978). Scope of silvipastral system in arid region. In *souvenir on arid zone research silver jublee,* CAZRI, Jodhpur, India.

Nair, P.K.R. (1993). An introduction to Agroforestry. Kulawar. Academic Publisher, The Netherland.

Naqvi, S.M.K. (2013). Present status of carpet wool production in India viz a viz world. Interactive meeting on Prospects in improving production, marketing and value addition of carpet wool. December 31, 2013. Arid Region Campus, CSWRI, Bikaner Page 1-7.

Narain Pratap and Kar Amal (2005). Drought in western Rajasthan impact, coping mechanism and management strategies, CAZRI, Jodhpur 45 pp.

Patel, A. K., Bohra, H. C. and Bhati, T. K. (2003). On Farm Trial of Multi Nutrient Block on Buffaloes in Arid Region. Proceeding of the 4th Asian Buffalo Congress on Buffalo for Food Security and Rural Employment, New Delhi, Feb. 25-28, 2003, Vol II Page No. 167.

Patel, A.K., Bohra, H.C and Rohilla, P.P. (2006). On-farm trial of Nutrient Mixture supplementation in lactating goats under semi-arid conditions

Indian Journal of Animal Production & Management 22 (3-4): 157-159.

Patel, A.K., Bohra, H.C., Patidar, M and Roy, M.M. (2012-14). Multi-nutrient feed block: a strategic alternative supplement for lactating buffaloes under field conditions in arid zone. Indian buffalo J. 10-12 (1 & 2): 40-43 (2012-14).

Pathak, P.S., Gupta, S.K. and Singh, P. (1996). IGFRI Approaches: Rehabilitation of degraded lands. IGFRI, Jhansi, 1-23.

Patidar, M. (2006). Strip cropping of grasses and legume in Silvipastoral system. In Abstract of National Symposium on Livelihood Security and Diversified Farming System for Arid Region. pp. 33. Jan. 14-17, 2006 CAZRI, Jodhpur.

Patidar, M. and Saxena Anurag (2007). Paryavarn sarankshan ke liye chara utpadhan. *Kheti* 60 (3): 18-22.

Patil, N. V., Mathur B. K., Patel A.K., Bohra H.C. and Khan M. S. (2006). Growth performance of arid breed kids weaned at different ages and received on complete feed block. Proceedings of Animal Nutrition Workers Association Conference, pp., 78. SKUAT, Jammu, September15-17, 2006.

Patil, N.V, Mathur B.K, Bohra H.C and Patel A.K. (2005). Relative preference index for arid forages for goat and sheep. National Seminar on *"Conservation, Processing and Utilization of Monsoon Herbage for Augmenting Animal Production"*, December 17-18, 2005. Organized by Indian Society for Sheep and Goat Production and Utilization, at RRS, CSWRI, Bikaner.pp.190.

Patil, N.V., Patel, A.K., Khan M.S. and Bhati T.K. (2009). Potential assessment of silivipasture system of arid Rajasthan based on its traditional use for rearing growing lambs and kids *Proceedings of world conference2009 on Animal nutrition Preparedness to combat challenges, New Delhi,* February 14-17, 2009 pp 278.

Rai, P., Solanki, K.R. and Rao, G.R. (1999). Silvipasture research in India. A review 1: 107-120.

Rohilla P.P., Patil N.V, Bohra H.C. and Patel A.K. (2009). Effect of supplemental feeding of urea molasses mineral block (UMMB) on growth performance of Marwari lambs. *Proceeding of National Symposium on "Organic livestock farming- global issues, trends & challenges" W B University of Animal & Fishery Sciences, Kolkata,* February 26-28, 2009. pp 87.

Sawal, R.K., Patel, A.K., Chopra, A., Narula, H.K. and Ayub, M. (2015). Growth performance of Magra lambs fed multi-nutrient blocks under field conditions of hot arid Rajasthan. *The Indian Journal of Small Ruminants* 21 (2) 336-337.

Sharma, S. C. and Roop Chand (2012). Research Contributions 1962 to 2012, Central Sheep & Wool Research Institute, Avikanagar (Celebration 50 Years Book: Edited by A.K. Shinde, C.P. Swarnkar and L.L.L. Prince)

Sharma S.K. and Diwakar G.D. (1989). Economic evaluation of horti-pastoral system in arid region of west Rajasthan. *Indian Journal of Range Management* 10: 119-122.

Sharma S.K. and Gupta, R.K. (1981). Effect of seasonal lopping on the top feed (long) production and growth of Prosopis cineraria (Linn.) Druce. *Indian J. Forestery* 4 (4): 253-255.

Shinde A.K., Karim, S.A., Singh N.P. and Patnayak B.C. (1995). Growth performance of weaner lambs and kids under intensive and semi intensive feeding management. *Indian Journal of Animal Sciences* 65: 630-633.

Venkatateswarlu, J., Anantharam, K., Purohit, M.L., Khan, M.S., Bohra, H.C., Singh, K.C. and Mathur, B.K. (1992). *Forage 2000 AD The Scenario for Arid Rajasthan.* CAZRI, Jodhpur. 32 p.

Yadav, G.L. and Poonia, T.C. (2003). Production potential of grasses between existing stock of trees under Silvipastoral system in semi arid watershed area. In *National Symposium on Sustainability Advancement and Future Thrust Area of Research on Forage* pp. 53. March 5-6, 2003. CCS, HAU, Hisar.

11

Optimizing Productivity through Watershed Management and Improving Livelihoods in Rainfed Areas

Mahesh K. Gaur and R.K. Goyal

Introduction

Watershed management is a process of formulating and carrying out a course of action involving manipulation of the natural system of a watershed to achieve objectives specific to the watershed such as control of soil erosion and land degradation, reclamation and rehabilitation of waste/degraded lands, land use changes consistent with land capability, or management of croplands, grasslands, and forests along with management of water resources (Wani *et al.*, 2011).

Effective use of land and water is fundamental to growth and sustainable development. The concept of watershed management has evolved to ensure effective use of both natural and social capitals. Thus, the watershed development programmes include land, water and human resources as essential components (Vaidyanathan, 2001). The watershed programme is primarily a land based programme, which is increasingly being focused on water, with its main objective being to enhance agricultural productivity through increased in situ moisture conservation and protective irrigation for socio-economic development of rural people (Joshi *et al.*, 2004, 2006). It has been essential in a country like India where majority of the population depends on agriculture and about 60 percent of total arable land (142 Mha) in the country is rain-fed. A large portion of the rain-fed areas (65% of arable land) in India is characterized by low productivity, high risk and uncertainty, low level of technological change and vulnerability to degradation of natural resources (Joshi *et al.*, 2004). Over the years, the sustainable use of land and water has received wider attention among policy makers, administrators, scientists and researchers (Tideman, 2000). Almost all major international developmental agencies like World Commission on Environment and Development (WCED), Food and Agricultural Organisation (FAO), and Consultative Group on International Agricultural Research (CGIAR) and its allied agencies amongst others have emphasized sustainable use of water and other natural resources (Vaidyanathan, 2006; Sharda *et al.*, 2008).

The watershed forms an appropriate unit which links all these three components and has a direct bearing on human lives. The watershed approach is a system-based approach that facilitates the holistic development of agriculture, forestry and allied activities in the proposed watershed (Vaidyanathan, 2001). It also forms an appropriate unit for analysing the development-linked resource problems, designing the appropriate solutions of identified problems and eventually testing the efficacy of the measures taken up (Bhathwal, 2003).

Watershed Development Programmes (WDPs) have been accorded high priority in India's development plans (Singh and Ballabh, 2008). These programmes have been initiated in India to improve and sustain productivity and the production potential of the dry and semi-arid regions of the country through the adoption of appropriate production and conservation techniques (Wani *et al.,* 2011). The WDP approach seeks to improve and develop all types of lands-government, forest, community and private lands- that fall within a particular watershed. It is a holistic approach to improve and develop the economic and natural resource base of dry and semiarid regions (Ninan and Lakshmikanthamma, 2001). The programmes have stressed upon improvement of wasteland, runoff reduction, water conservation and protective irrigation mechanism in all areas including desert prone areas and drought prone areas. Development programs, envisaged under its purview include almost every activity which concerns land, water and biomass production. Experiences have shown that watershed as a base is very effective in use and management of land and water resources (Vaidyanathan, 2001). With increasing awareness about the problems related to environment, use of watershed terminology is becoming popular and moreover in view of their potential for growth, improvement in income levels and augmenting the natural resource base of the disadvantaged regions of the country (Singh and Ballabh, 2008).

Significance of Watershed

Watershed management programmes have emerged as an appropriate strategy to manage natural resources (land, water and forests) and to provide sustainable livelihood to the rural poor. It is a holistic concept which tries to integrate several components like soil and water conservation, forestry development, agriculture, horticulture, livestock development etc. (Wani *et al.,* 2011). It has led to several dimensions of sustainable development, e.g. ecological sustainability (check in soil erosion, check in rate of silting, groundwater recharge etc.), economic sustainability (increase in crop intensity and crop productivity, milk production, etc.) and social sustainability, (equitable distribution of common property resources like water, forest produce and ensuring peoples' participation (Sharda *et al.,* 2008).

The watersheds essentially comprise various features but in varying degree. Because of this reason, there is a significant variation in their behaviour. The behaviour may be in terms of hydrological (rainfall-runoff-sediment yield), crop yield or in socio-economical aspects (Shah, 2001). Considerations of these behaviours are very important, because watershed provides all means for nature's creators. Watershed provides drinking water as well as water for recreation, irrigation and industrial activities. Watersheds are also significant for plants and animals as they provide food, shelter and water to them (Vaidyanathan, 2001).

As such, watershed management is not so much about managing natural resources, but about managing human activity as it affects these resources. The drainage area of the river provides the natural boundary for managing and mitigating human and environmental interactions (Shah, 2001). Because human activity includes actions by governments, municipalities, industries, and landowners, watershed management must be a co-operative effort. Effective watershed management can prevent community water shortages, poor water quality, flooding and erosion. The expense of undertaking watershed management is far less than the cost of future remediation (Reddy, 2000).

Objectives of Watershed Development Programmes (WDPs)

Rational utilization of land and water sources for optimum production causing minimum damage to the natural resources is known as watershed management. The main goal of Watershed Management is to implant the sustainable management of natural resources to improve the quality of living for the population. Watershed development aims to balance the conservation, regeneration and use by humans of land and water resources within a watershed. Common benefits from successful watershed development projects include improved agricultural yields and increased access to drinking water (Reddy, 2000; Vaidyanathan, 2001; Sharda *et al.,* 2008). The overall attributes of the watershed development approach, by and large, are three fold, viz. promoting economic development of the rural area, employment generation, and restoring ecological balance (Nath *et al.,* 2015). However, the multiple objectives include (Khalid *et al.,* 2004):

Environmental - For protecting vegetative cover throughout the year, to create ecological balance in the watershed area, protecting fertile top soil, utilizing the land based on its capabilities, in situ conservation of rain water, increasing ground water recharge, etc.

Economic - It draws attention for increase in cropping intensity through inter and sequence cropping, maximizing farm income through agricultural related activities such as dairy, poultry, sheep and goat farming, improved and sustained livelihood status of the watershed community with special emphasis on the poor/women.

Institutional - It includes formation of watershed committees and self-help groups, establishing sustainable community organization, etc.

Social - It includes alleviation of poverty, awareness generation, improving skills of the local community, capacity building activities, women's participation in decision-making process, empowerment of the community, etc.

Equity - To develop equitable distribution of the benefits of land and water resources development and the consequent biomass production, involvement of village communities in participatory planning, implementation, social and environmental arrangement, maintenance of assets and to operate in a more socially inclusive manner.

Agricultural land covers about one-third of the world's land area, with arable land representing less than one-third of agricultural land (about 10 percent of the world's land area). Agricultural land constitutes only a part of any country's total area, which can include areas not suitable for agriculture, such as forests, mountains, and inland water bodies. Agriculture is still a major sector in many economies, and agricultural activities provide developing countries with food and revenue. But agricultural activities also can degrade natural resources. Poor farming practices can cause soil erosion and loss of soil fertility. Efforts to increase productivity by using chemical fertilizers, pesticides, and intensive irrigation have environmental costs and health impacts. Excessive use of chemical fertilizers can alter the chemistry of soil. Pesticide poisoning is common in developing countries. And salinization of irrigated land diminishes soil fertility. Thus, inappropriate use of inputs for agricultural production has far-reaching effects. There is significant geographic variation in the availability of land considered suitable for agriculture. Increasing population and demand from other sectors place growing pressure on available resources. According to FAO, the world's cultivated area has grown by 12 percent over the last 50 years. The global irrigated area has doubled over the same period, accounting for most of the net increase in cultivated land. Agriculture already uses 11 percent of the world's land surface for crop production. It also makes use of 70 percent of all water withdrawn from aquifers, streams and lakes. Agricultural policies have primarily benefitted farmers with productive land and access to water, bypassing the majority of small-scale producers who are still locked in a poverty trap of high vulnerability, land degradation and climatic uncertainty.

Necessity of Watershed Programmes in India

Decrease in per capita arable land area: Substantial changes in land use pattern have taken place during the last few decades, mainly driven by biophysical factors and human needs. Arable land (% of land area) in India was reported at 52.62% (156.46 Mha) in 2016 whereas it accounted 54.8% (110.1 Mha) in 1950 (Tian et al. 2014). Arable land (% of land area) in India was 52.62 as of 2016. Its highest value over the past 55 years was 55.03 in 1984, while its lowest value was 52.40 in 1961. Agricultural land refers to the share of land area that is arable, under permanent crops, and under permanent pastures. Arable land (hectares per person) in India was reported at 0.11812 in 2016, according to the World Bank collection of development indicators, compiled from officially recognized sources. A noticeable decrease is seen in the total arable land area due to soil degradation, and conversion of prime agricultural land to industrial, urban, recreational, and other non-agricultural uses. Urbanization is a principal threat to prime agricultural land, especially in a densely populated country like India (Figure 1).

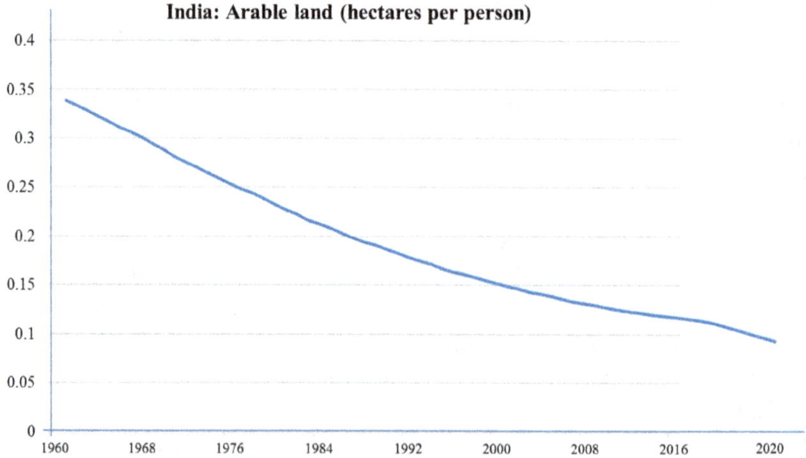

Data Source: https://www.macrotrends.net/countries/IND/india/arable-land (modified & updated by author)

Figure 1. India: Arable land (hectares per person).

In view of the ever-shrinking arable land resources, it is important to identify and implement strategies for restoration of degraded soils and intensification of existing prime agricultural land. The continuing decline of arable land per person is often cited as an indicator of impending problems (Bruinsma, 2018). The underlying cause for such problems is perceived to be an ever-increasing demand for agricultural products facing finite natural resources such as land, water and genetic potential. Scarcity of these resources would be compounded by competing demand for them, originating in urbanization, industrial uses and use in biofuel production, and by forces that change their availability, such as climate change and the need to preserve resources for future generations (environmentally responsible and sustainable use). Therefore, watershed management is an appropriate option to implement these strategies (Kerr, 2000, Tian *et al.,* 2014). The evidence shows that agricultural land conversion has become a serious issue in the country but the extent and intensity varies across different states. Between triennium ending (TE) 1991-92 and TE2011-12, net sown area in the country declined by about 1.8 Mha but it increased in some states, e.g. about 20 lakh ha in Rajasthan and 9.5 lakh ha in Gujarat. In contrast, Odisha lost over 17 lakh ha net sown area, Bihar (including Jharkhand) 12.4 lakh ha, Maharashtra (7.6 lakh ha), Tamil Nadu (7.1 lakh ha), Karnataka (3.1 lakh ha), Andhra Pradesh (2.7 lakh ha) and West Bengal (2.6 lakh ha). Contrary to general perception, Gujarat is the only state which has been able to add about 3 lakh ha to its total agricultural land during last two decades. Area under non-agricultural uses increased from 21.3 Mha in TE1991-92 to 26.3 Mha in TE2011-12 and almost all states witnessed an increase in area under non-agricultural uses (Sharma, 2015).

Natural resources degradation: Degradation of soil and water resources is a global threat. Land degradation, a decline in land quality caused by human activities, has been a major global issue during the 20th century and will remain high on the international agenda in the 21st century. The rapid economic growth based on rapid industrialization resulted in widespread degradation of natural resources. Serious environmental problems became visible in the 1960s, when many diseases occurred because of the environmental degradation. The main causes the depletion of our natural resources are:

- Over-population,
- Over-consumption and waste,
- Deforestation and the destruction of ecosystems leading to loss of biodiversity,
- Mining of minerals and oil,
- Technological and industrial development,
- Erosion, and
- Pollution and contamination of resources.

The importance of land degradation among global issues is enhanced because of its impact on world food security and quality of the environment. An estimated 950 Mha of salt-affected lands occur in arid and semi-arid regions, nearly 33% of the potentially arable land area of the world. Productivity of irrigated lands is severely threatened by build-up of salt in the root zone. In South Asia, annual economic loss is estimated at US$500 million from waterlogging, and US$1,500 million due to salinization (UNEP, 1994). Oldeman

(1994) reported that out of a total degraded land area of 1965 Mha, over 300 Mha (about the size of India) are strongly degraded on a world scale. These lands have lost their productive capacity, and restoration can only be achieved through major investments. Only about 3% of the global land surface can be considered as prime or Class I land and this is not found in the tropics. Another 8% of land is in Classes II and III. This 11% of land must feed the six billion people today and the 7.6 billion expected in 2020 (Eswaran *et al.*, 2001).

Table 1. Estimates of the global extent (in million km^2) of land degradation (Oldeman, 1994)

Type	Light	Moderate	Strong + Extreme	Total
Water erosion	3.43	5.27	2.24	10.94
Wind erosion	2.69	2.54	0.26	5.49
Chemical degradation	0.93	1.03	0.43	2.39
Physical degradation	0.44	0.27	0.12	0.83
Total	7.49	9.11	3.05	19.65

Problem of degradation of land in the rainfed or dry-land areas in the 1990s is likely to have proceeded at more than twice the rate observed in the 1980s, basically on account of soil erosion from runoff (GoI, 2007). Land and water degradation threaten food security for many of the poorest living in South Asia, Africa and Latin America and India is no exception (Gaur and Squires, 2018). The degradation also contributes to persistent poverty. Negative trends in resource degradation are a challenge that must be tackled to meet poverty alleviation.

About 144.43 Mha of the total land in India is under water and wind erosion. About 260 Mha of the land area is drought prone. Thus the total problem area is about 173.65 Mha.

***Watershed management and agriculture*:** It seeks to use water in a way that provides crops and animals the amount of water they need, enhances productivity, and conserves natural resources for the benefit of downstream users and ecosystem services. Agriculture is directly linked to many aspects of sustainable development including alleviation of poverty, sustainable consumption and production, management of natural resources, capacity building as well. Rainfed agriculture is subject to the vagaries of monsoon. Irrigation facility is one of the significant factors contributing to agriculture development in India. Despite various schemes of the Government to extend facilities for irrigation only around 40% of the net cultivable land has been irrigated so far while the remaining 60% is rainfed. In the absence of irrigation facilities, agriculture is nothing more than a gamble. In arid regions, irrigation is everything, water is more important than even land because when irrigated the productivity of soil rises almost six times. Hence the scope of Watershed Management Programme is tremendous and it has to be given the prime importance in the government programmes.

Watershed management programme in mitigating droughts: Drylands today are facing a landscape-scale water storage problem. Landscape losses of deep soils and vegetative cover in this region stem from historic overgrazing. Less summer precipitation and higher temperatures results in droughts. Agriculture traditionally and can again support this natural dynamic through watershed-scale systems of rain water harvesting. Drought is a creeping disaster, which brings large scale starvation, malnutrition, migration, unemployment and poverty. Overtime soil erosion has increased, ground water levels have gone down and consequently the severity of drought has increased leading to ecological degeneration. Droughts are becoming increasingly severe in the recent period. Drought connotes a situation of water shortage for human, cattle and agriculture consumption resulting in economic losses, primarily in agriculture sector. Drought leads to a situation of decreased food production, decreased availability of dairy and livestock products, loss to industries directly dependent on agricultural production, increased cost of water and other commodities, reduction of economic development and growth etc. Drought is a common phenomenon in one or other part of India. Arid and semi-arid regions, practicing largely rainfed agriculture are more prone to drought than the other climatic zones; Fifty four districts in the country together with parts of another 18 districts contagious to them are identified as drought prone.

The Drought Atlas for arid zone of western Rajasthan enables in the identification of the vulnerability of livelihoods to drought hazards by providing an effective tool to raise awareness on their exposure to drought. The Atlas identifies the variability of rainfall deficits zones, and allow visualizing how this climatic variability differs spatially within each zone, even at short distances. This is achieved by identifying the drought frequency, which is the probability of drought occurrence with a certain magnitude and duration at any given place in the area of interest. Specific deliverables focus on hazard maps for precipitation deficit (drought) and excess (floods) that allow identifying risks in vulnerable areas.

Evidently, the existing interventions for drought proofing have failed to counteract the overall processes of degradation of natural resources. To achieve transformation of the landscapes and the communities that rely upon them to a resilient state requires collaborations of integrated management systems and disciplines across the scales. To achieve such holistic land management requires developing common understandings of the system dynamics, the functional as well as the structural attributes of a system, which is often described as restoring the natural or biophysical functions and processes, or a process approach. Connectivity in watershed systems is emerging as a process that organizes dryland landscapes and is thus useful for revealing dryland dynamics. The dryland connectivity conceptual framework defines connectivity as the extent to which materials such as water, nutrients, and organisms can move, spread, or be redistributed from one place to another within the landscape. This also calls for evolving an overall policy frame-work that provides adequate incentive and opportunities for soil and moisture conservation, through watershed development.

An increase in lateral connectivity throughout a watershed increases the area of runoff sinks by spreading and slowing the flow, recharging soil storage capacity, and supporting vegetation productivity. Vegetation then provides the feedback function of increasing infiltration and water holding storage capacity through increasing surface roughness, introducing soil pathways, and increasing organic matter into the soil (Hudson, 1994).

Over exploitation of ground water: The importance of groundwater in Indian economy can be hardly overemphasized. Estimates state that groundwater sources account for as much as 70% to 80% of the value of farm produce attributable to irrigation. India's agriculture is closely tied to the availability of groundwater as we can see it from the increase in groundwater irrigation than canal irrigation. During drought years, groundwater happens to be the predominant source of irrigation. Groundwater provides the greatest measure of security on all the three fronts sought by farmers: timeliness, adequacy, and reliability. Besides, groundwater is the source of four-fifths of the domestic water supply in rural areas and around half that of urban industrial uses. Hence, groundwater's role as a catalyst of rural development as well as poverty alleviation cannot be denied.

The global scenario of groundwater overdraft indicates that over-exploitation of groundwater from the shallow aquifers has deteriorated its quality. Adverse effects of groundwater overdraft include, uneconomic pumping conditions, water quality degradation through induced intrusion of saline or poor quality groundwater, flow reduction in streams, wetlands and springs, land subsidence, interference with pre-existing water uses and water rights, and gradual depletion of groundwater storage (Khayat *et al.,* 2006). In India, over-development of groundwater resources resulted in water level depletion, shortage in water supply, worsening groundwater quality (Pophare, 2014). These factors have serious implications on the environment and socioeconomic conditions of the populace. About 85% of rural population and ~50% of urban population in India is directly dependent on groundwater for their drinking, domestic, and irrigation needs. Moreover, during the last few decades, the demand of groundwater as a reliable source for fulfilling these requirements has increased manifold (CGWB, 2007).

Groundwater, which is 38.5% of the available water sources of the country, plays a major role in irrigation, rural and urban drinking water supply and industrial development. Groundwater meets nearly 55% irrigation, 85% of rural and 50% of urban and industrial needs (Government of India, 2007). The use of groundwater in the agriculture sector has expanded rapidly because of the short gestation lags with which it can be developed, control over irrigation that it provides, free or subsidized availability of power in some states, water requirements for the crop production during critical growth stages caused due to erratic rainfall in dry land agriculture and paucity of surface irrigation (Wani *et al.,* 2009).

The average annual rainfall in the country is 1170 mm, which correspond to an annual precipitation of 4000 billion cubic meters (BCM). Out of this volume of precipitation, 1869 BCM appears as average annual flow in rivers. Due to various constraints, only 1123 BCM is assessed as the average annual utilizable water (690 BCM from surface water and 433 BCM from groundwater). The present total water use is 643 BCM of which 83% is for irrigation. This is projected to grow to 813 BCM by 2010, 1093 BCM by 2025 and 1447 BCM 2050, against utilizable quantum of 1123 BCM. As regards to use, the extent of extraction has increased significantly over the years due to steep increase in the number of wells (tube and open wells). The average rate of increase in number of wells per year in India was 2.3%. The number of tube and open wells increased at the rate of 6.3% and 2.4% per year, respectively. It is estimated that currently there are 19 million wells in the country, out of which 16 million wells are in use and drawing about 213 BCM of water (Government of India, 2007).

While groundwater development has had important implications for the economy, the overuse of groundwater has also been emerging as a major concern. Ground water is being used at a faster rate than the rate at which it is being recharged. Overexploitation of groundwater is a major and serious problem in states of Punjab, Haryana, Delhi, Tamilnadu, Rajasthan etc. At the country level, there is phenomenal increase in groundwater structures thus putting pressure on groundwater resources. The extent of extraction has increased significantly over the years as the number of wells and tube wells has gone up substantially.

Integrated watershed management through primarily water (surface and groundwater) conservation and management compounded with other improved practices have shown a significant increase in productivity, cropping intensity and income, while controlling degradation of natural resources. Compound growth rate (CGR) of productivity, net returns and benefit cost (B: C) ratio are mean of selected major crops (Wani *et al.,* 2009). In the Bundi watershed, Rajasthan, soil water conservation and rainwater harvesting interventions resulted in significant improvement in groundwater both in terms of duration of water available and the water yield from the wells. Before the watershed interventions, only 88 wells use to have water for 8 to 12 months in a year, whereas after the watershed interventions it increased to 187 wells. Before watershed interventions, 52 wells out of 227 were functional only for 1-4 months mainly during the rainy season, whereas after the watershed interventions particularly due to the construction of WHS, majority of the seasonally functional wells have become functional throughout the year. Similarly, the mean depth of water column in the wells before the watershed interventions was 4.5 m, compared to 9.5 m after the interventions (Wani *et al.,* 2009). So, the watershed management happens to be a useful technology for effectively recharging the groundwater by water and soil conservation methods.

Desertification: Desertification refers to a special type of land/soil degradation, where desert like conditions spread to the areas on the fringe of the desert in semiarid and arid regions. Desertification is the only and perhaps the most important hazard that is threatening the low rainfall areas of the world today. Vast areas of low rainfall zone are being inundated and lain waste by advancing deserts. Furthermore, many marginal lands of the arid world that otherwise could become productive to serve the needs of the people are now essentially wastelands as the result of exploitative human activities. Desertification implies decline in soil quality leading to reduced biological productivity and environmental moderating capacity of land in arid regions. Desertification may happen as a result of natural and anthropogenic or human-induced factors. The semi-arid to weakly arid areas of Africa are particularly vulnerable, as they have fragile soils, localized high population densities, and generally a low-input form of agriculture. Desertification is experienced on 33% (42 million km^2) of the global land surface and affects more than one billion people, half of whom live in Africa (Eswaran *et al.,* 2001). Desertification is an especially severe problem in dry regions in India. An appropriate strategy for desertification control may involve natural resource planning at the watershed level through watershed management programme.

The major physical manifestation of desertification in hot arid zone of India is wind erosion/deposition, followed by water erosion, as well as water logging and salinity (Ghose *et al.,* 1977). Wind erosion is a major problem in arid sandy tract of north-western India, especially in west Rajasthan. About 76 per cent area of western Rajasthan is affected by wind erosion, encompassing all major land uses, followed by vegetation degradation (3%), water erosion (2%), salinization (2%) and about 0.10 per cent area by mined spoiled degradation (Kar *et al.,* 2009). Desertification, manifested in different forms of degradation has been a challenging issue to the arid zone of India. The adoption of dryland conservation technologies, particularly the water resources conservation, can significantly increase productivity and profitability in these areas (Halvin and Schlegel, 1997).

In arid & semi-arid regions desertification, land degradation & drought affect more than 2 billion people - situation might worsen due to climate change. More than 2.6 billion people (44%) are affected by desertification and GDP in dryland areas is 50% lower than in non-dryland areas. Natural regeneration of vegetation cover & soils in arid areas takes 5-10 times longer than in favourable areas with greater and more regular rainfall.

The rainfed areas of western Rajasthan in India are characterized by very low and erratic rainfall, frequent droughts; and agricultural production is uncertain and risky on the degraded soils and depleted water resource. Due to watershed interventions, the annual runoff from the watershed is reduced by 52% and soil loss by 64%. This indicated that the watershed program is effective in conserving rainwater and controlling soil erosion, thereby minimizing land degradation. Over grazing, over exploitation and other anthropogenic factors resulted in severe problems of degradation of common lands. To combat this hazard and maintain ecological balance, watershed development program implemented a management system for the rehabilitation of common land and biodiversity development through silvipasture systems managed by the local communities. This system converted the degraded wasteland in to valuable and beneficial asset for community. The number of useful species of grasses and fodder increased (Pathak *et al.,* 2013).

Sustainable water management is thus crucial for economic development and livelihood of the people. In a country like India, where a lot of running water goes waste, it becomes very important to apply the technology of watershed management to solve recurring perennial problems of desertification.

Deforestation: Forests are important ecosystems because of their impact on global hydrological cycles, biodiversity, and numerous social, economic, and political issues, however, forests are rapidly dwindling. Global Forest Resources Assessment (FRA) reported a 3.16% decline in the global forest cover from 1990 to 2015, and the total forest cover stands at about 30.6% in the present time compared to 31.6% in 1990. The rate at which the forest cover is declining poses a direct threat in the near future if not checked. With an estimated annual loss of 18.7 million acres, it is evident that future demands on forest resources would certainly lead to immense competition among nations (Bradford, 2018)). The global annual rate of deforestation is estimated at 12.37 Mha out of a total remaining area of 1505 Mha or 0.82% yr^{-1}. Richards and Flint (1994) have reported that total forest area decreased from 100 Mha to 81 Mha while cropland area increased from 100 Mha to 120 Mha during 1880–1950. The temporal pattern of deforestation during 1880–2000 has a major control over temporal pattern of carbon emissions due to land use change (Chhabra and Dadhwal, 2004). A total of 26 Mha forest areas (from 89 Mha in 1880 to 63 Mha in 2010) and 20 Mha of grasslands/shrublands (from 45 Mha to 25 Mha) has decreased in India (Tian *et al.,* 2014). (Figure 2)

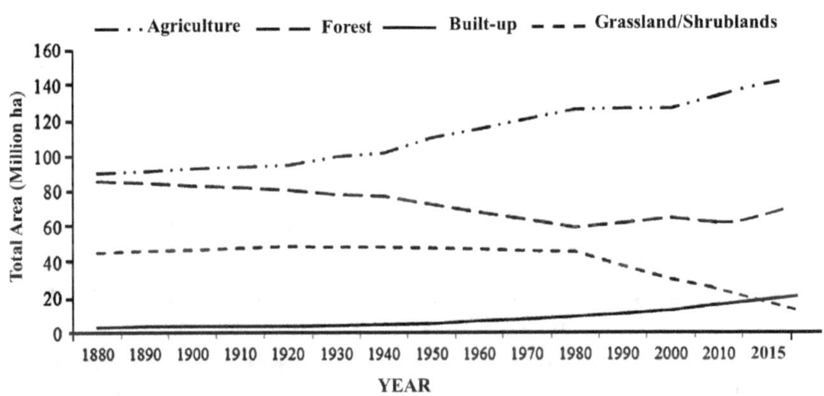

Figure 2. Temporal pattern of land cover and land use change during 1880–2010 (Adapted and Modified after Tian *et al.,* 2014 by the author).

The total forest cover (TFC) of India is 712,249 km² according to the biennial State of India's forest Report 2019 (ISFR, 2019). The TFC of India in 2019 is 21.67 per cent of the total geographical area (TGA) of the country as against 33% of National Forest Policy of 1988. It was 21.54 per cent (of TGA) in 2017 (Table 2). The forest cover shows a consistent increase in the area under the OF category, which includes commercial plantations. And this seems to be happening at the cost of the MDF category, which is normally the area close to human habitations.

Table 2. Forest cover in India (in km^2)

Year	2019	2017	2015	2013	2011
Very dense forests (VDFs)	99,278 (3.02%)	98,158 (2.99%)	85,904 (2.61%)	83,502 (2.54%)	83,471 (2.54%)
Moderately dense forests (MDFs)	308,472 (9.39%)	308,318 (9.38%)	315,374 (9.59%)	318,745 (9.70%)	320,736 (9.76%)
Open forests (OFs)	304,499 (9.26%)	301,797 (9.18%)	300,395 (9.14%)	295,651 (8.99%)	287,820 (8.75%)
Total Forest Cover	712,249 (21.67%)	708,273 (21.54%)	701,673 (21.34%)	697,898 (21.23%)	692,027 (21.05%)
Scrub	46,297 (1.41%)	45,979 (1.40%)	41,362 (1.26%)	41,383 (1.26%)	42,176 (1.28%)
Non-forest	25,28,923 (76.92%)	25,33,217 (77.06%)	25,44,2228 (77.40%)	25,47,982 ()77.51)	25,53,060 (77.67%)
Total Geographical area	32,87,469 (100%)	32,87,469 (100%)	32,87,469 (100%)	32,87,469 (100%)	32,87,469 (100%)

Source: State of Forest Reports (2011-2019), Forest Survey of India (it is in public domain).

However, the actual forest/tree cover is just about 19% of the total geographical area. The per capita forest area in the country is 0.08 ha as compared to the world average of 0.64 ha. A large part of these forests is degraded and productivity is very poor. The forest report also reveals that the forest area under the category "recorded forest area" (land notified as forest by the government) in tribal districts, which are home to about 60 percent of India's forests, is decreasing as well.

The rate of deforestation in country is very high, according to the data mapped by Global Forest Watch, a repository of forest data worldwide. The State-wise data shows that the loss of tree cover in north-east India amounts to over 70% of all tree loss in India that occurred between 2001 and 2018 (Table 3). According to Global Forest Watch (GFW) Water, watersheds lost more than half of their forests prior to 2000 in India. The watershed of Krishna, India was once covered by forests, but fewer than 3 percent of these trees remain today as a result of urbanization and cropland expansion. Communities in the area suffer from frequent droughts and floods, as well as high levels of water pollution from agricultural runoff. Sedimentation is a recurring challenge to reservoirs and dams. Planting new seedlings in deforested areas, enhancing natural forest generation, and integrating trees with crops and rangelands could help watersheds like Krishna.

Table 3. Primary forest loss and tree cover loss: India (ha)

Year	Primary forest	Tree cover (30%)
2001		62,441
2002	11,718	53,035
2003	10,846	47,782
2005	15,148	74,217
2007	17,208	73,976
2009	17,112	79,339
2011	16,287	88,585
2013	14,399	80,943
2015	20,997	116,374
2018	19,310	132,429
Primary loss 2002-2018	310,624	
Tree cover loss (2001-2018)		1,625,397
Primary loss/ Total tree cover loss	19.1%	
Percent loss (2001-2018)	3.0%	
Loss/year (2001/2-2010)	15,376.23	66,993
Loss/year (2011-2018)	21,529.71	127,239

The rate of deforestation in country between 1970 to 1990 was up to 1.5 Mha year^{-1}. India faces a growing crisis of land degradation: Nearly 30% of its land area, as much as the area of Rajasthan, Madhya Pradesh and Maharashtra put together, has been degraded through deforestation, over-cultivation, soil erosion and depletion of wetlands. India has pledged to restore 5 million hectares of degraded land by 2030. But this is just 1.5% of the country's geographical area, 28.5 percentage points less than the total land left degraded. India need to stop land-use change, work on forest conservation and step up land restoration on priority, to halt land degradation. Watershed Management can play a crucial role through better land management for a judicious management of forest ecosystems, and in restoration of degraded soils.

***Water scarcity*:** The water relationships of arid zones are perhaps more critical to a greater number of people on earth than those of more humid regions. Water is always in critical balance with arid ecosystems and this balance is presently being upset by man and his animals at alarming rates over vast acreages of the earth. Water crisis is usually viewed in terms of an increasing imbalance between water supply and demand. The increase in population at geometrical rate and over use of water for various purposes are considered to be the major reasons for decreasing the water resources in the world. When the annual per capita availability of renewable fresh water in a country or a region falls below 1700 m^3, it is held to be a situation of water stress. If the availability is below 1000 m^3, the situation is labelled as that of water scarcity. And when the per capita availability falls below 500 m^3, it is said to be a situation of absolute scarcity (Engleman and Roy, 1993).

The per capita availability of water in the country is 1545 m^3 as per the 2011 census. The average annual per capita availability of water in the country, taking into consideration the population of the country as per the 2001 census, was 1816 m^3 which reduced to 1545 m^3 as per the 2011 census. So, the per capita water availability in the country is reducing progressively due to rapid increase in population. Also there are reports that demand for water in India would rise dramatically to about 833 km^3 in 2025 and 899 km^3 in 2050 due to increase in population, rapid urbanization and a growing GDP with significant changes in the style of living. It is estimated that by the year 2025, as much as two-thirds of the world's population-estimated to have expanded by an additional 2.6 billion people will be living in conditions of serious water shortage and one-third will be living in conditions of absolute water scarcity.

Africa, which is one of the world's driest continents, is facing a very severe water crisis. Over 90% of Sub-Saharan Africa agriculture is rain-fed, and mainly under smallholder management (Batino and Waswa, 2011). As per the WHO/UNICEF report more than one billion people still use unsafe drinking water. The African river system identifies three river systems (de Wit and Stankiewicz, 2006) i.e. the areas receiving very low rainfall have virtually no perennial drainage (dry regime), then the areas with an intermediate range in which drainage density increases with increasing rainfall (intermediate rainfall regime) and the areas of high rainfall (high rainfall regime). The dry regime covers the largest area of the African continent i.e. approximately 41%, but most important is the intermediate rainfall regime which covers approximately 25% because this is the area where changes to precipitation would result in serious changes in drainage supply.

Integrated water management is based on a multi-sectoral approach to water management which seeks to optimize the use of all sources of water for the following uses; supply and sanitation, agriculture, industry, urban development, hydropower generation, inland fisheries, transportation, recreation, etc. It advocates the integration of water management efforts with that of other natural resources such as land and forests within the framework of the national economic and social policy. Integrated water management should be carried out at the catchment or sub-basin level and would involve optimal water allocation for competing uses; promotion of efficiency in use; integration of environmental and resource quality considerations; and flood and drought management. These would be effected through pricing mechanisms, regulatory measures,; development of databases, forecasting models, environmental impact assessment methods, public awareness, involvement of local communities and authorities in water management and promotion of international scientific research cooperation. Effective management of problems related to growing water scarcity in arid/ semi-arid regions calls for suitable adaptation and mitigation measures. So, the crisis can be averted by improving water use and management.

Watershed technology can address the problem of water scarcity to a certain extent. In a water scarce region such as western Rajasthan, water availability has a direct impact on livelihood security such as food availability, animal husbandry and forest produce. To effectively cope with water scarcity, adaptation measures are needed to be taken at the individual/ community level to develop the water resources in a sustainable manner in order to mitigate the adverse impacts of droughts, stabilize agricultural production as well as reduce the community's vulnerability to water shortage. The core components of the required approach suitable for western Rajasthan comprise the following:

a. Soil and water conservation measures mainly in agricultural land such as check dams and vegetative barriers, stone bunds and terraces.
b. Water harvesting structures (WHSs) such as anicuts and percolation tanks for efficient utilization of available rainwater.
c. Alternate land uses for non-arable lands such as afforestation and pasture development supported by rainwater conservation measures such as contour trenches and vegetative barriers.
d. Water saving measures such as efficient irrigation practices.
e. Promotion and strengthening of Self Help Groups (SHGs) as village institutions and capacity building of participating households/ beneficiaries with support from local Non-Governmental Organizations (NGOs), in order to ensure effective utilization, maintenance and sustainable operation of WHSs and soil and water conservation practices even after the withdrawal of support from NGO.

Importance of watershed management to food security: Rainfed lands account for more than 80 percent of global crop area and 60 percent of global food output but are especially susceptible to the impacts of climate change, more so in the arid and semi-arid regions. In general, the productivity of rainfed agriculture in developing countries is considerably lower than in more secure irrigated conditions, where the use of other factor inputs is generally higher. In the northern hemisphere intensive, mechanized and highly productive rainfed agriculture is the norm, benefiting from reliable and well-distributed rainfall (Gaur 2019). Agriculture of western hot arid Rajasthan desert is predominantly rainfed, with low and erratic rainfalls over a large portion of the continent and is dominated by small subsistence farms.

Water and food security are the important challenges under climate change as both are highly vulnerable to continuously changing climatic patterns. Climate change brings additional threats arising from stresses and shocks caused by higher temperatures and a lack of rainfall (Zougmoré *et al.,* 2014). According to Lobell *et al.,* (2011), the potential yield loss due to the climate change impact is about 5% for each degree Celsius of global warming. Some of the prominent climate change impacts are, growing deserts, and increase in the magnitude of floods and droughts. An extreme decline in crop yields in arid and semi-arid areas globally has caused food shortages and a manifold increase in food inflation. Countries of Africa, Middle East, Arab and Asia have close economic ties with natural resource and climate-dependent sectors such as forestry, agriculture, water, and fisheries. As per the IPCC (2007) report, the maximum growth in the emission of greenhouse gases (GHG) has occurred between 1970 and 2004, i.e. 145% increase from energy supply sector, 120% from transport, 65% from industry, 40% from change in land use patterns and during this period global population increases by 69%.

India's principal issue of the 21st century is food security and environment quality. Despite the phenomenal advances made in agricultural technology, India's food production has not kept pace with the increase in population. India is ranked at number 67, way below neighbouring countries like China and Pakistan, in 2010 Global Hunger Index by the International Food Policy Research Institute. Degradation of soil and water resources and lack of appropriate technology to address the basic issue of resource management are the primary factors responsible for low agricultural productivity. Important among environmental issues are poor water quality and the accelerated greenhouse effect. Managing water is one of the greatest and more urgent challenges, which requires an accurate assessment of the available and renewable water resources. However, such an assessment is especially difficult in arid regions, since the science base is limited, data are scarce and humid zone experience is inappropriate. Therefore, improved scientific understanding, co-operation and data sharing are fundamental to facilitate ways of better water governance and to reach integrated management of water resources in arid and semi-arid regions (Squires and Gaur, 2019).

The problem of food security needs to be addressed at immediately. As one of the world's largest emitters of greenhouse gases India contributes significantly to global warming. Watershed Management programme decreases greenhouse effect and increases food security.

Global warming: Another serious issue is that of global warming. The atmospheric concentration of CO_2 and other greenhouse gases is rapidly increasing (Lal *et al.*, 1995ab; 1998ab). Emission of greenhouse gases from soil to the atmosphere is accentuated by soil degradation and desertification processes are driven by deforestation, land misuse, and soil mismanagement. Adoption of improved watershed management technologies has a potential to sequester C in soil, enhance soil quality, improve productivity, and mitigate the greenhouse effect.

Mitigating the migration: Migration of labourers, especially rural to urban is on the increase. A typical case is the drought-prone Barmer district in Rajasthan, which has had high migration rates for several decades. It is now well known for the legendary Rajasthani labourers who work in construction all over India. A common understanding is that deteriorating agriculture leads to out migration and improving the natural resource base and generating employment in rural areas can mitigate migration.

Watershed projects has the strong potential of employment generation that varies across regions depending on cropping intensity and labour intensive crops grown in that region. This resulted in minimizing migration of landless and other labour in the villages.

Poverty and Watershed Management: The watershed programmes has increasingly become more poverty focused. There has been a shift from assessing the impact of watershed management on regeneration of natural resource base, health of the environment and agriculture productivity to enhance to overall impacts on poverty and livelihood security in a sustainable manner. Natural Resources Management and Poverty alleviation should be considered as two-sides of the same coin. Watershed management per se does not reduce poverty. But watershed programme conserves natural resources which lead to improved agricultural productivity. Agriculture is a key driver of development and poverty reduction in rural areas. Growth in agriculture usually generates the greatest improvements for the poorest people.

References

Bationo A., B. Waswa, J.M. Okeyo, F. Maina, J. Kihara (Eds.) (2011). Innovations as Key to the Green Revolution in Africa: Exploring the Scientific Facts, Vol. 1, Springer, Dordrecht, The Netherlands, pp. 3-20.

Bhathwal, C.P. (Editor). (2003). Good Governance in India, Deep and Deep publication Pvt. Ltd, New Delhi.

Bradford, A. (2018). Deforestation: Facts, Causes and Effects. Retrieved from: https://www.livescience.com/27692-deforestation.html; Accessed on: 28/01/2019

Bruinsma, Jelle (2018). Looking Ahead in World Food and Agriculture: Perspectives to 2050. http://www.fao.org/3/i2280e/i2280e06.pdf (accessed on 4th March 2020).

CGWB (2007). Manual on artificial recharge of groundwater; Central Ground Water Board, Ministry of Water Resources, Govt. of India, 185p.

Chhabra, A., Dadhwal, V.K. (2004). Assessment of major pools and fluxes of carbon in Indian forests. *Clim. Chang.* 64 (3), pp. 341–360.

de Wit, Maarten, Jacek Stankiewicz (2006). Changes in surface water supply across Africa with predicted climate change, *Science*, 311, pp. 1917-1921.

Engelman R and Roy, P. (1993). Sustaining water: population and the future of renewable water supplies Washington, DC: Population and Environment Program, Population Action International. 56 pp.

Eswaran, H., R. Lal and P.F. Reich (2001). Land degradation: an overview. In: Bridges, E.M., I.D. Hannam, L.R. Oldeman, F.W.T. Pening de Vries, S.J. Scherr, and S. Sompatpanit (eds.). Responses to Land Degradation. Proc. 2nd International Conference on Land Degradation and Desertification, Khon Kaen, Thailand. Oxford Press, New Delhi, India. (Accessed on 3rd March 2020).

Gaur, Mahesh K. and Squires, Victor R. (Eds.) (2018). Climate Variability, Land Use change and Impact on Livelihoods in the Arid Lands. Springer. ISBN: 978-3-319-56680-1.

Gaur, Mahesh K. and Squires, Victor R. (2019). Changes in Agricultural Land Use and Food Security: Challenges for North West, India. In: Food Security and Land Use Change under Conditions of Climatic Variability (Eds. Victor R. Squires and Mahesh K. Gaur), Springer, pp. 257 – 280. https://doi.org/10.1007/978-3-030-36762-6_14

Ghose, B., B. Singh and Amal Kar (1977). Desertification around the Thar – A Geomorphological Interpretation. Annals of Arid Zone 16: 290-301.

Government of India (2007). Report of the Working Group on Natural Resources Management: Eleventh Five Year Plan (2007–2012), Planning Commission, New Delhi.

Government of India (2007). Report of the expert group on "Groundwater management and ownership" submitted to Planning Commission, September 2007. Government of India, Planning Commission, Yojana Bhavan, Parliamentary Street, New Delhi. 61 pages.

Halvin, J.L. and A.J. Schlegel (1997). Dryland conservation technologies: Enhancing agricultural profitability and sustainability. *Annals of Arid Zone* 36(3): 291-303.

Hudson, B.D. (1994). Soil organic matter and available water capacity. *J. Soil Water Conserv.* 1994, 49, 189–194.

IPCC (2007). Fresh water resources and their management climate change 2007; Impact, adaptation and vulnerability. Contribution of working group-II to the fourth assessment report of the intergovernmental panel on climate change. Cambridge University Press: Cambridge, UK.

Joshi, *et al.,* (2004). Socio economic and Policy Research on watershed Management in India: Synthesis of Past experiences and needs for future research, ICRISAT, Hyderabad. (Available at *SAT eJournal,* August 2006, Vol. 2, Issue 1.

Kar, Amal, P.C. Moharana, P. Raina, Mahesh Kumar, M.L. Soni, P. Santra, Ajai, A.S. Arya and P.S. Dhinwa (2009). Desertfication and its control measures. In: Trends in Arid Zone Research in India (Eds. Amal Kar, B.K.Garg, M.P.Singh and S.Kathju), Central Arid Zone Research Institute, Jodhpur, pp.1-47.

Kerr, J. *et al.,* (2000). *An Evaluation of Dry Land Watershed Development Projects in India*, Environment and Production Technology Division, International Food Policy Research Institute 2033, K Street, N.W Washington, D.C. 20006 U.S.A.

Khayat S, Heinz H, Stefan G and Wasim, A. (2006). Hydrochemical investigation of water from the Pleistocene wells and springs, Jericho area, Palestine; *Hydrogeol. J.* 14:192–202.

Khalid, M.A. *et al.,* (2004). *Impact Assessment Study of the Watershed Development Programme, a Compendium*, The Energy and Resources Institute, New Delhi.

Lobell DB, Schlenker WS, Costa-Roberts J. (2011). Climate trends and global crop production since 1980. *Science,* 333:616–620. doi:10.1126/science.1204531

Nath, Reshmita, Luan, Yibo, Yang, Wangming, Yang, Chen, Chen, Wen, Li, Qian and Cui, Xuefeng (2015). Changes in Arable Land Demand for Food in India and China: A Potential Threat to Food Security. *Sustainability* 2015, 7, 5371-5397; doi:10.3390/su7055371 (accessed on 4th March 2020)

Ninan, K.N. and S. Lakshmikanthanthamma (2001). Social Cost-Benefit Analysis of a Watershed Development Project in Karnataka, Vol.30 No.3, *Royal Swedish Academy of Sciences.*

Oldeman, L.R. (1994). The global extent of land degradation. In: *Land Resilience and Sustainable Land Use,* eds. D.J. Greenland and I. Szabolcs, 99–118. Wallingford: CABI.

Pathak, Prabhakar, Chourasia, Anil Kumar, Wani, Suhas P. and Raghavendra Sudi (2013). Multiple Impact of Integrated Watershed Management in Low Rainfall Semi-Arid Region: A Case Study from Eastern Rajasthan, India. *Journal of Water Resource and Protection,*

5, 27-36. doi:10.4236/jwarp.2013.51004 (accessed on 5th March 2020).

Pophare, Anil M.; Lamsoge, Bhushan R.; Katpatal, Yashwant B.; Nawale, Vijay P. (2014). Impact of over-exploitation on groundwater quality: A case study from WR-2 Watershed, India. *J. Earth Syst. Sci.* 123, No. 7, October 2014, pp. 1541–1566. https://doi.org/10.1007/s12040-014-0478-0.

Reddy, V. Ratna (2000). Sustainable Watershed Management Institutional Approach, *Economic and Political Weekly*, September issue.
Richards, J.F., Flint, E.P. (1994). Historic land use and carbon estimates for South and Southeast Asia 1880–1980. In: Daniel, R.C. (Ed.), ORNL/CDIAC-61, NDP-046. Oak Ridge National Laboratory, Tennessee, U.S.A. 326 pp.

Shah A. (2001). Who Benefits from Participatory Watershed Development? Lessons from Gujarat, India, IIED *Gatekeeper Series 97* IIED, London.

Sharda, V.N., Juyal, G.P., and Naik, B.S. (2008). *Watershed Development in India: Status and Perspective,* Central Soil & Water Conservation Research &Training Institute, Dehradun, Uttarakhand.

Sharma, Vijay Paul (2015). Dynamics of Land Use Competition in India: Perceptions and Realities. W.P. No. 2015-06-02. IIMA. Pp. 40. https://web.iima.ac.in/assets/snippets/workingpaperpdf/2079964 8232015-06-02.pdf (accessed on 4th March 2020).

Singh K. and Ballabh V. (2008). Incidence, Impacts and Management of Droughts in India: An Overview. In Jairath J. and Ballabh V. (eds.) Droughts and Integrated Water Resource Management in South Asia Sage: New Delhi, India.

Squires, Victor R. and Gaur, Mahesh K. (2019). Food Security and Land Use Change under Conditions of Climate Variability: Synthesis and Uniting Perspectives. In: Food Security and Land Use Change under Conditions of Climatic Variability (Eds. Victor R. Squires and Mahesh K. Gaur), Springer, pp. 331 – 345. https://doi.org/10.1007/978-3-030-36762-6_17.

Tian, Hanqin, Banger, Kamaljit, Bo, Tao and Dadhwal, Vinay K. (2014). History of land use in India during 1880–2010: Large-scale land transformations reconstructed from satellite data and historical archives, *Global and Planetary Change,* 121:78-88. http://dx.doi.org/10.1016/j.gloplacha.2014.07.005

Tideman, E. M. (2000). *Watershed Management: Guidelines for Indian Conditions*, Omega Scientific, New Delhi.

UNEP (1992). *World Atlas of Desertification.* London: Edward Arnold.

UNEP (1993). Good news in the fight against desertification. *Desertification Control Bulletin,* 22, 3.

UNEP (1994). *Land Degradation in South Asia: Its Severity, Causes and Effects upon the People.* INDP/UNEP/FAO. World Soil Resources Report 78. Rome: FAO.

Vaidyanathan, A. (2001). Watershed Development: Reflections on recent developments" in Nair, K.N., Chattopadhyay & Srikumar (eds.): 'Watershed Management for Sustainable Development Field experiences And Issues', Kerala Research Programme on Local Level Development, Trivandrum.

Vaidyanathan, A. (2006). India's Water Resources: Contemporary Issues on Irrigation, Oxford University Press, New Delhi.

Wani, Suhas P., Rockström, Johan, and Sahrawat, Kanwar Lal (2011). Integrated Watershed Management in Rainfed Agriculture. CRC Press, The Netherlands. Pp. 488.

Wani, Suhas P., Raghavendra Sudi and P. Pathak (2009). Sustainable Groundwater Development through Integrated Watershed Management for Food Security. *Bhujal News Quarterly Journal*, Oct-Dec, 2009.

Zougmoré, Robert, Jalloh, Abdulai and Andre Tioro (2014). Climate-smart soil water and nutrient management options in semiarid West Africa: a review of evidence and analysis of stone bunds and zaï techniques, *Agriculture & Food Security,* 3:16. http://www.agricultureandfoodsecurity.com/content/3/1/16.

12

Irrigation Scheduling for Efficient Water Management in Watershed

Ajita Gupta and Chetankumar P. Sawant

Introduction

Irrigated agriculture is the biggest water consumer in the world and always competes with industrial and urban sectors. Estimates indicate that more than 56 % of the world food grain production comes from irrigated agriculture. In arid and semi-arid regions with limited and irregular rainfall, the use of supplemental irrigation is necessary for enhancing crop production. Moreover, enhancing water productivity (WP) of different crops is one of the major activities to ensure water saving in agriculture sector (Gupta *et al.,* 2018). The misuse of water due to improper scheduling of irrigation and low irrigation water use efficiency would result in higher production costs and adverse environmental impacts. Matching the water supply and demand is essential for enhancing productivity and attaining sustainability under irrigated environment which can be obtained from proper irrigation scheduling. Moreover, knowledge of crop-water requirements is crucial for management and planning of water resources in watersheds under changing climate scenarios.

Judicious irrigation scheduling is simply applying the right amount of water at the right time. Scheduling considers physical factors such as water holding capacity of the soil, climatic parameters, and plant characteristics (e.g. water demand, root characters and sensitivity to water stress). But, scheduling also involves human factors, technical knowledge of irrigators, labour constraints, facilities available (e.g. electricity supply, equipment and water pricing and rostering), irrigation system type and management practices, and a host of other human considerations. Judicious irrigation scheduling besides other crop management practices plays a vital role in enhancing water productivity as well as effective and efficient irrigation water management.

Irrigation Scheduling

Irrigation scheduling is a decision-making process which enables an irrigator to take decision on when to irrigate and how much water to be applied per irrigation. This knowledge helps estimating the earliest date at which the next irrigation should be applied, before plants suffer water stress. The goal of an effective irrigation scheduling is to supply the plants with sufficient amount of water while minimizing losses through deep percolation or runoff. The water productivity of crops is generally low, only 40 to 60% of the applied water is effectively used by the crops and the rest of water is lost in irrigation system or in farm either through evaporation, runoff, or by deep percolation into the groundwater. Proper irrigation management practices require a sound knowledge of irrigation scheduling techniques which ranges from personal experience to following neighbours' practices and use of modern tools and techniques.

Criteria for Irrigation Scheduling

Three main criteria to schedule quantity and timing of irrigation are soil water (moisture) content, meteorological parameters and plant stresses. However, conventional scheduling methods use soil water content or meteorological parameters to calculate the crop water requirement. The most common irrigation criteria are soil moisture content or soil moisture tension, where the soil moisture status in terms of water content or potential is measured directly to determine the need for irrigation. Similarly, detailed methods for estimating evapotranspiration and crop water requirements for different crops and different climates, are used for irrigation scheduling. The third approach of irrigation scheduling is based on monitoring plant stress. For this approach irrigation scheduling decisions are based on plant responses rather than on direct measurements of soil water status or meteorological parameters. Plant stress responses can provide the most visual way to identify plants water demand. However, it should be noted, they do not provide a direct volumetric measure of water required. Whichever method used, each has its merits and limitations. Hence modern scientific irrigation scheduling practise uses a combined approach based on weather, soil or plant indicators. A number of devices, techniques, and computer methods are available to assist irrigators in determining time and the amount of water required. Currently, only few numbers of irrigators schedule their crops with either computer programmes or soil moisture/plant sensors because of the complexity involved in understanding and operation of the system. Therefore, successful irrigation scheduling methods must combine together scientific, economic and social considerations. To understand how to manage irrigation water by scheduling, one must understand some key words given below:

Effective root depth: Effective root depth indicates the active root development of an irrigated crop with no restrictive soil layers. This is the soil depth from surface, mostly the upper part where the root system is denser from which the plants take nearly 80-90% of their total water need. The rooting depth of plants depend on plant physiology, soil type, availability of hard pan below soil surface, and water availability (Table 1). General water extraction pattern of plants from the soil is as follow (Figure 1): 40% of their water need from the first quarter (topmost layer) of root zone, 30% from the second quarter, 20% from the third quarter and 10% from the fourth quarter (bottom-most layer) of their effective root zone depth.

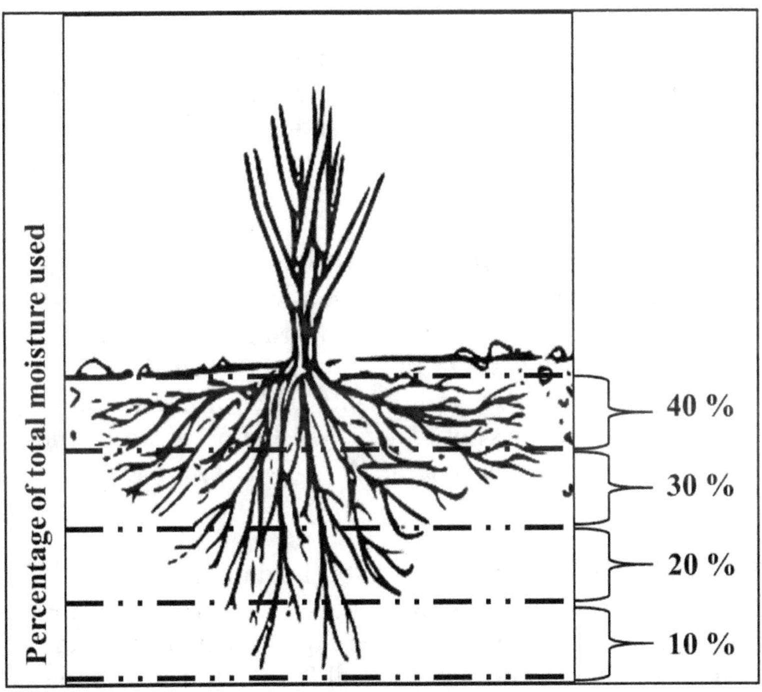

Figure 1. Soil moisture extraction pattern of plant.

Table 1. Effective root zone depth of different crops for moisture extractions (Doorenbos and Pruitt, 1977; Martin et al., 1990; and Allen *et al.,* 1998)

Shallow rooted crop *(Root depth upto 60 cm)*	**Moderately deep rooted crop** *(Root depth upto 90 cm)*	**Deep rooted crop** *(Root depth upto 120 cm)*	**Very deep rooted crop** *(Root depth upto 180 cm)*
Potato, onion, cucumber rice, groundnut, cauliflower, cabbage, most vegetable crops	Wheat, castor, pea, tobacco, carrot, chillies, muskmelon, french bean	Cotton, maize, bajra, jawar, soybean, pearl millet, grass(pasture), watermelon, sugar beet, citrus and fruit tree	Sugarcane, safflower, alfa-alfa, lucerene, apple, grape wine, coffee

***Field Capacity (FC)*:** The term field capacity (FC) denotes the quantity of water stored in a soil after drainage of gravitational water from soil pores. After heavy rainfall or irrigation above than the infiltration capacity of soil, soil becomes saturated. Once the soil has drained by gravity for about 24-48 hours, water in the pore spaces is held by the force of surface tension, and little further drainage takes place. This condition is called field capacity, and at this time the soil is holding as much water as it, as well as enough air to supply oxygen needed to plant roots.

***Permanent wilting point (PWP)*:** As plants extract water from the soil, a point reaches where root system can no longer obtain water from the soil. This point is called the wilting point for that soil. The soil is not absolutely dry, but the remaining water is held so tightly that the roots cannot extract it. Plants suffer stress at wilting point and will die if moisture is not replenished. For better crop production irrigated crops should never be allowed to approach the wilting point.

***Available water*:** The amount of water held in the soil between field capacity and permanent wilting point is known as available water to plant. The available soil water can be expressed as a percentage, or in millimeters of water per meter of soil (Table 2). For example, a soil with 17 percentage water has 170 mm of water per meter of soil. The available water-holding capacity of the soils vary according to soil texture, soil structure and organic matter content. A soil with better soil structure and organic matter will have more available water than poorly structured soils. In irrigation practice, only a percentage of AWC (Available water content) is allowed to be depleted, called management allowable depletion (MAD, %) (Table 3) because plants start to experience water stress even before soil water content reaches close to PWP. However, MAD varies with the crop type and the amount of stress a crop can withstand. Therefore, MAD (Table 3), must be specified for selected crops.

Table 2. Soil physical properties (average values) (Hillel, 1980)

Soil properties (average values)	Type of soil		
	Light textured (coarse) soil	**Medium textured soil**	**Fine textured soil**
Field Capacity (%) by weight	8-10	18-26	32- 42
Wilting point (%) by weight	4-5	10-14	20-24
Available soil moisture (mm/m depth of soil)	Upto 65	120	160-200
Bulk density (g/cm^3)	1.4-1.6	1.2-1.4	1.0-1.2
Infiltration rate (mm/h)	25-75	8-16	2-6
Time required from saturation to field capacity (h)	18-24	24-36	36-48

Table 3. Range of management allowed depletion (MAD) levels in major root zone activity of different crops (Doorenbos and Pruitt, 1977; Martin *et al.,* 1990; and Allen *et al.,* 1998)

Crops	Allowable soil water depletion (%)
Major crops, wheat, maize, sugar beet, cotton, pasture, cucumber, grass seeds, nursery stock, tree fruits, groundnut, citrus	50

Broccoli, cabbage, cauliflower, asparagus, grapes	45
Beans, cauliflower, mint, small grains, watermelon, Tomato	40
Potato, pumpkin, tobacco, spinach, carrot, banana	35
Sugarcane, peas, squash (winter), cotton, sweet potato, olives	65
Garlic, onion, lettuce, capsicum, berry	30

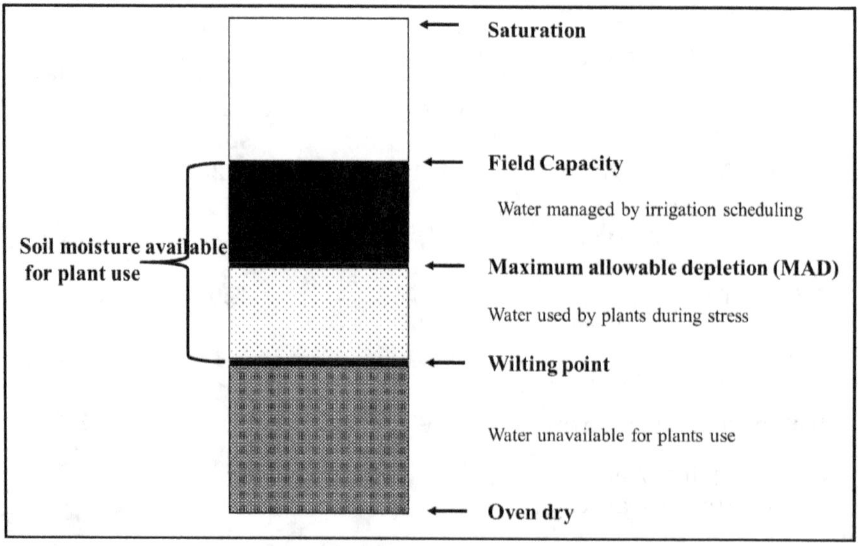

Figure 2. Soil moisture reservoir.

***Net irrigation requirement*:** The amount of water to be applied in the root zone to bring soil moisture at field capacity is estimated as the net irrigation requirement (d_{net}). Based on soil moisture content values and soil properties, net irrigation requirement can be calculated as:

$$d_{net} = \sum \frac{W_{fc} - W_i}{100} . BD. R_z$$

where d_{net} is net depth of irrigation required; w_{fc} is moisture content (%) of soil at field capacity; w_i is moisture content (%) of soil just before irrigation; BD is bulk density of soil; R_z is Effective root zone depth; and $\sum$ stands for the summation of soil layers from soil surface up to the effective root zone depth

However, during the irrigation process, a considerable amount of water lost through evaporation, seepage, deep percolation, etc. The amount of water lost depends on the efficiency of system and irrigation methods (Table 4). Hence, these irrigation efficiencies should incorporate to calculate actual amount of water given to field, this is known as gross irrigation requirement and can be calculated as follows:

Gross irrigation requirement = Net irrigation requirement/ Ea (%) * 100

where, Ea is irrigation water application efficiency, percentage.

Table 4. Irrigation system application efficiencies (Burt *et al.,* 2000 and Kansas CC, 2005)

Method	Efficiency Range (%)
Basin	75-80
Border strips	75-85
Low Gradient/blocked	80-90
Contour Border	70 – 80
Continuous Flood	80-85
Sprinkler Irrigation	65-85
Hand-move, End-tow, Side Roll Traveling gun, boom	60-75
Micro irrigation	80 – 93
Sub-surface micro irrigation	95 -98
LEPA (Low Energy Precise Irrigation)	80-93

Irrigation interval or Frequency: This is the number of days between two consecutive irrigations.

$$f = \frac{d}{CU}$$

where, f is irrigation interval or frequency (days), d is depth of irrigation to be applied (mm) and CU is daily crop consumptive water use (mm/day).

Irrigation cycle will be short at the period of peak water use. If there is rain in between consecutive irrigation, the depth and the frequency have to be corrected according to the most recent moisture contents of the soil.

Leaching requirements: An excess amount of water, 10–15 percent, is applied during the irrigation to flush out or leach down the soluble salts, where necessary. In this process a portion of applied irrigation water percolates through and below the root zone carrying with it accumulated soluble salts.

Different Methods for Irrigation Scheduling

Irrigation scheduling consists of two major activities: (1) applying the right amount of water and (2) at the right time. But these two variables are inversely co-related, so an irrigator has always a choice to adjust either amount of water per application or time interval between irrigation. Less amount of water can be applied at frequent time interval or an adequate amount of water can be applied at definite interval. Irrigation scheduling is generally governed by (1) measuring soil water status (soil based), or (2) by measuring plant water status (plant based), (3) measuring or estimating crop water demand (climate based).

The first two methods involve taking physical measurement of either soil or plant parameters. Both soil-based as well as plant-based methods of irrigation scheduling are rational approaches since they provide direct measurement of inadequacy of water. However, the climate-based method uses different weather parameters to estimate crop water demand over a certain period. The climate-based method to schedule irrigation is much easier and less expensive than soil- and plant-based methods, especially today when current weather data can be found readily available through internet and weather stations. Since this method is an estimation of crop water demand, it is important to periodically check the validity by measuring a tangible parameter, such as the soil moisture or plant water status.

Soil-based methods of irrigation scheduling

Soil-based irrigation scheduling methods are the most reliable and accurate because these measure available water and current moisture content in the soil at different depths. As a plant uses water, the root zone soil moisture reservoir is depleted and when soil moisture level reaches a critically low value, irrigation is applied to replenish the reservoir. Many devices/instruments are available in the market that directly provide information about the soil water status. These include tensiometers, resistance devices (gypsum and granular matrix blocks), neutron meters, and dielectric constant based measurement devices such as TDR (Time Domain Reflectometry) and FDR (Frequency Domain Reflectometry). Soil properties such as resistance, capacitance, and dielectric constant provide an indirect estimate of moisture present in the soil, only oven dry method provide direct measurement of moisture present in the soil. In addition, many irrigators still rely on the age old "hand feel method" to determine soil moisture.

***Hand feel method*:** Many irrigators still determine soil moisture by "feel method" as it is simple, fast and does not require any technical knowledge and costly instruments. The moisture status is determined by firmly squeezing a handful of soil samples to form a ball, this is bounced three times lightly in the palm and also some of the sample ribbons out between thumb and forefinger. The relative soil moisture for different soil type can be determined by using Table 5. Soil samples can be taken out with a soil probe, auger, and a shovel or simply by hand. This method is not recommended as a sole method for making irrigation scheduling decisions, but should be used as verification of other methods.

Disadvantages:

1) It is non-quantitative and non-standard method,
2) It does not give any information about amount of water present in the soil,

Table 5. The relative soil moisture for different soil type (Van der Gulik (2006)

Available water	Feel or appearance of soil			
	Coarse texture **fine sands to loamy fine sands**	*Moderately coarse texture* **sandy loam to fine sandy loam**	*Medium texture* **sandy clay loam, loam, & silt loam**	*Fine texture* **clay loam to silty clay loam**
	Available water (mm/m)			
	50-100	***108-142***	***125-175***	***108-200***
> 100 %	Wet, Forms a weak ball, Free water appears when soil is bounced in hand	Wet, Forms a soft ball, Free water is released with kneading	Free water can be squeezed out	Puddles; free water forms on surface
100 %	Upon squeezing, no free water appears on soil, but wet outline of ball is left on hand	Appears very dark. Upon squeezing, no free water appears on soil, but wet outline of ball is left on hand. Makes short ribbon.	Appears very dark. Upon squeezing, free water appears on soil, but wet outline of ball is left on hand. Will ribbon about 2.5 cm.	Appears very dark. Upon squeezing, no free water appears on soil, but wet outline of ball is left on hand. Will ribbon about 5cm.
75 - 100 %	Tends to stick together slightly sometimes forms a weak ball with pressure.	Quite dark. Forms weak ball, breaks easily. Will not slick.	Dark coloured. Forms a ball, is very pliable, slicks readily if high in clay.	Dark coloured. Easily ribbons our between fingers, has slick feeling
50 – 75 %	Appears to be dry, will not form a	Fairly dark. Tends to form ball	Fairly dark. Forms a ball somewhat plastic, will	Fairly dark. Forms a ball, ribbons out be-tween

	ball with pressure.	with pressure but seldom holds together.	sometimes slick slightly with pressure.	thumb and forefinger.
25 - 50 %	Appears to be dry, will not form a ball with pressure.	Light coloured. Appears to be dry, will not form a ball	Lightly coloured. Somewhat crumbly, but holds together with pressure.	Slightly dark. Some-what pliable, will ball under pressure.
0 - 25 %	Dry, loose, single-grained, flows through fingers.	Very slightly coloured. Dry loose, flows through fingers.	Slightly coloured. Powdery, dry sometimes slightly crusted, but easily broken down into powdery condition	Slightly coloured. Hard, baked, cracked, sometimes has loose crumbs on surface.

3) It requires knowledge and experience.

Gravimetric soil moisture sampling method: This is the standard method of soil moisture estimation and used to calibrate other devices and instruments. Soil samples are collected from different depths and locations using a soil probe or soil auger. Sample is kept in an oven at 105° C for 24 hours to become bone-dry then weighing is done before and after drying. Percentage of total soil-water content on a dry weight basis is computed. The gravimetric moisture content of the soil is computed as follows:

$$\theta = \frac{w_w - w_d}{w_d} \, .100$$

where, θ is the gravimetric moisture content (%), w_w is the weight of the soil sample at field condition (g), w_d is the weight of the oven dry soil sample (g). To calculate volumetric moisture content, the gravimetric moisture content is multiplied by the soil bulk density and also available soil water is calculated by subtracting current soil moisture content from soil moisture content at wilting point. This is the most accurate method for determining soil water content and widely used for research purposes, but is time consuming and physically impractical at field level.

***Tensiometers*:** A tensiometer is an instrument used to measure the soil water potential (metric suction or metric potential, which is directly related to soil moisture status. The measured value indicates the energy exerted by the plants to extract water from the soil. Generally, soil water tension increases with decreased soil water content, this means high readings for dry soils and low for wet soils. A tensiometer consists of a porous cup, connected through a transparent plastic tube and a vacuum gauge. Tube is filled with distilled water. The porous cup is normally made up of ceramic material due to its good strength and permeability. A Bourdon tube type vacuum gauge is commonly used, but where greater precision is required mercury manometer or pressure transducer can be used for automatically and continuously data recording.

Working principle: Tensiometers are installed in the plant root zone with their porous end in contact with surrounding soil. Water moves through porous upto create an equilibrium condition with the soil water. As the moisture level in the root zone depletes, the suction or negative pressure extract water from the plastic column through the porous tip, thus creating a vacuum in the tube equal of the soil moisture tension. This value is noted on the vacuum gauge. The tensiometric values are commonly measured in units of bars (and centibars (1/100th of a bar) in the English unit) or kilopascals (in metric units). One bar is approximately equal to one atmosphere pressure and one centibar is equal to one kilopascal. The volumetric water content corresponding to tensiometric values is pre-determined for different soil types and provided by the manufacturer, but calibration is necessary before field installation. The range of operation of a tensiometer is limited to 0 to about 85 centibars, which cover the entire range of soil water tension in sandy soils. Hence, tensiometers are best suited in sandy soils for irrigation scheduling. However, irrigation scheduling with tensiometer can be done by using trigger values, for different types of irrigation systems and soil types.

Advantages:

1) They are simple, low cost and easy to install.
2) They can be operated for long periods of time if maintained properly.

3) They can be easily adapted for automatic data recording and transmission system.

Disadvantages:

1) Tensiometers are not suited for heavy soils, where large amounts of available water occur outside its operational range.
2) They do not work properly if air enters the water column.
3) Tensiometers measure soil water tension rather than soil water content, soil moisture characteristic curve for specific soil is required to determine water content.
4) They require careful handling and regular maintenance otherwise breakage may occur during installation or field activities.
5) Tensiometers do not record the osmotic potential arises due to dissolved salts and poor irrigation water.

Electrical-resistance instruments: The basic principle of electrical resistance based instrument is that, a change in moisture content produces a change in electrical property of the soil. These are constructed of blocked shape porous materials usually plaster of Paris, nylon, fiberglass, gypsum, or combinations of these materials. Two electrodes embedded in the blocks and attached by wires. A resistance or conductance meter measures changes in electrical resistance in the blocks. Blocks are buried in the soil to the root zone depth at which the water content to be monitored. These instruments make close contact with soil particles and create equilibrium with surrounding soil. Since the amount of moisture in the blocks corresponds to electrical resistance, thus any change in moisture content provide change in the resistance values which can be measured using resistance meters. Nylon and fiberglass units are more sensitive in the higher ranges of soil moisture than plaster of Paris blocks. Plaster of Paris blocks function more effectively at a tension between 1 to 15 atmospheres, and fiberglass units operate satisfactorily over the entire range of available moisture. A combination of fibre glass and plaster of Paris provides sensitivity in both the wet and dry range and provides good contact with the soil. Electrical-resistance instruments are sensitive to salts and fertilizer concentrations and do not work properly in salt affected soil. The gypsum blocks are the most commonly used electrical resistance-based sensors and works well in salt affected soil due to its buffer action. Generally gypsum blocks have a one-year life. However, they might operate well for up to two years in acidic soils, and three to five years in neutral or alkaline soils.

Advantages:

1) They are inexpensive and made up of low cost materials.
2) They cause only minor disturbance to the soil profile.
3) They allow irrigators to monitor soil moisture in more locations and depths, compared to tensiometers.
4) They do not create any hindrance in farming operations.

Disadvantages:

1) They become unresponsive at the higher levels of moisture content.
2) They dissolve in the presence of salts and fertilizers, thus creating a poor soil surface contact.

***Neutron probe*:** The neutron probe is a very accurate, fast and capable of providing real time soil water content measurement, when calibrated properly. It is calibrated to give the total volume of water per unit soil depth. Neutron probes are sophisticated instrument that can provide repeated soil water measurements at the same location in different time intervals.

Working principle: The neutron probe works by emitting neutrons into the soil and reading the number of neutrons bounce back. This device consists of two main components, an emitter and a detector. The emitter consists of a radioactive source of fast neutrons, and the detector monitors the slow neutrons scattered back to the instrument. An access tube made up of aluminium or plastic which is transparent to neutrons, is inserted into the ground. Readings are taken by lowering the probe down to the access tube to the desired depths, and the backscattered slow neutrons are measured by a detector. The flux of backscattered slow neutrons is directly proportional to the density of hydrogen atoms present in the soil volume near to the circle of influence of probe. The number of scattered neutrons by hydrogen molecules, increases with the amount of water present in the soil. Three main processes involved in the measurement are: 1) emission off a neutrons from a radioactive source; 2) Collisions of neutrons in the soil medium and backscattering towards the instrument by hydrogen molecules; 3) Detection and counting of neutrons close to the source through detector.

Two major neutron sources used with this technique are, americium-beryllium (Am-Be) and radium-beryllium (Ra-Be). Am-Be is the most common type of commercially available neutron probe that is used for soil water measurement. They can be used for a number of years without an appreciable change in radiation flux.

Advantages:

1) Accurate readings and large sphere of influence.
2) Working does not affected by temperature, soil type or ph.
3) The system is portable and can be carried to multiple locations for monitoring spatial and temporal variation of soil water.

Disadvantages:

1) Cost is relatively high, labour intensive and needs sound technical knowledge before application.
2) Requires proper handling and care to avoid radiation hazard.
3) Proper calibration is needed at frequent time intervals.
4) They have accuracy problems at shallow depths (at near surface, less than 15 cm).
5) Slight delay occurs over the period of time causes error in receiving readings.

TDR (Time Domain Reflectometry): Time domain reflectometry (TDR) is a relatively new method for volumetric field measurement of soil water content. It is an indirect measure of soil water content based on the electrical conductivity and permittivity (dielectric constant) of the material (Jones et al., 2002).The propagation of electrical signals in the soil material is influenced by soil water content and dielectric constant of material. The dielectric constant of a non-conducting material is a measure of the transmission capacity of electromagnetic pulses through it. The dielectric constant of dry soil ranges from 2 to 5 and for water the value is 80 at the frequency of 30 MHz - 1 GHz.

Working principle: TDR consists of 2-3 parallel metal rods, acting as transmission lines (TL), pulses travel along a transmission line that is built in a coaxial cable and reflected back to the sensor for analysis. Travel time for electromagnetic waves to traverse the probe length is determined to calculate dielectric constant, from which the water content is inferred. The speed or velocity of the travelling pulse along the rod is related to the dielectric constant of the substrate. In relatively wet soil the velocity of the pulse is slower than in drier soil. The TDR probes are inserted directly into the soil for in situ soil water measurement at desire point. The measurement volume of TDR sensors depends on the length of the rods which ranges from 8 cm to 100 cm.

Advantages:

1) This technique is fast, accurate, non- destructive and easy use.
2) Soil specific-calibration is usually not required.
3) Relatively insensitive to normal salinity levels.
4) TDR measurements can be easily automated using a computer or data logger.

Disadvantages:

1) Relatively high cost equipment due to complex electronic circuit.
2) Limited applicability under highly saline conditions or heavy clay soils.
3) Soil-specific calibration required for soils having large amounts of bound water (i.e. highly porous soil, soil with high organic content, volcanic soils etc.).

Plant-based methods

Water deficit affects all physiological process and ultimately reduces growth and yield of a plant. Various plant physiological traits can provide an excellent way to measure the plant water status and consequently, irrigation timing. These traits include sap flow, visual symptoms, stem shrinkage, stomatal conductance or photosynthesis, leaf water potential, Leaf turgor pressure, chlorophyll content and canopy temperature etc. Monitoring the crop water stress through the direct visual inspections is the most common and feasible method of measurement (Jones, 2004). An experienced agronomist or farmer can detect early symptoms of water deficiency by visually inspecting the crop canopy, especially during the period of peak water use. However, the most comprehensive way is the measurements of total plant transpiration and photosynthesis. Various sensors have been developed so far to measure plant water stress, which relates integral components of the plant-soil-atmosphere complex. For example presence of nematodes, salinity, diseases, climatic conditions and compacted soil layers have combined effect on water transpired through the plants. Monitoring the crop canopy temperature with an infrared thermometer is widely used for calculating the water stress and crop water stress indices. With modern tools and technique nowadays remote sensing and GIS (Geographical Information System) have been emerged as an ultimate way to measure plant water stress at large scale.

However, critical growth stages (Table 6) to be found more practical than other methods. Soil moisture stress at these sensitive stages reduces growth and yield of crop. Therefore, adequate amount of water should be supplied during these stages to maintain crop transpiration. But the real number of irrigations will depend on agro-climatic conditions of the region, rainfall, and soil types. Thus, measurement of plant water status provides a fundamental approach to integrate both soil water available to the plants and the climatic conditions, but none of the technique developed so far is applicable to field conditions directly. Some of the plant based measurement techniques are discussed below:

Stomatal/leaf conductance: Stomata are small pores on the top and or bottom of a leaf that are responsible for exchanging CO_2 and water vapour in the surrounding atmosphere. Therefore stomatal conductance (mol m^{-2} s^{-1}) estimates the rate of gas exchange (i.e., carbon dioxide uptake) and transpiration (i.e., water loss) through the leaf stomata. It is a measure of the degree of stomatal opening that can be used as an indicator of plant water status. Stomatal opening changes constantly with changes in soil water status and atmospheric conditions. Measurements of stomatal conductance are done by using porometers and infrared gas analysers (IRGAs). The porometer measures resistance to water vapour movement while the IRGA measures resistance to both water and CO_2 exchange. Both devices consist of a chamber in which the whole leaf, or part of it, is enclosed during the measurement. In case of permeable cuticle, the leaf conductance to water vapour and CO_2 is measured. However, impermeable cuticle yields the stomatal conductance. This method is considered one of the most effective water stress indicators in terms of earliness and sensitivity. However the measurements are done manually, which limits its use for precision irrigation where automatic data sensing and recording is necessary.

Table 6. Critical growth stages of crops to supply water (FAO, 2002)

Crop	Critical growth stages
Wheat	Crop root initiation, tillering, flowering and grain development
Rice(up-land)	Boot, flowering, and milk
Rice (wet)	Tillering, panicle initiation, flowering and milk
Maize	Tasselling and cob development
Sorghum	Flowering and grain filling
Pearl millet	Flowering and grain filling
Gram	Pre- flowering and pod development
Green gram	Flowering and pod development
Black gram	Flowering and pod development
Lentil	Pre- flowering and pod development
Pea	Pre- flowering and pod development
Groundnut	Flowering and pod development
Linseed	Flowering and pod development
Soybean	Flowering and seed filling
Sugarcane	Sprouting and tillering
Potato	Tuberisation and tuber elongation
Tomato	Flowering and fruit development
Cauliflower	Early seeding and head development
Onion	Bulbing
Tobacco	Suckering and leaf maturity

Leaf water potential : The plant tissue water potential is measured as the difference between osmotic potential of the inner cell solution (ψ_o), which is created by the concentration of ions, and the counter pressure of the cell walls, the turgor pressure (P_t). Thus the leaf water potential (ψ_l) is the negative pressure or tension found in plant leaves, neglecting other minor potentials, can be expressed as:

$$\psi_l = P_t - \psi_o$$

The value LWP (English units are bars and SI units are mega Pascals [MPa]) for well-watered plants is higher (i.e., less negative in value) than that of plants under water stress. It can serve 2 purposes:

- To provide an indirect estimate of soil water potential. Since plant transpiration rate at night is very small or reaches nearly zero, at this time the leaf water potential comes into equilibrium with the water potential of the soil in the root zone. Therefore, leaf water potential values are at its peak when measured before dawn and will provide a reasonable estimate of the soil water potential.
- To provide current leaf water status which is extremely dependant on environmental conditions.

The pressure chamber, or pressure bomb as it is sometimes referred to, is one of the commonly used device that measures the leaf water potential of a plant according to the technique described by Scholander (1965). The pressure chamber measures "turgor pressure or xylem pressure potential," because the "osmotic potential" is usually negligible or small. The less free water there is in plant, the greater pressure is required to cause it to exude. Measurement is done by detaching a leaf from the plant and inserting into a chamber with its petiole projecting to the exterior through the sealing port. An increasing pressure is applied by allowing gaseous nitrogen to enter into the pressure chamber. When the internal pressure of the chamber become equilibrium with xylem pressure, small water bubbles appear and can be seen on the cutting edge of the petiole, and this pressure is recorded as the LWP value. However, application of this method is limited to solar noon when transpiration is at its peak and environmental conditions change slowly. Similarly, leaves must be sealed in plastic bags or wrapped with aluminium foil immediately after excision to reduce transpiration losses prior to the measurements.

Table 7. Critical leaf water potential (ψ_l) for selected crops at which stomatal regulation of transpiration will begin to reduce transpiration rates (Slabbers, 1980)

Crop	Ψ_l (MPa)
Alfalfa	-1.4
Barley, bean, cotton	-0.9 to -2.4 (average -1.4)
Cotton	-1.1 and -1.3
Corn	-1.7, -1.1 to -1.9†
Potato	-0.35 to -0.4
Soybean, Tomato, Grasses	-1.0
Sorghum	-0.9 to -1.6† (average -1.2), -2.0
Pea	-0.8
Sugar beet	-0.5
Sunflower	-0.75
Wheat	-1.0, -0.7 to -1.9‡

†depends on growth stage; ‡depends on leaf position in canopy

Canopy temperature: Infrared (IR) thermometry is an ideal method to monitor plant water stress since it is continuous, non-destructive, scalable from single plants to whole fields, and is less expensive than many alternative methods (Jones *et al.,* 2002).Plants cool themselves with transpiration and maintain a cooler body temperature, then ambient air during the daytime if sufficient moisture is available. As the soil moisture decreases transpiration also decreases, and temperature of plant leaves rises. Thus, measurement of canopy temperature can be used as a good irrigation scheduling tool for getting information about current soil moisture status. Water deficiency results in a considerable increase in canopy temperature. Various hand held infrared thermometer sensors are available that can measure the canopy temperature quickly and accurately. However, in order to make sound decisions, average of a large number of repeated values, at the different height, angle, location with respect to the sun, etc., should be calculated. Values obtained from IR thermometers can be used to calculate the crop water stress index (CWSI) (Idso *et al.,* 1981) which provide a linear relationship between the canopy-air temperature difference (T_c - T_a) and the vapour pressure deficit (VPD). At the wilted baseline, value of CWSI becomes 1 and at the well-watered baseline its value becomes 0.IR instruments are generally not robust enough for on-farm use, since they require proper maintenance and calibration. Full canopy has to be achieved before they can be employed for the measurements, and, thus, the early season irrigation scheduling is not possible by this method. Finally, this method is dependent on cloud free conditions and will not go well for humid regions.

***Sap flow measurement*:** In this technique, flow of the transpiration within the stem is measured by applying heat around the circumference of the stems and measuring the time of its advance (Cohen *et al.,* 1981; Fernandez *et al.,* 2001) or the time of its dissipation (Weibel and Devos, 1994). A series of small heat pulse is applied by a miniature heater imbedded radially into the trunk (typically a thin heater strip surrounding the plant stem) and the rate of sap flow is determined from the time taken by the pulse to reach a series of thermistors arranged as differential thermopiles, located at a short distance above the heater. Heater strips are more applicable for stem diameters less than 50 mm, but can be used successfully up to 150 mm stem diameter. However, the larger diameters require more number of thermocouples and become more expensive. The applied heater power varies between 0.2 to 10 W depending upon the crop species and transpiration rate to be measured. Mass flow of sap is measured by balancing energy into and out of the heated section and given as:

$$F = {(P - Q_v - Q_r)}/{C_p} * dT$$

where: F = Flow rate per unit of time, P= Power supplied (W), Q_v = Vertical or Axial flow (W), Q_r = Radial heat flow (W), C_P = Specific heat of water (4.186 J $g^{-1}C^{-1}$), dT = Temperature increase (C). However, the major limitations of this method are that only a few number of plants (4 - 32) can be monitored economically in a field (Howell and Meron, 2006) because it requires large wiring needed to connect sap flow meters to data logger. Also, instruments are of high cost and sophisticated.

***Dendrometer*:** As water status of the plant changes, a slight change also occurs in its diameter over a certain time period (usually from morning to noon). Therefore, measurement of daily stem contraction may indicate the water balance within the plants and need for irrigation. Both seasonal tree growth and water storage fluctuations over the year can be extracted from dendrometer measurements. This procedure measures a more comprehensive way of conditions being experienced by the entire plant.

Dendrometer is a device, used for measuring small changes in diameter. Dendrometer uses strain transducers mainly differential variable reluctance transducers (DVRT) fixed to a steel arm, which can detect a small arm movement and recording the form of electrical signals that are subsequently converted to micrometrics values. A wide variety of dendrometers are available in the market differing in accuracy, precision, cost, operational simplicity and robustness. With high-resolution dendrometers, diurnal changes in stem size can be detected. The trunk radius changes show a sinusoidal waveform over a period of 24 hours with a maximum value before sunrise, at the end of the increment phase. Subsequently, a decrease in stem radius will occur after sunrise, reaching a minimum value approximately two hours before sunset. The difference between these two critical points, is known as the maximum daily shrinkage and is used as an indicator for drought stress measurement (Fernández and Cuevas, 2010).

Climate based methods

Climate-based approach of irrigation scheduling uses weather information (meteorological parameters) to estimate the amount of water required by plant to meet its evaporative demand, which varies over time and thus set the irrigation amount accordingly (Hillel, 1990). Inevitably, during crop growth period the water is depleted primarily due to plant evapotranspiration, which declines available soil moisture pool. When sufficient moisture is available at plant root zone, the evapotranspiration is governed by atmospheric demand of water, but as soil moisture reaches to a critical value, at that time rate of evapotranspiration govern by soil moisture available to plant root zone (Jensen *et al.,* 1970). It is therefore fields are irrigated to replace depleted water and the soil water reservoir before plants suffer water stress. The general approach of climate-based methods is to maintain the water balance of cropped area by keeping record of the evapotranspiration losses, irrigation applied and precipitation amount. This procedure is analogous to maintaining the balance in a bank account, it is therefore commonly referred to as the "checkbook" method of scheduling.

The water requirements differ from one crop to another. However, the rate of crop water use mainly depends on (1) type of crop, (2) crop growth stage/degree of maturity and (3) atmospheric conditions, such as temperature, wind, and humidity. During initial growth stage plant water need is generally low, but it increases rapidly during the maximum growth stage. Whereas, during the later stages, water use decreases as the crops reach to ripening stage.

Evapotranspiration definition: Evapotranspiration (ET) is the sum of evaporation from soil surface, canopy interception, and water bodies plus plant transpiration through stomata. Evapotranspiration plays an important role in the hydrologic cycle and water balance parameters. Sun is the source of energy behind evapotranspiration process. Driving force of evaporation process is the difference between water vapour pressure at evaporating surface and vapour pressure of surrounding atmosphere. Since evaporation requires energy to change the water molecules from liquid to vapour state, the source of this energy is solar radiation and temperature. Whereas, in case of transpiration the driving force is the difference between water vapour inside the leaf and the atmosphere (Allen *et al.,* 1998).Its value is largely determined by climate factors, such as solar radiation, temperature, humidity and wind, and by the environment. Out of the total evapotranspiration, evaporation accounts for about 10 percent and plant transpiration for the remaining 90 percent.

Factors affecting evapotranspiration

Solar radiation: It provides the energy that drive photosynthesis activity of plants and evaporation from soil surface. The energy received is more in high solar radiation area, subsequently water used rate is also high in these areas as compared to area receiving low solar intensity.

Humidity: It is easier for water to evaporate into dryer air than into more saturated air, because the surrounding air is less able to remove the water vapour coming out of the plant's leaves and soil surface.

Wind speed: Crop evapotranspiration rate increases with increased movement of the air around a plant. Wind serves as a conductive medium, if there is no wind, humidity of the air around the leaf would rise. As wind moves around leaves, the more saturated air close to the leaf is replaced by drier air.

Temperature: Transpiration rates go up as the temperature goes up, especially during the growing season.

Soil-moisture availability: When sufficient moisture is available at plant root zone, the crop evapotranspiration is govern by atmospheric parameters, but as soil moisture depletes, the rate of evapotranspiration govern by soil moisture available to plant root zone.

Crop factors: The crop type, variety, developmental stages, crop roughness, stomata type and crop rooting characteristics result in different ET levels for different types of crop under the identical environmental conditions.

Besides this some factors such as poor land fertility, soil salinity, presence of hard pan below the ground surface, pest and disease management, plant density, cover crops, tillage practices and use of fertilisers have a significant effect on actual crop water use

The crop evapotranspiration rate is normally expressed in millimetres (mm) per unit time. It expresses the amount of water lost in terms of depth from a cropped surface per time unit that can be an hour, day, month and even entire growing period or year.

For example, loss of 1 mm of water from one hectare land (10 000 m^2) corresponds to a loss of 10 m^3 of water per hectare. In other words, 1 mm day^{-1} is equivalent to 10 m^3 ha^{-1} day^{-1}. Water depths can also be expressed in terms of energy received to vaporize free water (known as the latent heat of vaporization (λ)) per unit area. For example, at 20°C, λ is about 2.45 MJ kg^{-1}. In other words, an energy input of 2.45 MJ per m^2 is able to vaporize 0.001 m or 1 mm of water, and therefore 1 mm of water is equivalent to 2.45 MJ m^{-2}.

Determination of evapotranspiration (ET)

Crop evapotranspiration (ET_c) is calculated under standard conditions and non-standard conditions. ET_c under standard conditions refers to evapotranspiration that occurred from excellently managed, well-watered crop, growing over a large area that achieve its full production under given climatic conditions (FAO-56).But poor management practices and sub-optimal environmental conditions affect evapotranspiration. Therefore, ET_c under such non-standard conditions generally requires adjustment or correction for all types of stresses. There are different methods of determining evapotranspiration: direct measurement using weighing type or drainage type field Lysimeters and indirect methods from weather data and soil water balance. Some of these methods of crop evapotranspiration estimation are discussed below:

Field water balance study: Evapotranspiration can be determined by using field water balance method by measuring various input and output components of field water balance (Heermann, 1985). In this method basically dynamics of water on, into, out of and below the soil surface is calculated. The equation can be written as;

$$\mathrm{ET} = \mathrm{I} + \mathrm{P} + \mathrm{U} - \mathrm{D} \pm \Delta \mathrm{S_f} \pm \Delta \mathrm{S_w}$$

where, I is the irrigation water supplied, P is the rainfall, Upward capillary rise into root zone, the capillary rise is important in case shallow water table, R is surface runoff, D is the deep percolation beyond root zone, Δ S_f is subsurface flow in or out of the root zone and Δ S_w is a change in the root zone soil moisture storage. In this approach evapotranspiration is determined as a residual term and, all the other components have to be either measured or estimated. It is relatively easy to measure amount of water added to the field by irrigation and rainfall. However, the most difficult parameter to measure is deep percolation D. Generally, this method yields better result for short-term evapotranspiration calculation.

Lysimeters are the device that works on the water balance principle to calculate crop evapotranspiration. Lysimeters are typically tanks or containers installed in the field that defines a specific boundary and contain isolated and undisturbed soil volume in which vegetation is grown. The amount of water lost by evapotranspiration can be worked out by calculating the change in weight of soil tank at predefined intervals. However, Lysimeters are expensive and need well trained research personnel which make this method inappropriate for daily field measurements.

Reference evapotranspiration: The calculation of reference evapotranspiration (ET_0) is a very common method to calculate the crop water requirement for efficient irrigation scheduling. It refers to the rate of evapotranspiration from a uniform green cover crop surface such as grass, not short of water. The FAO-56 defined ET_0 as the evapotranspiration rate from a reference surface (hypothetical grass or alfalfa), with an assumed crop height of 0.12 m, with a fixed surface resistance of 70 sm^{-1} and an albedo of 0.23 is closely resembling the evapotranspiration from an extensive surface of green grass of uniform height, actively growing, well-watered, and completely shading the ground (Allen *et al.,* 1998). ET_0 describe the atmospheric demand of water and independent of crop type, growth stage and management practices. The actual evapotranspiration ET_c for different crops is calculated by multiplying reference evapotranspiration with crop coefficient Kc as is shown in next equation.

$$ET_c = ET_0 * K_c$$

where, ET_c actual crop evapotranspiration (mm), ET_0 reference evapotranspiration (mm) and K_c crop coefficient (crop characteristic based specific coefficient)

Calculation of reference evapotranspiration: Different reference evapotranspiration estimation methods have been developed over the years. Most of these methods require some climate data and geographic information as well to calculate ET_0 i.e. (a) temperature based (Thornthwaite, 1948; Doorenbos and Pruitt, 1977), (b) radiation based (Doorenbos and Pruitt, 1977; Hargreaves and Samani, 1985), and (c) combination methods (FAO-56 Penman-Monteith) (Allen *et al.,* 1998). Numerous studies worldwide have shown that the FAO-56 Penman-Monteith (PM) method is the most accurate method under various climatic conditions and declared as standard method for calculating reference evapotranspiration by FAO.

FAO-Penman-Monteith method: Penman-Monteith equation is a combination equation that has generally been accepted as a scientifically sound formulation for estimation of reference ET_0. This equation is expressed as combined function of radiation, maximum and minimum temperature, vapour pressure, and wind speed. The Penman-Monteith method is considered to offer the best results with minimum possible error in relation to a living grass reference crop. This method introduces resistance (both aerodynamic resistance and surface resistance) factors. Aerodynamic resistance describes resistance from the vegetation upward and involves friction from air blowing over vegetative surfaces. Surface resistance describes resistance of vapour flow through stomata opening, total leaf area and soil surface. The FAO Penman Monteith equation is written as the following:

$$ET_0 = \frac{0.408\Delta(R_n - G) + \gamma\frac{900}{T+273}U_2(e_a - e_d)}{\Delta + \gamma(1 + 0.34\,U_2)}$$

where: ET_0 is reference evapotranspiration (mm day^{-1}); Rn is net radiation at the crop surface (MJ m^{-2} day^{-1}); G is soil heat flux density (MJ m^{-2} day^{-1}); T is mean daily air temperature at 2 m height (^oC); U_2 is wind speed at 2 m height (m s^{-1}); e_a is saturation vapour pressure (kPa); e_d is actual vapor pressure (kPa); (e_a – e_d) is saturation vapour pressure deficit (kPa); Δ is the slope vapor pressure curve (kPa $^oC^{-1}$) and γ is psychometric constant (kPa $^oC^{-1}$).

Reference evapotranspiration ET_0 can be calculated using a spreadsheet or computer programs eg. CROPWAT (Smith, 1992) and ET_0 Calculator (Raes, 2009).The equation uses standard climatological records of solar radiation (sunshine), air temperature, humidity and wind speed. To ensure accuracy, measurement of weather data should be made at 2 m (or converted to that height) above an extensive surface.

However, the major drawback of FAO-56 Penman-Monteith method is that it requires numerous weather data viz. maximum and minimum air temperature, maximum and minimum relative humidity, atmospheric pressure, wind speed, wet bulb and dry bulb temperature, dew point temperature, daily net radiation, daylight hours etc. which are not easily available in many meteorological stations, especially in hilly regions or remote places. Therefore, application of alternative ET_0 equations that require a few meteorological parameters is necessary for locations where limited weather parameters are available. Some of these methods are discussed below:

***Penman method*:** This equation was developed by Penman (1948) to compute evaporation from open water surfaces based on the principles of energy budget and mass transfer. This combination type of equation is given as:

$$E_p = \frac{\Delta}{(\Delta + \gamma)}E_r + \frac{\gamma}{(\Delta + \gamma)}E_a$$

where: E_p is evaporation (mm) calculated by the Penman model; Er is the net radiation expressed in terms of equivalent evaporation (mm/day); Ea is evaporation by aerodynamic method (mm/day); Δ is the slope of the saturation vapour pressure curve at given air temperature (kPa $^{o}C^{-1}$) and γ is the psychrometric constant (kPa $^{o}C^{-1}$).

Priestley-Taylor method: Priestley and Taylor (1972) proposed an equation to calculate potential evaporation. Aerodynamic component of Penmen equation was deleted and energy component was multiplied by a short wave reflectance coefficient (albedo), α = 1.26. The Priestley–Taylor approach for estimating evaporation under conditions of minimum advection and neglecting heat flux into ground for daily interval is described as:

$$EPT = \alpha \frac{\Delta}{(\Delta + \gamma)} E_r$$

where: EPT is evaporation (mm) by the Priestley-Taylor model; α is a constant and E_r is the net radiation expressed in terms of equivalent evaporation (mm/day); Δ is the slope of the saturation vapour pressure curve at given air temperature (kPa $^{o}C^{-1}$) and γ is the psychrometric constant (kPa $^{o}C^{-1}$).

Stephens and Stewart method: Stephens and Stewart (1963) proposed an empirical linear equation that requires only radiation and temperature data.

$$Ess = SR(a + bT_{mean})$$

where: E_{ss} is evaporation in mm by Stephens and Stewart model; SR is solar radiation expressed as equivalent water evaporation (mm/day); T_{mean} is mean air temperature (^{o}C), 'a' and 'b' are fitting constants. Solar radiation is calculated with the Angstrom formula, which relates solar radiation to extra-terrestrial radiation and relative sunshine duration (Allen *et al.,* 1998).

Blaney-Criddle method: The Blaney-Criddle (1950) procedure is one of the most popular methods in the western U.S.A. region and requires mean temperature data only. The usual form of the Blaney-Criddle equation in metric unit is written as:

$$ET = kp(0.46T_a + 8.13)$$

where: ET = potential evapotranspiration from a reference crop (mm); T_a = mean temperature in °C; p = percentage of total daytime hours for a given period (daily or monthly) out of total daytime hours of the year; k = monthly consumptive use coefficient, depending on vegetation type, growth stage, location and season.

Energy balance method: The process of evapotranspiration is governed by energy exchange at vegetation surface and is limited by the amount of energy available. Therefore, the principle of energy conservation can be applied to predict evapotranspiration rate. Energy received at the surface must be equal to energy leaving the surface for a given duration.

$$Rn - G - \lambda ET - H = 0$$

where: ET is evapotranspiration (mm); λ is latent heat of vaporization (kJ/kg); ρw is water density (kg/m^3); Rn is net radiation flux (W/m^2); H is sensible heat flux (W/m^2) and G is ground heat flux (W/m^2).

Mass-transfer based methods

This is one of the oldest methods and is still an attractive method for estimating free water surface evaporation because of its simplicity and reasonable accuracy. The mass-transfer methods are based on the Dalton equation (1802) and can be written as:

$$E = C(e_s - e_a)$$

where: E is free water-surface evaporation (mm); e_s is saturation vapour pressure at the temperature of water surface (mmHg); e_a is actual vapour pressure in the air (mmHg); C is an empirical constant depends on wind characteristics.

Many variations of the Dalton equation have been proposed. For example Rohwer (1931) proposed the given equation.

$$E = 0.44(1 + 0.27u2)(e_s - e_a)$$

where: E is in mm d^{-1}; u2 is the wind speed at 2 m height in m s^{-1}; e_s and e_a are as described above.

Penman (1948) concluded the best form of equation for practical use is

$$E = 0.35(1 + 0.98/100u2(e_s - e_a))$$

These empirical equations provide adequate and reliable estimates of evapotranspiration for the most engineering purposes where limited meteorological data are available. Their use does not require much skill, time and effort. Radiation based empirical equations generally provide more reliable estimates than those based on air temperature as primary variable. However, the major limitation using empirical equations it that these equations are region specific and cannot be used in other regions without proper calibration and validation.

Crop coefficient (K_c): Once the reference evapotranspiration (ET_0) has been determined, it should be adjusted for local conditions and weather. The actual evapotranspiration is then determined by multiplying adjust ET_0 value with a crop specific coefficient called "crop coefficient". The value of crop coefficient differs from one crop to another, depending on their characteristics which includes crop type and crop development for that specific crop. To calculate ET_c, crop growth stages, their lengths, and corresponding (K_c) coefficient for each stage must be identified.

Allen *et al.,* (1998) (FAO-56) divided growing season into four stages. The length of each of these stages depends on the climate, altitude and planting date.

Stages	Identification
Initial	From planting date to approximately 10% ground cover
Development	This stage runs from 10% ground cover to effective full cover which is 70 or 80 percent ground cover
Mid- season	Effective full cover to start of maturity
Late- season	Start of maturation to full senescence or harvest

K_c, is basically ratio of the crop ET_c to the reference ET_0, and it represents integrated effects of four primary characteristics (crop height, albedo of crop-soil surface, canopy resistance and evaporation from soil surface) that distinguish crop from reference grass.

Main factors affecting K_c

Some important factors affecting crop coefficient at each stage, are described below:

Crop type: Different crops have different properties, which makes crop coefficient different from one crop to another. The most important crop properties, affecting K_c are crop height, aerodynamic properties, albedo (reflectance) of crop-soil surface, leaf resistance, leaf and stomata properties. For example, some crops close their stomata during day, like pineapples have very small crop coefficients.

Climate: The main climatic parameters that affect crop coefficient K_c values are wind speed, air temperature and relative humidity. As wind speed increases and relative humidity decreases, value of K_c will increase. Similarly as temperature increases the value of K_c will also increase.

Soil evaporation: The soil evaporation has a big effect at initial stage period. At initial stage, crop is small so most of the ET occurs as evaporation from soil surface, when soil is wet by irrigation or rain, evaporation from soil surface will be considerable and K_c increases accordingly.

Crop growth stage: At each growing stage each crop exhibit different characteristics, which makes the crop coefficient values different. At initial stage when crop is small, the evapotranspiration predominately occurs in the form of soil evaporation. During the crop development stage, K_c value varies according to ground cover and plant growth. At mid-season stage crop starts to mature and fully covers the ground, K_c reaches its maximum value. At late season stage, Kc value starts decreasing due to senescence and affected by crop and water management practices. The K_c end value is high if the crop is frequently irrigated until harvested fresh.

Estimating K_c for different stages

Allen *et al.,* (1998) has given lists of typical K_c values for K_c ini, $K_c$mid, and K_c end for various agriculture crops and trees (Table 9) in FAO-56 paper. These values are average K_c under a standard climate condition, which is defined as a sub-humid climate with average daytime minimum relative humidity 45 % and having calm to moderate wind speeds averaging 2 m/s. Kc values must be modified for other climate condition differing from standard condition. Differences in evaporation and transpiration between field crops and the reference grass surfaces can be estimated using single crop coefficient (K_c) approach. Whereas, in the dual crop coefficient approach (K_{cdual}), effects of crop transpiration and soil evaporation are bifurcated as the basal crop coefficient (K_{cb}) to describe plant transpiration, and the soil water evaporation coefficient (K_e) to describe evaporation from the soil surface. Thus single K_c coefficient is replaced by: $K_{cdual}= K_{cb} + K_e$. Selection of approach depends on the purpose of calculation, accuracy required, climatic data available and the time-step.

Table 8. General selection criteria for the single and dual crop coefficient approaches (Allen *et al.,* 1998)

Criterion	Single crop coefficient Kc	Dual crop coefficient Kcb + Ke
Purpose of calculation	irrigation planning and design irrigation management basic irrigation schedules real time irrigation scheduling for non-frequent water applications (surface and sprinkler irrigation)	research, real time irrigation scheduling irrigation scheduling for high frequency water application(micro irrigation and automated sprinkler irrigation) supplemental irrigation detailed soil and hydrologic water balance studies
Time step	daily, 10-day, monthly (data and calculation)	daily (data and calculation)
Solution method	graphical computer, pocket calculator	computer

Crop coefficient curve

The curve coefficient curve represents changes in the crop coefficient over the length of the growing season. It allows determination of K_c values for any period during the growing season To construct crop coefficient curve total growing period is divided into four general growth stages (initial, crop development, mid-stage and late season). Then length of each growth stage and crop coefficient corresponding to each stage must be determined. Initial stage K_c value must be adjusted by multiplying it with fraction of soil surface wetted depending on irrigation methods or precipitation.

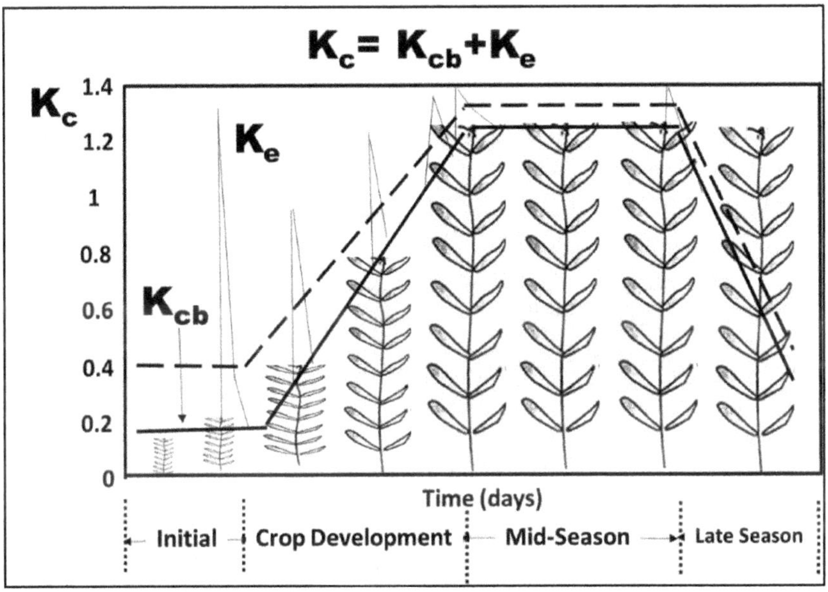

Figure 3. A typical crop coefficient curve showing K_c, K_{cb} and K_e values, adopted from FAO-56 paper (www.fao.org/docrep) (Redrawn).

Steps to schedule the irrigation process using the crop coefficient method:

Step 1. Collect meteorological data to calculate reference evapotranspiration (ET_0) value.
Step 2. Determine K_c values based on crop and stage of growth (Table 9).
Sub-Step 2a (Optional): Modify the amount of ET_c if soil moisture is limited.
Sub-Step 2b (Optional): Separate evaporation (E) and transpiration (T) to obtain more accuracy.
Sub-Step 2c (Optional): Adjust default K_c values to better fit in local conditions.
Sub-Step 2d (Optional): Modify the time scale used.

Step 3. Calculate daily crop water use (ET_c) by multiplying ET_0 with K_c values.
Step 4. Balance the checkbook by subtracting out daily ET_c and adding effective rainfall or other inputs.
Step 5. Irrigate when the soil checkbook balance reaches to a predefined level.

Table 9. Single (time averaged) crop coefficient K_c, Values for non-stressed, well managed crops in sub-humid climate (Allen *et al.*, 1998)

Crop	K_{cini}	K_{cmid}	K_{cend}
Groundnut	0.4	1.15	0.4
Soybean	0.5	1.15	0.5
Chick Pea	0.4	0.00	0.35
Rapeseed, Canola	0.35	1.0-1.15	0.35
Cotton	0.35	1.15-1.20	0.7-0.5
Wheat	0.7	1.15	0.25-0.40
Maize	0.3	1.20	0.60
Sorghum	0.3	1.00-1.20	0.55
Rice	1.05	1.20	0.90
Barley/Oats	0.30	1.15	0.25
Sugarcane	0.40	1.25	0.75
Onion	0.70	1.05	0.75
Garlic	0.70	1.00	0.70

Tomato	0.60	1.15	0.70-0.90
Cucumber	0.60	1.00	0.75
Potato	0.50	1.15	0.75

***Pan evaporation method*:** This is the most common and practical method for determining evapotranspiration. This approach combines the effects of temperature, humidity, wind speed and sunshine. Evaporation from the pan closely resembles evapotranspiration from grass surface that can be taken as an index of ET_0 for calculating actual evapotranspiration ET_c. The pan readings (E_{pan}) are related to ET_0 with the aid of pan coefficient (K_{pan}), which depends on type of pan, its location (surrounding with or without ground cover vegetation) and climatic conditions.

$$ET_0 = Epan \times Kpan$$

From ET_0 we can calculate crop water requirement (ET_c) by determining the specific pan crop coefficient (K_c). There are different types of pan used to determine evaporation rate. The most common types are circular pan or square pan. Descriptions of commonly used evaporation pans (Howell and Meron, 2007) are given below:

Name (Reference)	Description
Colorado Sunken Pan	This pans having square shape with 0.92 m sides and 0.46 m deep, made of 18-gage galvanized iron. Pan is buried in the ground with its rim 5 cm above the ground and water level is maintained at about ground level.
U.S. Weather Bureau Class A Pan	This is one of the most widely used pans. The Class A Evaporation Pan is circular in shape, having 121.5 cm inside diameter, 25 cm deep with a stilling well. A vertical pointer is fixed in stilling well, which shows level of water in the pan. The water level is maintained 50 to 75 mm from pan rim. It is usually made of 22-gage galvanized iron or 0.8-mm thick Monel metal, placed on a wooden frame 10 mm above the levelled surface so that air may circulate beneath the pan. Daily

	evaporation rate is recorded by fall of water level in the stilling well during 24-hour period.

***Pan coefficient (K_{pan})*:** A pan coefficient accounts for differences between characteristics of the pan and natural evaporating surfaces such as a lake or a cropped field. This coefficient is used to adjust pan measured evaporation values. Pan coefficients depend on type of pan, size, and state of upwind buffer zone (fetch), ground cover, wind and humidity conditions. Pan coefficient for Class A evaporation pans and Colorado sunken pans can be calculated using regression equation given in table 10 under different conditions.

Table 10. Pan coefficients (K_p): regression equations adapted from FAO-56 paper

Pan type	Regression Equation
Class A pan with green fetch	$K_p = 0.108 - 0.0286\, u_2 + 0.0422 \ln(FET) + 0.1434 \ln(RH_{mean}) - 0.000631\, [\ln(FET)]^2 \ln(RH_{mean})$
Class A pan with dry fetch	$K_p = 0.61 + 0.00341\, RH_{mean} - 0.000162\, u_2\, RH_{mean} - 0.00000959\, u_2\, FET + 0.00327\, u_2 \ln(FET) - 0.00289\, u_2 \ln(86.4\, u_2) - 0.0106 \ln(86.4\, u_2) \ln(FET) + 0.00063\, [\ln(FET)]^2 \ln(86.4\, u_2)$
Colorado sunken pan with green	$K_p = 0.87 + 0.119 \ln(FET) - 0.0157 [\ln(86.4\, u_2)]^2 \ln(RH_{mean}) - 0.000053 \ln(86.4 u_2) \ln(FET) RH_{mean}$
Colorado sunken pan with dry fetch	$K_p = 1.145 - 0.080\, u_2 + 0.000903 (u_2)^2 \ln(RH_{mean}) - 0.0964 \ln(FET) + 0.0031\, u_2 \ln(FET) + 0.0015 [\ln(FET)]^2 \ln(RH_{mean})$
Coefficients and parameters	K_p pan coefficient, u_2 average daily wind speed at 2 m height (m s^{-1}) RH_{mean} average daily relative humidity (%) = (RH_{max} + RH_{min})/2 FET fetch, or distance of the identified surface type (grass or short green agricultural crop for case A, dry crop or bare soil for case B upwind of the evaporation pan)
Range for variables	$1 \leq FET \geq 1000$ m (these limits must be observed) $30\% \leq RH_{mean} \geq 84\%$ $1\ m\ s^{-1} \leq u_2 \geq 8\ m\ s^{-1}$

Checkbook method

Total available soil water in crop root zone represents the bank account whereas amount of usable soil water represents checkbook balance. Three important components of this checkbook balance are (1) water withdrawals (evapotranspiration and deep percolation), (2) water deposits (rain and irrigation), and (3) critical water balance level (the point of irrigation), referred to variously as application amount, irrigation trigger point, irrigation deficit etc. The water withdrawals occur via crop evapotranspiration (ET_c) and deep percolation. ET_c is the most dynamic part of accounting system, changing with weather and stage of crop growth. Once soil water balance account is defined and current water balance is determined, irrigations are scheduled when available soil water drops to a predetermined level by accounting field capacity, application rate and irrigation efficiencies (Werner, 1978).This amount is equivalent to gross irrigation depth. This method requires keeping the records of net amount of water applied at each irrigation, rainfall occurred between the irrigations, total water available, usable soil water in crop root zone and rate at which crop extracts water from the soil. These factors are balanced against each other to ensure root zone soil water is not depleted too much. If water is a limiting factor, then different irrigation strategies (Table 11) can be adopted to minimize the losses.

Table 11. The most-used deficit irrigation strategies (Fernández, 2017)

Strategy	Definition
Supplementary or complementary irrigation	A single irrigation event is applied when a fixed threshold for water stress is achieved. Just 2 or 3 irrigation events in the irrigation season. This method of irrigation is used when water for irrigation is very scarce or in temperate regions with low evapotranspiration rates or high rainfall. It can lead to substantial increases in crop performance, as compared to dry-farming, if irrigation events are applied at the right time.
Low frequency deficit irrigation (LFDI)	The soil is left to dry until the readily available water is consumed. Irrigation is then applied, until field capacity. This is repeated several times along the irrigation season.

Sustained deficit irrigation (SDI)	A fixed fraction of the crop water needs is applied all throughout the irrigation season. Irrigation is applied daily or 2–4 times per week.
Partial root zone drying (PRD)	Similar to SDI, but water is applied to half of the root zone, switching to the other half every 2–3 weeks.
Regulated deficit irrigation (RDI)	Irrigation amounts equal or close to the crop water needs is supplied at the phenological stages most sensitive to drought. Irrigation during these periods is applied daily, or at least several times per week. For the rest of the crop cycle, irrigation is drastically reduced (not only the dose is reduced but also the frequency, to one or two irrigation events per week) or even withheld.

Computer Models

The complexity of soil-water-crop inter-relationships makes it difficult for farmers to develop their own irrigation scheduling plan, but with the availability of low cost personal computers, various models/expert system shave been developed to assist in taking irrigation scheduling decisions. Simple models perform checkbook type record keeping or water-balance approach. Whereas sophisticated models require weather data inputs (either manually or automatically from in-field weather stations).Some models also incorporate various alternate scenarios for a particular situation as well as other management decisions such as weeds, insects and disease management. Such types of models are called expert systems. Sensor systems installed in the fields monitor soil-water status and weather conditions, and send signals/data directly to the expert system. Computer models can be very precise in decision making, save time, money and human efforts. However, the statement "Garbage in – Garbage out" is very applicable to computer models. These models should be calibrated properly for geographical conditions and the input data should be validated otherwise results could be inappropriate or misleading.

Accounting for System Efficiencies

Performance of any irrigation systems must be determined by making field observations and by evaluating net irrigation depth applied, uniformity in application, and different efficiencies. This result in money saving by conserving water and energy, reducing nutrient loss, improving crop yields and ultimately improve working of irrigation system and water management objectives. Since part of the water applied by irrigation systems is lost due to poor conveyance, application and storage efficiencies, which result in run-off, deep percolation and seepage losses. The average application efficiency (AE) for various types of irrigation system is given in Table 4. These efficiencies are used to determine the gross irrigation amount, to be applied. The actual application efficiency of any given irrigation system or irrigation event can vary beyond the given ranges depending on soil, environmental and human factors.

Advantages of Irrigation Scheduling

A proper irrigation scheduling method offers several advantages:

- It enables farmer to schedule a proper water rotation among the various fields to minimize water stress during cropping season and maximize yields.
- It reduces the pumping cost of water and labour, by making maximum use of available soil moisture storage.
- It lowers nutrient losses and fertilizer costs by reducing surface runoff and deep percolation (leaching) thus reduces non-point pollution.
- It increases net income of farmers by increasing crop yields and crop quality.
- It controls root zone salinity, secondary salinization problems, water-borne diseases and water-logging problems by reducing the drainage requirements.
- It saves water which subsequently can be applied to other crops for life saving irrigations during water-short periods and thus increases net income.

Difficulties on Applying Irrigation Scheduling at Farm Level

In spite of the variety of methods and tools developed to schedule the irrigation, adoption of these techniques by farmers is still limited. Identification of these limitations by farmers and managers is important in the selection of the appropriate scheduling methods. Some of the limitations and difficulties on applying irrigation scheduling tools at farm level are listed below.

1. Irrigation scheduling becomes particularly sensitive under scarce water situations, where availability of water defines the timing of water applications in order to minimise yield reductions. Similarly, under saline conditions, scheduling requires proper knowledge of salt-water balance and salt tolerance levels.
2. Options, special requirements and limitations of irrigation scheduling under variable rainfall conditions need to be considered.
3. Management of deficit irrigation should be proper through suitable and reliable water stress indicators.
4. Irrigation efficiencies as measured in terms of adequacy and application efficiency, as well as design criteria of irrigation method, needs to be considered in the selection and operation of irrigation scheduling method.
5. Costs of labour, equipment and technologies, capital investment capacity of farmers, pricing of water, and profitability of irrigation decides the irrigation method to be applied.
6. Some farmers do not fully understand hydrology and water budgets and prefer to irrigate as usual. The farmers can lack the appropriate technical skills to conduct proper scheduling.

Conclusions

Judicious irrigation scheduling is basic element of proper management of irrigation water by applying the right amount of water at the right time to meet crop water requirement and to achieve best possible yield under prevailing conditions. A proper irrigation scheduling can avoid the common irrigation problems like salinity, water logging, nutrient leaching, increased level of water table and yield loss. Any irrigation scheduling technique depends on the type of crop, its stage of development, its sensitivity to drought stress, soil and weather conditions. Irrigation scheduling is more efficient when an appropriate irrigation method is selected, this helps in controlling the flow rate and distribution of water in the field. When the irrigation scheduling is designed either by climate data or estimated by computer program, it is important to assess field conditions for measuring actual soil water status and water deficiency to make sure that the estimation is right, because an inundation of water is as harmful as a dearth of water.

References

Allen, R. G., Pereira, L. S., Raes, D. and Smith, M. (1998). Crop evapotranspiration: guidelines for computing crop water requirements. Irrigation and Drainage Paper 56, UN-FAO, Rome, Italy.

Blaney, H. F. and Criddle, W. D. (1950). Determining Water Requirements in Irrigated Area from Climatological Irrigation Data, US Department of Agriculture, Soil Conservation Service, Technical Paper No. 96, 48 pp.

Burt, C.M., A.J. Clemmons, R. Bliesner, J.L. Merriam, and L. Hardy. (2000). Selection of Irrigation Methods for Agriculture. American Society of Civil Engineers: Reston, Virginia.

Cohen, Y., Fuchs, M., Green, G.C. (1981). Improvement of the heat pulse method for determining sap flow in trees. *Plant, Cell and Environment,* 4, 391–397.

Dalton, J. (1802). 'Experimental Essays on the Constitution of Mixed Gases: On the Force of Steam or Vapor from Water or Other Liquids in Different Temperatures, Both in a Torricelli Vacuum and in Air; on Evaporation; and on Expansion of Gases by Heat', Manchester Literary Philosophical Society Memoirs Proceedings 5, 536–602.

Doorenbos, J. and W. O. Pruitt. (1977). Guidelines for Predicting Crop Water Requirements. Irrigation and Drainage Paper No. 24, Second edition. Food and Agricultural Organization, United Nations, Rome, Italy. 156 pp.

FAO (Food and Agricultural Organization of the United Nations, Land and Water Division). (2002). Crop Water Information. Crop Water Information. Rome. From http://www.fao.org/ag/AGL/AGLW/cropwater/cwinform.stm Crop Water Information.

Fernández, J.E. (2017). Plant-Based Methods for Irrigation Scheduling of Woody Crops. *Horticulturae,* 3, 35.

Fernández, J.E. and Cuevas, M.V. (2010). Irrigation scheduling from stem diameter variations: A review. *Agricultural and Forest Meteorology,*150:135-151.

Fernandez, J.E., Palomo, M.J., Diaz-Espejo, A., Clothier, B.E., Green, S.R., Giron, I.F., Moreno, F. (2001). Heat-pulse measurements of sap flow in olives for automating irrigation: tests, root flow and diagnostics of water stress. *Agricultural Water Management,* 51:99–123.

Gupta, A., Sarangi, A. and Singh, D. K. (2018). Determination of crop evapotranspiration and partitioning of bare soil evaporation of mustard using weighing type field lysimeters. *Journal of Soil and Water Conservation,* 17(2), 169-175.

Hargreaves, G. L. and Samani Z. A. (1985). "Reference crop evapotranspiration from temperature." *Applied Engineering Agriculture Transaction,* ASAE, 1(2):96-99.

Heermann, D. F. (1985). Evapotranspiration in irrigation management. In Proceedings of the National Conference on Advances in Evapotranspiration. ASAE Publication 14-85. p 323-334.

Hillel, D. (1980). *Fundamentals of soil physics.* New York: Academic Press. ISBN-13: 978-0-12-348560-1.

Hillel, D. (1990). Role of irrigation in agricultural systems. In: *Stewart, B. A., & Nielsen, D. R. (editors), Irrigation of Agricultural Crops*: Agronomy 30. American Society of Agronomy. Madison.

Howell, T.A. and Meron, M. (2006). Irrigation scheduling. *In Micro irrigation for crop production: Design, Operation, and Management* (1st ed., pp.61-130). Elsevier B.V.: F.R. Lamm, J.E. Ayars and F.S. Nakayama.

Idso, S.B., Jackson, R.D., Pinter, P.J., Reginato, R.J., Hatfield, J.L. (1981). Normalizing the stress-degree-day parameter for environmental variability. *Agricultural Meteorology,* 24:45–55.

Jensen, M. E., Robb, D. C. N. and Franzoy, C. E. (1970). Scheduling irrigations using climate crop- soil data. *ASCE Journal of Irrigation and Drainage,* 93:25-28.

Jones H.G., Stoll, M., Santos, T., de Sousa, C., Chaves, M.M. and Grant, O.M. (2002). Use of infrared thermography for monitoring stomatal closure in the field: application to grapevine. *Journal of Experimental*

Botany, 53:2249–2260.

Jones, H.G. (2004). Irrigation scheduling: advantages and pitfalls of plant-based methods, Water-Saving Agriculture Special Issue. *Journal of Experimental Botany*, 55(407):2427–2436.

Jones, S. B., Wraith J. M. and Or, D. (2002). Time domain reflectometry measurement principles and applications, *Hydrological Processes,* 16:141–153.

Kansas Conservation Commission (2004). Irrigation Initiative. (http://www.accesskansas.org/kscc/irrigation.html).

Martin, D. L., Stegman, E. C. and Fereres. E. (1990). Irrigation scheduling principles. In: Management of Farm Irrigation Systems, ASAE Mono. G. J. Hoffman, T. A. Howell, and K.H. Solomon, (Eds.), ASAE, St. Joseph, Michigan. pp.155-203.

Penman, H. L. (1948). Natural evaporation from open water, bare soil and grass. Proceedings. Royal Society, London, United Kingdom. Series A193:120-146.

Priestley, C.H.B. and R.J. Taylor (1972). On the assessment of surface heat flux and evaporation using large-scale parameters. *Monthly Weather Review,* 100:81-82.

Raes, D., Steduto, P., Hsiao, T.C., & Fereres, E. (2009). Aquacrop - The FAO Crop Model to Simulate Yield Response to Water: II. Main Algorithms and Software Description. *Agronomy Journal*, 101:438–447.

Rohwer, C. (1931). 'Evaporation from Free Water Surface', USDA Tech. Null., 217:1–96.

Sadler, E.J., Camp, C.R., Evans, D.E., Millen, J.A. (2002). Corn canopy temperatures measured with a moving infrared thermometer array. *Transactions of the American Society of Agricultural Engineers,* 45:581–591.

Scholander, P. F., Bradstreet, E. D. and Hemminger. E. A. (1965). Sap pressure in vascular plants. *Science*, 148:339-346.

Slabbers, P. J. (1980). Practical prediction of actual evapotranspiration. *Irrigation Science,* 1:185-196.

Smith, M. (1992). CROPWAT: A Computer Program for Irrigation Planning and Management (FAO Irrigation & Drainage Paper). FAO, Rome.

Stephens, J.C. and Stewart, E.H. (1963). A comparison of procedures for computing evaporation and evapotranspiration. Publication 62, International Association of Scientific Hydrology. International Union of Geodesy and Geophysics, California, USA, pp.123–133.

Thornthwaite C.W. (1948). An approach towards a rational classification of climate. Geogr. Revue, 38.

Van der Gulik, T. (2006). Irrigation scheduling techniques. British Columbia, Ministry of Agriculture and Lands. Water Conservation Factsheet. from: http://www.agf.gov.bc.ca/resmgmt/publist/500series/577100-1.pdf.

Weibel, F. P. and J. A. Devos. (1994). Transpiration measurements on apple trees with an improved stem heat balance method. *Plant and Soil,* 166:203-219.

Werner, H. D. (1978). Irrigation Scheduling – Checkbook Method. Agriculture Extension Services M-160. University of Minnesota., St. Paul, Minnesota. 12 pp.

13

QGIS-Based Drastic Modeling for Assessing Groundwater Vulnerability in Dewas Watershed (Madhya Pradesh)

Pramod Dubey, D.C. Gupta, J.P. Shukla and Shailesh Chaure

Introduction

Groundwater has been considered as an important source of water supply due to its relatively low susceptibility to pollution in comparison to surface water, and its large storage capacity (US EPA, 1985). However, there are significant sources of diffuse and point pollution of groundwater from land use activities, particularly agricultural practices. The concept of groundwater pollution vulnerability was first introduced in France by the end of the 1960s to create awareness of groundwater contamination (Vrba and Zoporozec, 1994). Dewas is an industrial city of the Madhya Pradesh. It has many medium-sized and small industries. For the present work, a small watershed (sub-watershed of Kshipra Watershed Code 09) covering Dewas town was taken and using the various geological, hydrological and geomorphological parameters a groundwater. The study area is bounded by Latitude 22°57'35.112" - 22°57'31.112"N and Longitude 75°56'26.901" - 76°05'4.129"E, on Survey of India topographical map sheet No. 46N/13 and 55B/1. Total geographical area of the watershed is 67.71 km^2 shown in Figure 1a.

Material and Methodology

The DRASTIC method (Aller *et al.,* 1987) developed by U.S. Environmental Protection Agency, has been widely used to identify the area where groundwater supplies are most susceptible to contamination. The DRASTIC method has been widely used by various researchers in the United State including Hearne *et al.,* (1992), Atkinson and Thomlinson (1994). GIS-based groundwater pollution hazard assessment by the DRASTIC model has been conducted by Merchant (1994). The method has been used for assessment of aquifer vulnerability (Napolitano *et al.,* 1996).

A DRASTIC model applied in a GIS environment was used to evaluate the vulnerability of the Dewas city watershed. The DRASTIC model based on the concept of the hydro-geological setting that is defined as "a composite description of all the major geologic and hydrologic factors that affect and control the groundwater movement into, through and out of an area" (Aller *et al.,* 1987). In DRASTIC model seven parameters used in the model which are-Depth to water (D), Recharge (R), Aquifer media (A), Soil media (S), Topography(T), Impact of vadose zone (I) and Hydraulic Conductivity(H) shown in Table 1. The DRASTIC model yields a numerical index that is derived from ratings and weights assigned to the seven parameters of model. The significant media types of each parameter represent the ranges, which are rated from one to ten based on their relative effect on the aquifer vulnerability. These seven parameters are then assigned weights ranging from one to five reflecting their relative importance. The pollution potential (DRASTIC Index) was computed by sum of all the seven parameters using following equation:

$$\textbf{Drastic Index } (D_i) = D_rD_w + R_rR_w + A_rA_w + S_rS_w + T_rT_w + I_rI_w + C_rC_w$$

where D, R, A, S, T, I, and C are the seven parameters and the subscripts r and w are the corresponding ratings and weights, respectively.

The DRASTIC equations, weights and technique are well defined and are used worldwide for the data analyses. The model implementations were performed using the QGIS software. QGIS is free and Open Source GIS software. The drastic index calculation was done using QGIS. All the vector layers were converted into raster layers using Rasterize option. The summing up of all the layers along with their correspond weightage and rating was done using 'Raster Calculator', generate final groundwater pollution vulnerability map.

In the present study are a range of thematic layers specifically Depth to water, net Recharge, Aquifer media, Soil media, Topography, Impact of vadose zone and hydraulic Conductivity. The DRASTIC model yields a numerical index that is derived from ratings and weights assigned to the 7DRASTIC model parameters. The significant media types or classes of each parameter represent the ranges, which are rated from 1 to 10 based on their relative effect on the aquifer vulnerability. The seven parameters are then assigned weights ranging from 1 to 5 reflecting their relative importance, the Rainfall data collected from Land Record department Dewas and groundwater samples data from Irrigation department Dewas. A Raster calculator was used to prepare Groundwater potential zone map using QGIS.

The depth to water table (D) is a significant factor controlling the ability of pollutants to reach the aquifer. It affects the contamination of chemical and biological reactions such as dispersion, oxidation, natural attenuation, etc. A low depth of water table will show the way to a high vulnerability ranking. Depth to water table map (Figure 1b) has been estimated based on post monsoon (2014) water table data collected from network stations. The depth to water table was obtained by subtracting the water table level (msl) from surface contours (msl).

The depth to water table map was then classified into ranges defined by the DRASTIC model and assigned rates ranging from 1 (minimum impact on vulnerability) to 10 (maximum impact on vulnerability). The deeper the groundwater, the smaller the rating value. Rating values for vulnerable to pollution is based on depths of water table.

Recharge (R) describes the amount of water available at the surface that infiltrates into the soil and then continues to percolate through the vadose zone into the aquifer. Recharge represents the primary contaminant transport mechanism into the aquifer and depends on the soil characteristics. A sand or loamy sand soil will have the maximum infiltration capacity whereas compacted clay loam may allow very small amounts of infiltration. In urban areas where the soil is covered by an impermeable layer no recharge would occur (Ckakraborty Surajit. *et al.*, 2007). The recharge map was constructed from the rainfall data according to the following formula:

Net recharge = (rainfall - evapotranspiration) x recharge rate

As per the records of Water Resources Department, Madhya Pradesh, the total annual average rainfall is 982 mm and average evapotranspiration for the study area is 154.75 mm. The recharge rate was assumed to be 20%, for the urbanized area, 30% for sloppy areas and 85% for the rest of the study area. The recharge map was then classified into ranges and assigned ratings between 1to10. High recharge rates were assigned high numerical rates. Ratings were assigned based on the fact that sandy soil with higher recharge rate and more vulnerable. Distribution of recharge rate of the study area is shown in Figure 1c.

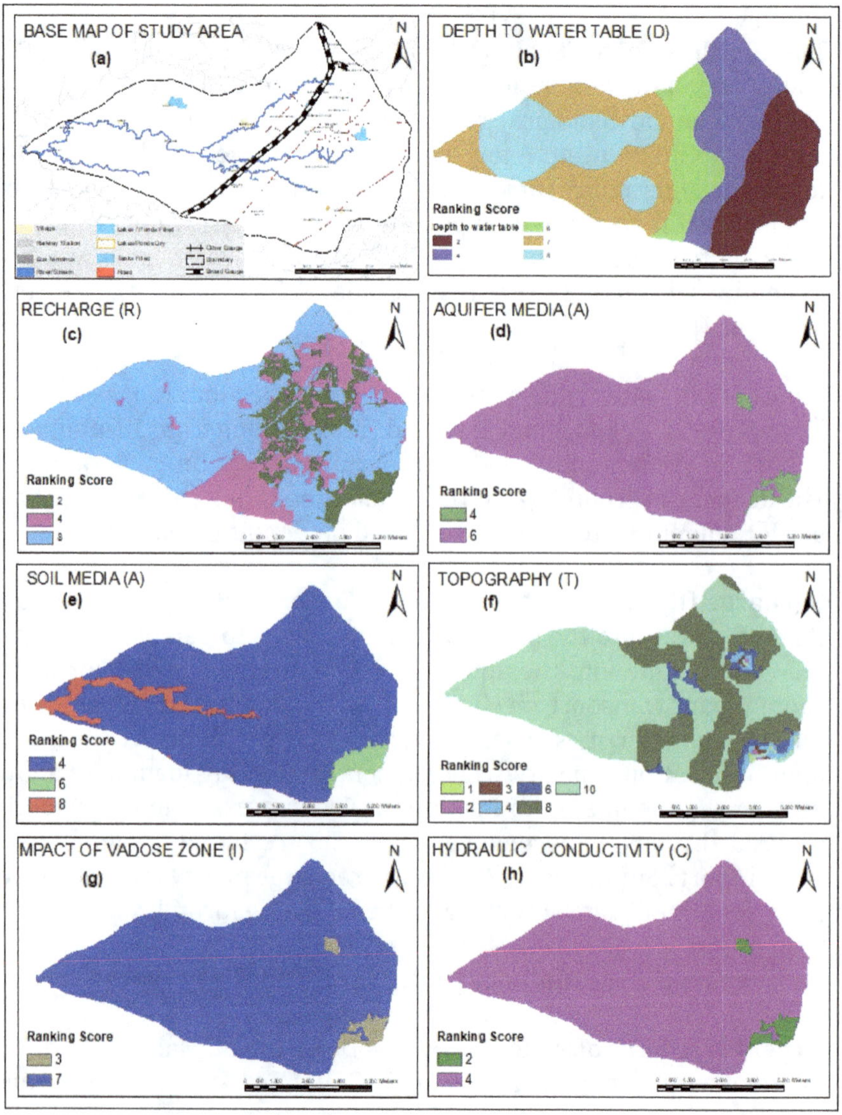

Figure 1. Seven DRASTIC Model layers (a. Base map b. Depth to water, c. Recharge, d. Aquifer media, e. Soil media, f. Topography, g .Impact of vadose zone and h. Hydraulic Conductivity)

Aquifer media (A) has been recognized from the field mapping and borehole data. Aquifer in the study area consists of fine to medium massive basalt and weathered basalt. The aquifer media has been grouped under 2 categories covering massive basalt and weather basalt. Aquifer Media map is shown in Figure 1d.

Table 1. Weightage assignment of various thematic maps for DRASTIC Model

Criteria	Ranges & Classes	Ranking
Depth to water table Weightage (D)- 05		
1. Depth to water table (D)	15-20	8
	20-30	7
	30-40	6
	40-50	4
	50-65	2
Recharge (R) Weightage – 04		
2. Recharge (R)	High recharge zone	8
	Medium recharge zone	4
	Poor recharge zone	2
Aquifer media (A) Weightage – 03		
3. Aquifer media (A)	Massive Basalt	4
	weathered basalt	6
Soil media(S) Weightage - 02		
4. Soil media (A)	Alluvial Soil	8
	Mixed soil	6
	Black Cotton Soil	4
Topography (T)Weightage – 01		
5. Topography (T)	Plain	10
	Slightly Sloping	8
	Gently Sloping	6
	Strongly Inclined	4
	Steep	3
	Very Steep	2
	Precipitous	1
Impact of Vadose Zone (I) Weightage - 05		

<table>
<tr><td rowspan="2">6. Impact of Vadose Zone (I)</td><td colspan="2">Hard</td><td>3</td></tr>
<tr><td colspan="2">Weathered and unconsolidated</td><td>7</td></tr>
<tr><td colspan="3">Hydraulic Conductivity (C)- 03</td><td></td></tr>
<tr><td rowspan="2">7. Hydraulic Conductivity (C)</td><td>2×10^{-11} to 4.2×10^{-7}</td><td>Massive Basalt</td><td>2</td></tr>
<tr><td>4×10^{-7} to 2×10^{-2}</td><td>weathered basalt</td><td>4</td></tr>
</table>

Soil media (S) is the portion of the unsaturated zone characterized by significant biological activity. The characteristics of the soil influence the amount of recharge infiltrating into the groundwater body, the amount of pollutant dispersion, the purifying process of contaminants etc. (Ckakraborty Surajit *et al.,* 2007).Soil types are strongly depend and controlled by the surface deposits and geomorphology. Rating value was assigned to define soil classes based mainly on the texture of the soils. Texture is based mainly on the proportion of sand: silt: clay. More the finer particles present less is the vulnerability of the soil media to pollution. Soil media map is shown in Figure 1e.

Topography (T) refers to the slope of land area and is expressed as percent slope. Areas with low slope tend to retain water longer. This allows a greater infiltration of recharge water and a greater potential for contaminant migration (Bassam-kattaa *et al.,* 2010). Where Areas is steep slope, have large amount of run-off and smaller amount of infiltration, are less vulnerable to groundwater contamination. Slope categories have been marked (Figure 1 f) with the help of LPS software and cartosat-1 stereo image.

The vadose zone (**I**) (Figure 1 g) is the unsaturated horizon above the water table. If vadose zone is highly permeable, then this will lead to a high vulnerable rating (Corwin, *et al.,* 1997). Vadose zone map has been generated from the geological map with 2 classes I and II. Class I contains thin vadoze zone and class II contains thick vadoze zone and rating has been assigned accordingly.

Geologic material (C), composed of coarser particles, such as sand and gravels have high conductivity values. The aquifer vulnerability is related to the hydraulic conductivity through the aquifer's capacity to transport the pollutants away from the point at which they enter the aquifer. The greater is the hydraulic conductivity, the further the contaminants will travel and potentially contaminate greater volume of groundwater (Ckakraborty Surajit *et al.,* 2007). The different hydraulic conductivity zones in the area were defined and assigned ratings according to DRASTIC and Conductivity map is shown in Fig. 1 h.

Results and Discussion

Aquifer vulnerability assessment The DRASTIC vulnerability index was computed according to Eq. (1). In order to understand the vulnerability index, it is necessary to choose a representation method which can expose the aquifer vulnerability in an appropriate fashion and simultaneously allows comparability between different areas (Aller *et al.,* 1987).All the maps were summed up as per the Drastic index equation and final aquifer vulnerability map was prepared and shown as Figure 2. The aquifer vulnerability values ranges from 64 to 163.

The values were categorized into four classes. They are low (64-88), medium (89-117), high (118-139), and very high (140-163) groundwater vulnerability. Figure 2 and Table 2 indicate that in the central part of the study area, the vulnerability to pollution is ranges between high and low. In the western parts, the vulnerability ranges high.

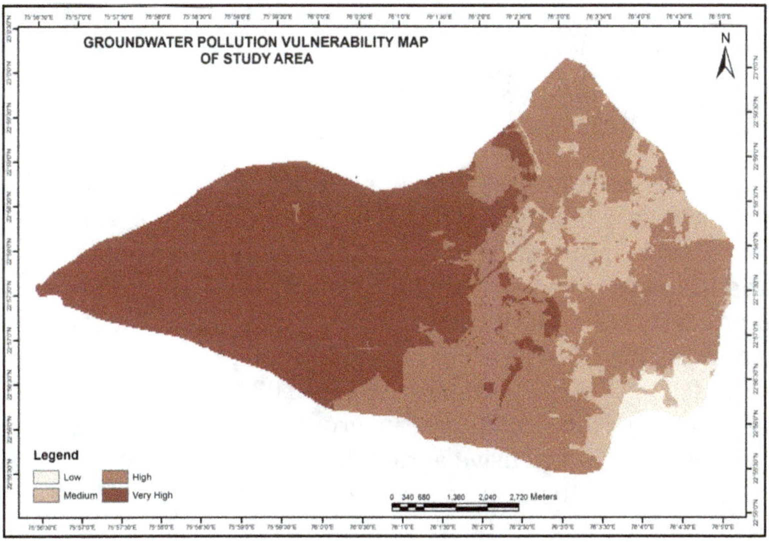

Figure 2. Groundwater Pollution Vulnerability map of study area.

Assessment of the aquifer vulnerability of the study area has been done employing the empirical index DRASTIC model of the U.S. Environmental Protection Agency (EPA). Seven environmental parameters were used to represent the natural hydrogeological setting of the study area aquifer; Depth to water, net Recharge, Aquifer media, Soil media, Topography, Impact of vadose zone, and hydraulic Conductivity.

Table 2. Aquifer Vulnerability classes of the study area

S. N.	Aquifer Vulnerability Class	Area (sq. km)
1	Low Vulnerability	1.42
2	Medium Vulnerability	8.19
3	High Vulnerability	26.88
4	Very high Vulnerability	31.23

Validation of the Result

The nitrate concentration in the groundwater was taken as the main parameter to validate the results. Nitrate is not naturally found in groundwater, but it usually enters via the surface. Leaching of nitrate occurs because of recharges from the surface over a long period of time (Feola *et al.,* 2015; Guler and Ali, 2013; Hamza *et al.,* 2014; Neshat and Pradhan, 2015). Therefore, the nitrate concentration is the indicator of the vulnerability index to reflect the actual situation in the study area.

To validate the results the locations of the water samples was plotted on the map and the concentration of the nitrate in the water sample was classified into 4 classes as shown in Figure 3. The nitrate concentration is increasing towards the west and is higher in the very high vulnerability zone. The nitrate values were classified into four classes based on its permissible (limit) concentration as follows: 0.00-0.0140 mg/L; 0.0141 -0.0270 mg/L, 0.0271-0.0340 and 0.0341-0.0540 mg/L. The High levels of nitrates in groundwater can be attributed to the extensive and intensive agricultural activities in the study area, which contribute to nitrate pollution as result of the hydrogeological condition.

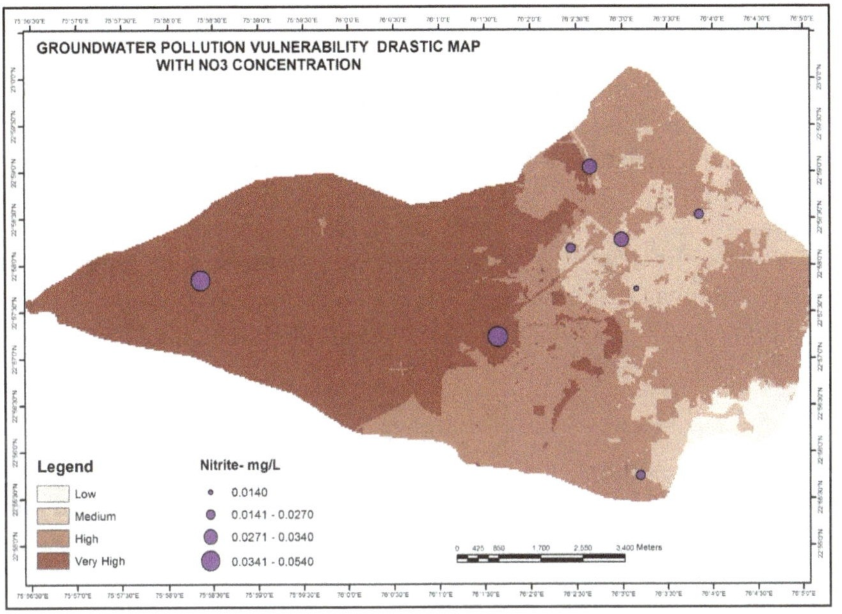

Figure 3. Groundwater Pollution Vulnerability DRASTIC map with NO_3 Concentration.

Conclusion

This work reveals that the value of drastic index in the study area ranges from 64 to 163. Based on the drastic index values the entire area was divided into 4 classes namely low, medium, high, and very high vulnerability. About 85% part of the study area falls under the high and very high vulnerability class. The high and very high vulnerability areas are mostly agricultural lands with gentle (02^{0}-05^{0}) slope. The nitrate concentration in groundwater samples correlates well with the groundwater pollution vulnerability classes. The nitrate concentration is low in the eastern part of the area and is increasing towards the west; the same trend can be seen in the vulnerability classes. Areas with very high vulnerability show also elevated actual nitrate concentration in the range of 0.0341 mg/L to 0.0540 mg/L.

References

Aller, L., T. Bennett, Lehr, J.H., Petty, R., and Hackett, G. (1987). DRASTIC: A standardized system for evaluating groundwater pollution potential using hydrogeologic settings. USEPA 622p.USEPA, Cincinnati, OH.

Bassam-kattaa *et al.,*(2010). Groundwater Vulnerability Assessment For The Banyas Catchment of The Syrian Coastal Area Using GIS And The Riske Method, Journal of Environmental Management 91 (2010) 1103–1110

Ckakraborty Surajit *et al.,* (2007). Assessing aquifer vulnerability to arsenic pollution using DRASTIC and GIS of North Bengal Plain: A case study of English Bazar Block, Malda District, West Bengal, India, Journal of Spatial Hydrology Vol.7, No.1, spring.

Corwin, D.L., Vaughan, P.L., and Loague, K. (1997). Modelling non-point source pollutants in the vadose zone with GIS. Environmental Science and Technology, 31, 2157-2175.

Feola, G., Lerner *et al.,* (2015). Researching farmer behavior in climate change adaptation and sustainable agriculture: Lessons learned from five case studies, 39, 74-84.

Guler, C., Ali, M., (2013). Ocean & Coastal Management Assessment of groundwater vulnerability to nonpoint source pollution in a Mediterranean coastal zone Mersin, Turkey under conflicting land use practices, 71, 141-152.dio:10.1016/j.ocecoaman.2012.10010.

Hamza, S.M., *et al.,* (2014). Accomplishment and subjectivity of GIS-based DRASTIC groundwater vulnerability assessment method: a review. Doi: 10.1007/s12665-014-3601-2.

Hearne, G. M., *et al.,* (1992). Vulnerability of the uppermost groundwater to contamination in the Greater Denver Area, Colorado. USGS water-resources investigations report 92-4143, 241pp.

http://www.mpwrd.gov.in/documents/18/0e5001ad-d37a-4731-befb-1e6c5b114434.

Merchant, J.W. (1994). GIS-Based Groundwater Pollution Hazard Assessment: A Critical Review of the DRASTIC Model. Photogrammetric Engineer & Remote Sensing, 60(9), 1117-1127.

Napolitano P, Fabbri A.G. (1996). Single-parameter sensitivity analysis for aquifer vulnerability assessment using DRASTIC and SINTACS HydroGIS 96: application of geographical information systems in hydrology and water resources management. Proceedings of Vienna Conference. IAHS Pub, vol. 235, 1996. p. 559–66.

Neshat, A., Pradhan, Biswajeet. (2015). An integrated DRASTIC model using frequency ratio and two new hybrid methods for groundwater vulnerability assessment. Nat Hazards 76:543-563 DOI 1007/s11069-014-1503-y.

Shahid, S. (2000). A study of groundwater pollution vulnerability using DRASTIC/GIS, West Bengal, India. Journal of Environmental Hydrology, 8, paper1.

US Environmental Protection Agency (1985). DRASTIC: a standard system for evaluating groundwater potential using hydrogeological settings, Ada, Oklahoma WA/EPA Series; 1985. 163.

Vrba, J., and Zaporotec, A. (1994). Guidebook on Mapping Ground Water Vulnerability. IAH Int. Contribution for Hydrogeology, 16(94), Heise, Hannover, 131.

14

Watershed Management Using Geospatial Technology: A Case Study of Uttarakhand

N. Negi, T.P. Kumar, S. Karwariya and Suraj K. Singh

Introduction

Water is one among the foremost very important necessities for economic and social development. Indian population and economy is increasing thereby increasing the demand for water for domestic, agricultural and industrial use. Reduction of surface runoff are often achieved by constructing appropriate structures, that mechanically helps to manage the opposite natural resources like soil and vegetation because the watershed condition have an effect on these recourses (Zhang *et al.,* 2015). For the economical management, one must take appropriate unit of management. Here, by taking watershed as a unit the water, soil, and vegetation's are often managed with efficiency, together, and at the same time. The matter of water and sensitive land is actually acute and harsh, that demands special strategies of management and attributable to increasing population and development of agriculture, water from numerous sources is consumed at such a quick rate that it's planning to cause a grave water crises in close to future (Singh 1995). Besides, within the areas wherever irrigation water is created obtainable unforeseen issues like water work, pH scale and salinity of the soil and numerous water borne diseases and social downside have appeared within the absence of any scientific management of watershed (Pakhmode *et al.,* 2003; Chandra 1990). Hence, this analysis has been planned to provide thought to those issues and to counsel ways in which and suggests that through rain water harvest home and correct watershed management. GIS has evolved as an extremely refined knowledge management system to place along and store the voluminous knowledge usually needed for such studies (Santhanam *et al.,* 1982; Pandey *et al.,* 2007). Thus, remote sensing and GIS along offer data base for economical management of land and water resources. The synoptic read provided by satellite remote sensing and therefore the analysis capability provided by GIS provide a technologically acceptable technique for learning these resources.

Watershed may be a natural geo-hydrological entity wherever excess water collected through precipitation drains to a typical purpose like an outlined drain, stream, channel or a stream within the variety of surface run-off once infiltration into the highest porous layers of the planet (Chow *et al.,* 1988; Agarwal 1998). Development of the watershed wants higher understanding regarding the assorted natural resources their relations with one another and their relations with resource of the stakeholders, property land and water program is that the primary objective of any watershed programme (Anil *et al.,* 2003; Nag and Chakraborty, 2003). This can promote socio-economic development through optimum utilization of natural resources and can mitigate the adverse effects of drought. From the hydrological purpose of read, the various phases of hydrological cycle during a watershed are keen about the assorted natural options and human activities (Choi *et al.,* 2005).

Watershed Management is a reiterative method of integrated decision-making relating to uses and modifications of lands and waters among a watershed (Sreedevi *et al.,* 2005; Singh and Frevert, 2006). This method provides an opportunity for stakeholders to balance numerous goals and uses for environmental resources, and to think about however their additive actions could have an effect on semi-permanent property of those resources. Watershed management needs use of the social, ecological, and economic sciences. Common goals for land and water resources should be developed among folks of numerous social backgrounds and values. The choice method should additionally weigh the economic advantages and prices of other actions, and mix current market dynamics with issues of semi-permanent property of the scheme (Gajbhiye *et al.,* 2014).This approach counsel the combination of technologies among the natural boundaries of a catchment area for optimum development of land, water, and plant resources to fulfill the essential wants of individuals and animals during a property manner. This approach aims to boost the quality of living of people by increasing his earning capability by giving all facilities needed for optimum production (Singh and Kanga, 2017; Singh *et al.,* 2017). So as to realize its objective, integrated watershed management suggests to adopt land and conservation practices, water harvest home in ponds and recharging of groundwater for increasing water resources potential and use of improved sort of seeds, integrated nutrient management and integrated persecutor management practices (Strahler, 1964). Another necessary impact of watershed development was its impact on dominant wearing. Several studies have disclosed that watershed development intervention were productive in dominant wearing. During this study, a number of the parameters, together with flow direction, flow accumulation, slope of the watershed were studied. Performance criteria and therefore the level of study at that they operate (Singh *et al.,* 2018).

The aim and objectives of the analysis work is to show the watershed designing through the readying of GIS techniques it are often helpful for decision-makers for numerous strategic designing for numerous body units. to get the land use/ land cowl map for the employment of land and water resources improvement to extend the crop productivity and to organize the Natural Resources thematic maps (LULC, spring water Prospects map, Soil and Slope) for the Study space on 1:25,000 scale. The study space emphasizes to formulate a model by victimization remote sensing representational process for distinctive areas for appropriate watershed treatment plans and eventually to get implementable action plans for watershed supported remote sensing and GIS techniques.

Study Area

Tanakpur is situated in the Champawat district of Uttarakhand, in the Terai-Bhabar region of Kumaun. Tanakpur is a calm and small town located on riverside of Sharda River. It is located at the foothills and is the last plain area on the road to Kumaun zone of Uttarakhand. It acts as a junction for the Kumaon District's mountainous part. It is also the first point in the Kailash Manasarovar Pilgrimage. Gradually benefitting from the nationwide progress, being on the Indo-Nepal Border Tanakpur is very important on the security perspective of India as a nation.

Tanakpur is located at 80° 3'47.97"E to 80° 8'5.00"E and 28°59'2.97"N 29° 5'44.90"N located in the northwestern part of the country. It has an average elevation of 837 feet above sea level. Uttarakhand has a total geographic area of 53,483 km^2, of which 93% is mountainous and 64% is covered by forest. As of 2001 India census, Tanakpur had a population of 15,810. Males constitute 53% of the population and females 47%. Tanakpur has an average literacy rate of 65%, higher than the national average of 59.5%: male literacy is 71%, and female literacy is 58%. In Tanakpur, 15% of the population is under 6 years of age. The town of Tanakpur experiences a pleasant climate throughout the year, and the type of climate can be classified as sub-tropical. The basic seasons in Tanakpur are those of summer, winter and monsoon. Summers in Tanakpur are not too hot, and the weather will normally be ideal for holidaying.

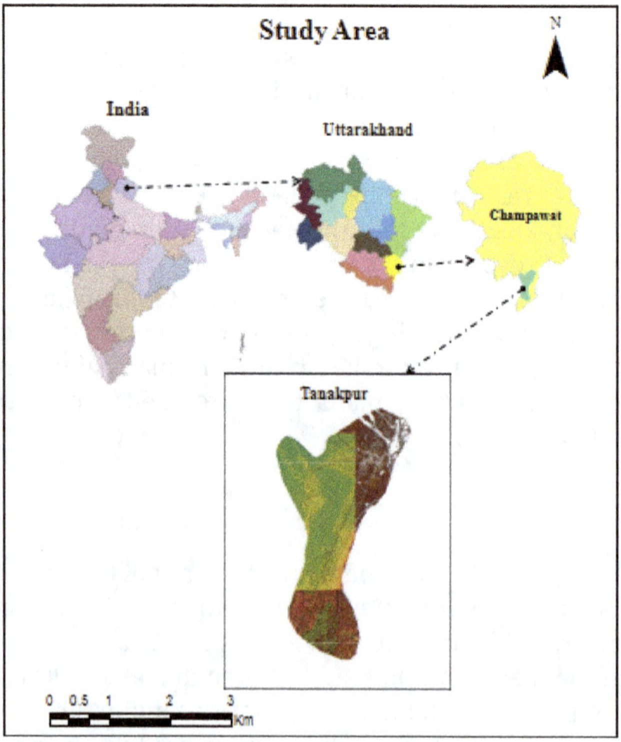

Figure 1. Location map of the study area.

Data Used and Methodology

In this area, datasets and examine approach has been clarified individually. In the main subsection, different datasets, which have been utilized, is clarified. Distinctive sort of Data sets has been utilized in this venture. Information are given by National foundation of provincial advancement (NIRD). NIRD has given us Base guide, Toposheet, soil disintegration information of Tanakpur town. The chose investigation territory clutch Survey of India Toposheet of scale 1:25,000. Than the Toposheet number are 62 C/4/SW. Curve Hydrology has been intended to speak to information from hydrography and hydrology, consequently making a reason for getting a more profound comprehension of surface water frameworks. The depiction, study and outlining of waterways, for example, streams, lakes and oceans, is called hydrography.

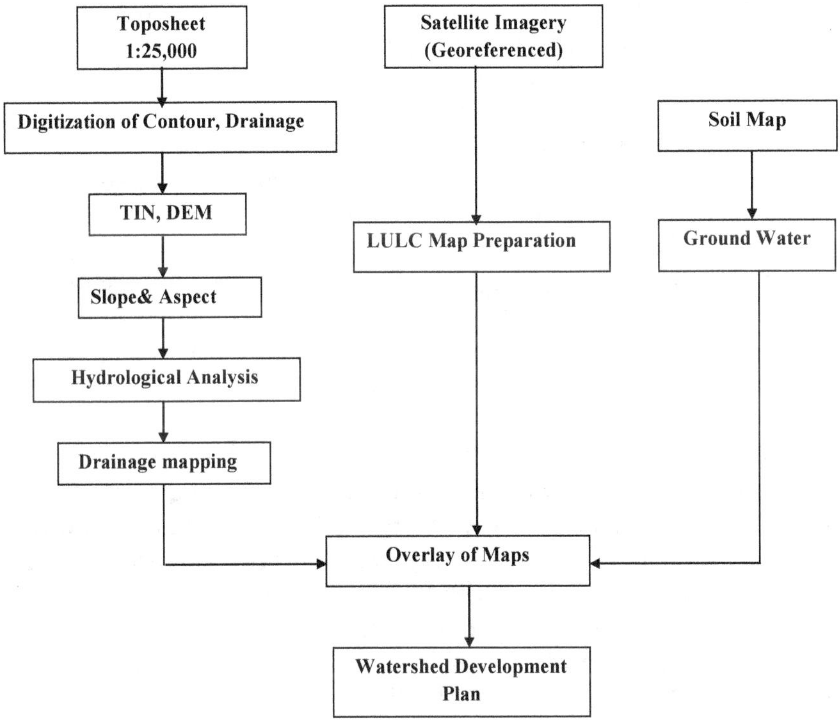

Figure 2. Flow chart of methodology.

Results and Discussion

Base guide of the examination region are set up from the overview of India Toposheet on 1:25000 scale. The readiness of the base guide of the investigation region is initial phase in the examination of land use/land spread. Different perpetual highlights like woodland, settlement, cropland, and ranch or some other land based component were move to the base guide. Filtered maps ought to be enlisted as for framework base utilizing ERDAS Imagine programming so as to have certifiable directions. The 16 converging purposes of longitudes and scopes ought to be considered as GCPs (Ground control focuses) to build up a GCP model with mean remaining mistake. Each Toposheet on 1:50000 scale Toposheet ought to be enrolled independently. After enlistment of individual Toposheet a mosaic ought to be set up for the whole investigation territory. Distinctive topical maps to be specific, waste, land use/land spread, soil disintegration and slant can be set up by embracing standard translation systems utilizing Survey of India Toposheet and Satellite Imagery. Essentially, updation of waste should be possible with the assistance of Satellite Imagery. Transformation of point, line and polygon credit information from raster to vector structure must be done through digitization procedure utilizing ArcInfo GIS programming. After digitization the mistakes, for example, pseudo hubs/dangle hubs, undershoots, overshoots ought to be adjusted. Names ought to be entered in all polygons and furthermore to various circular segments (curve ids) of the highlights. Projection is the procedure of change of three dimensional space onto the two dimensional procedure. The geodatabase in this manner created from corrected toposheets through digitization procedure are to be anticipated into true arranges by receiving standard projection methodology with the end goal of incorporation during spatial investigation. A control point is an enrollment point or geographic control point which speaks to x and y facilitates (longitudes and scopes) of genuine arrange framework. In any event four control focuses are required for projection. For all toposheets tics ought to be set up for change process. In the wake of anticipating inclusions to genuine arranges by standard projection system, all highlights like polygons and lines ought to be changed to certifiable facilitates by change process. The serious issue with the enlistment of cadastral maps is because of absence of directions. So as to take care of this issue these maps ought to be changed on to the toposheets by recognizing surely understood reference focuses on the toposheets in the region of intrigue. The hydrological examination procedure is one of the viable strategies as far as expense and time in proposing different water collecting structures. This procedure manages evaluating different hydrological qualities of a surface. The fundamental parameter which controls the surface water stream (run-off) is the state of the surface (landscape). Slant and viewpoint assume an essential job fit as a fiddle of a surface. The essential information sources required to create a hydrological model for a district are slant, angle, stream bearing, stream aggregation, pour focuses and a conceivable stream organize which requires a TIN (Triangulated Irregular Network) and DEM (Digital Elevation

Model). The entire hydrological procedure can be extensively separated into 2 stages for example (1) Surface investigation and (2) Hydrological examination. Surface investigation procedure includes production of rise surfaces, for example, Triangulated Irregular Network (TIN) and Digital Elevation Model (DEM) from rise information accessible either as spot statures or forms. To create TIN and DEM shape information which is accessible in the vector structure the Survey of India Toposheet on 1:25000 scale with 10 m interim can be utilized. Right off the bat, Survey of India Toposheet on 1:25000 scale ought to be checked utilizing AO size level bed scanner and changed over to Tiff raster design. These maps ought to be anticipated to genuine arrange framework. Shapes (poly lines of equivalent height) are to be digitized utilizing Arc Info GIS programming. After digitization altering ought to be done to expel pseudo hubs, overshoots and undershoots. Bend topology has been made by 'Clean' direction. The inclusion must be anticipated to polyconic projection in the wake of building up tics to create TIN which is a basic parameter in surface displaying.

So as to build up stream collection raster and conceivable stream organize a rise raster must be made. For this shape information consequently created in the vector configuration must be utilized to produce TIN utilizing ArcMap 3D investigator capacities. Later this TIN must be changed over to DEM (Elevation raster) raster structure by utilizing spatial investigator capacities. Slant guide can be produced from the height raster by surface examination apparatuses under spatial investigator capacities. Thus, viewpoint guide can likewise be set up from the slant map in a similar strategy referenced. Incline and viewpoint maps assume an essential job notwithstanding stream heading and stream aggregation in hydrological displaying. Slant is one of the significant territory parameters which can be clarified by flat dispersing of the forms. By and large, in the vector structure firmly separated shapes speak to more extreme slants and meager forms display delicate slant while in the rise yield raster each cell has a slant esteem. Here, the lower incline esteems show the compliment territory (delicate slant) and higher slant esteems compare to more extreme slant of the landscape. In the height raster incline is estimated by the distinguishing proof of most extreme pace of progress in an incentive from every phone to neighboring cells. The slant esteems can be determined either in rate or degrees in both vector and raster structures. Note that when the slant point equivalents to 45°

the ascent and run are equivalent and the slant edge is equivalent to 100%. The slant map in this manner produced from rise raster displays diverse slant classes depending up on the slant of the landscape. The yield cell size assumes a significant job in creating yield incline raster. Incline guide can be readied physically according to AIS&LUS (All India Soil and Land Use Survey) and furthermore by utilizing 'Slant' choice under Surface examination work in Arc Map.

These slant classes can be renamed according to our necessity. As a rule slant guide includes seven expansive classes according to AIS&LUS (All India Soil and Land Use Survey). The general slant classes are appeared in Table 1.

Table 1. Slope classification

Class	Percentage	Slope Category
1	0-1	Nearly level
2	3-Jan	Very gently sloping
3	5-Mar	Gently sloping
4	10-May	Moderately sloping
5	15-Oct	Strongly sloping
6	15-35	Moderately steep to steep sloping
7	>35	Very steep sloping

Sinks are the intersection highlights which draw stream from the parts of the bargains according to the general definition in hydrology, however the sinks that show up in the height raster information are because of mistakes like examining impacts and adjusting of rises to whole number numbers. Sinks in this model have indistinct bearings of one of the eight substantial qualities in a stream course raster (for example 1 to 128). Sinks by and large happen in rise raster because of neighboring cells are higher than the handling cells. The height raster information ought to be free from sinks to make an exact portrayal of stream course and gathered stream. 'Fill sinks' alternative must be utilized under hydrology capacity to expel sinks from the height raster information (DEM). The stream bearing shows the course of surface stream which is a whole number raster worth extents from 1 to 255. In a rise raster, in case a phone is lower than its neighboring cells, the heading of the stream will be towards that phone. In some rise rasters, the point at which various neighbors have the least qualities then the stream will be characterized by sifting through one cell sinks. Now and again if a cell has a similar change in 'Z' esteem in numerous headings the came about stream course will be total of those bearings. The stream heading can be dictated by finding steepest plunge from every cell which can be determined from this condition. For example, Change in Z esteem/Distance *100. The height raster consequently produced without sinks must be utilized to create the stream course utilizing 'Stream bearing' choice in hydrology examination work. The yield stream course raster indicates eight (8) potential headings in which the conceivable surface would be.

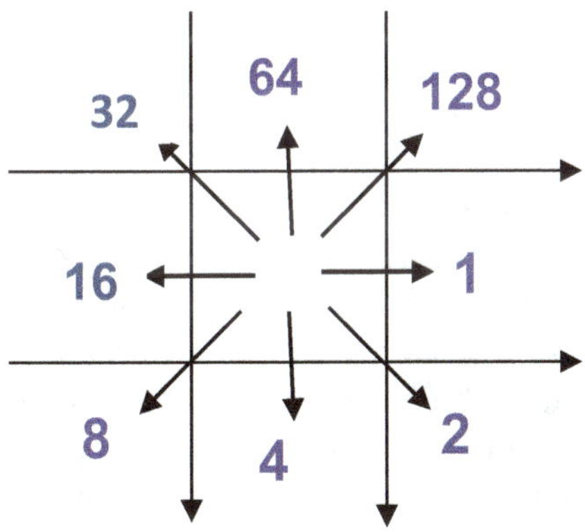

Flow direction

Stream gathering will be created from the blunder free height raster information. The cells of unclear stream headings other than (1 to 8) will just get stream aggregation. The gathered stream in the yield raster will be determined dependent on the quantity of cells streaming towards every cell. The high stream zones in the yield raster are the territories of concentrated stream which are essential to distinguish conceivable stream channels. So also, those territories with stream gathering worth zero (low) are the zones of geographically high like edges. A stream system can be made by utilizing the aftereffects of the high gathered stream. Also, this stream system can be utilized as contribution to produce stream request, stream line and stream connect.

0	0	0	0	0	0
0	1	1	2	2	0
0	3	7	5	4	0
0	0	0	20	0	1
0	0	0	1	2	0
0	2	4	7	3	1

Flow accumulation

Getting stream request is one of the significant parameters in hydrology displaying. Stream request raster can be created in ArcMap utilizing hydrology device. Stream request capacity allots a numeric request to the sections of a raster which speaks to parts of a straight stream arrange. Stream amassing raster can be utilized as info surface to create a raster stream arrange by applying a limit an incentive to choose cells with a high gathered stream. The info stream raster straight system ought to be spoken to as qualities more prominent than or equivalent to one on a foundation of no information in getting yield stream request. Strahler stream request technique must be embraced for relegating stream request. The yield stream request accordingly created is a whole number raster. The accompanying condition can be assessed utilizing raster number cruncher in spatial examiner to separate raster stream request from stream collection raster. To make digitization through SOI Toposheet and update with Google earth picture. Waste example, otherwise called waterway frameworks, are the examples shaped by the streams, streams, and lakes in a specific seepage bowl. They are administered by the geology of the land, regardless of whether a specific locale is ruled by hard or delicate rocks, and the slope of the land. A seepage bowl is the topographic district from which a stream gets spillover, through stream, and groundwater stream. The number, size, and state of the waste bowls found in a territory differ and the bigger the topographic guide. Soil disintegration is a normally happening procedure that influences all landforms. In agribusiness, soil disintegration alludes to wearing a method for a field's topsoil by the regular physical powers of water and wind or through powers related with cultivating exercises, for example, culturing.

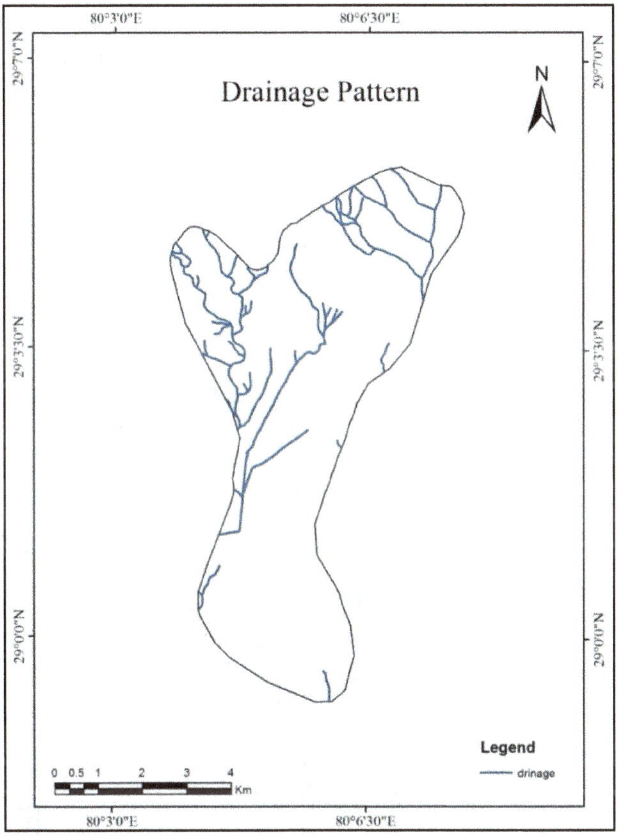

Figure 3. Drainage pattern.

Disintegration, regardless of whether it is by water, wind or culturing, includes three unmistakable activities – soil separation, development and statement. Topsoil, which is high in natural issue, richness and soil life, is moved somewhere else "on location" where it develops after some time or is conveyed "off-site" where it fills in seepage channels. Soil disintegration lessens cropland efficiency and adds to the contamination of nearby conduits, wetlands and lakes. Soil disintegration can be a moderate procedure that proceeds generally unnoticed or can happen at a disturbing rate, causing genuine loss of topsoil. Soil compaction, low natural issue, loss of soil structure, poor inner seepage, salinization and soil acidity issues are different genuine soil debasement conditions that can quicken the dirt disintegration process. Water asset the executives assumes a fundamental job in practical improvement of watershed which is conceivable just through the usage of different water gathering systems. The proficient path for sub-surface water stockpiling, soil dampness protection or ground water revive advances ought to be received appropriately submerged asset improvement plan. Correspondingly, creating of badlands by proper measures can improve water assets on uplands in the watershed.

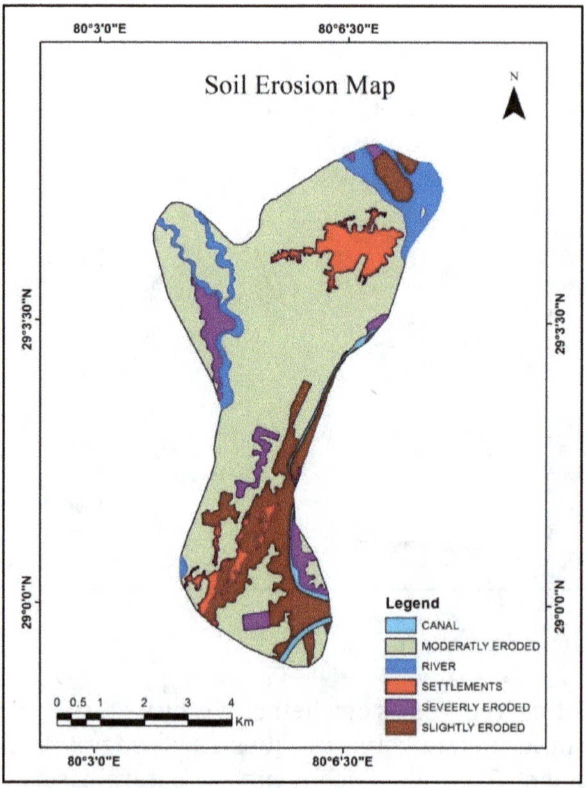

Figure 4. Soil erosion map.

The aftereffect of this venture can be talked about with land use land spread guide task is demonstrating that larger part of land is spread by deciduous woodland, crop land, squander land, settlement, estate and land which demonstrates that lion's share of populace is rely upon harvest generation for their living. Land use land spread guide of Table 2 demonstrates the aftereffect of LULC in forbidden structure which is straightforward.

Table 2. Land use/land covers statistics

LU/LC Classes	Class Name	Area in sq.km
1.	Settlement	3.10
2.	Waste land	1.69
3.	Canal	0.82
4.	Crop land	10.09
5.	Forest	11.65
6.	Plantation	7.39
7.	Water body	3.67

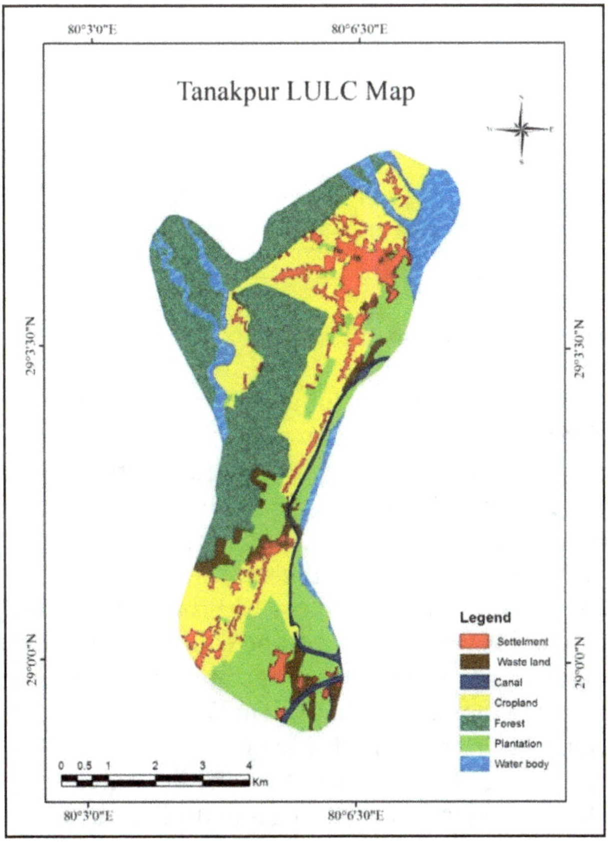

Figure 5. Land use/land cover map.

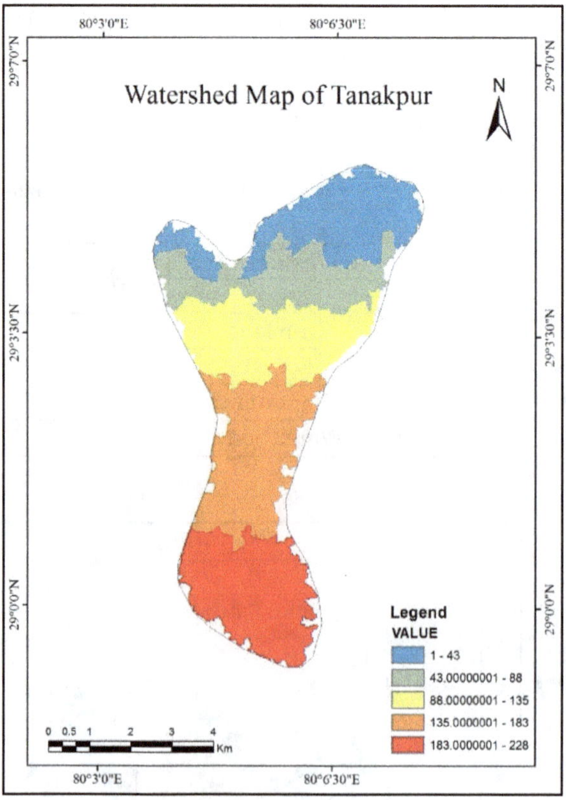

Figure 6. Watershed map.

The idea of watershed development is that the improvement and the executives of the assets in the watershed ought to be taken up, for example, to accomplish higher generation that can be supported without causing any crumbling in the asset base or causing no biological irregular characteristics. The point is to anticipate watershed corruption that outcomes from the cooperation of physiographic highlights, wipe out informal land use by fitting editing design, soil disintegration in this way improving and continuing profitability of assets prompting higher salary and expectation for everyday comforts for the occupants in the watershed zone. Watershed as way to deal with improvement was first embraced in quite a while in Tennessee Valley for diminishing soil disintegration in the valley territories during mid-1920s. The watershed being a hydrological unit, any mediation to store water or moderate the overflow reacts all the more positively when the program is sorted out on a watershed premise. A watershed is a zone from which spillover, coming about because of precipitation streams past a solitary point into a huge stream, waterway, lake or a sea. Watershed guide is set up by outlining watersheds utilizing the waste guide following Ridge to Valley idea. The all India Soil and Land use reviews grouping and coding is received in this guide. Watershed coding is done based on hydrological units at various phases of outline. Watershed code is alphanumerical number speaking to all the Watershed units viz., Region, Basin, Catchment, Sub catchment, Watershed, Mini-watershed, Micro watershed, and Nano watershed. Generally, watershed limits are drawn physically onto a topographic guide. So the limits use includes on the guide to figure out where a gap is found. Outline of watershed or sub watershed can occur at various spatial scales. A huge scale watershed spread may tributaries in the stream framework. The computerized strategy for outline watershed pursues an arrangement with a filled DEM. These arrangement includes with both morphometric and hydrological units.

A bund is the man-made earthen bank of a tank. The bunds are built over the inclines of the land or surface or as indicated by the geography. The initial phase in bund-development is to study the geology and water stream. In some spot, bunds can likewise be developed in plain land, if the water stream proper and enable ranchers to get the water for rural land. Be that as it may, sandy territories are unsatisfactory for bund development since their water stockpiling limit is low. The bund size relies on the capacity limit of the tank and kind of soil.

LBS (Lower Bund Structures)

Water from the homestead lakes can be used for water system crops just as for drinking purposes. Slow dissipation over significant lots of the year changes the small scale atmosphere and builds soil inset populace. Guaranteed of water assets ranchers will remain longer on their property. Grub accessibility will be helpful for creature farming projects. 1123 Farm lakes will be exhumed in the task zone.

ENBS (Earthen Nala Bund Structures)

Earthen bunds of appropriate stature and width, contingent on the incline and size of the plot, are all around tried and attempted as basic measures to limit top soil misfortunes. The bund is furnished with an outlet for abundance water to stream out. This measure likewise improves the ground water energize. The outlet has been found where the incline is steepest or the stream is greatest.

CNBS (Cement Nala Bund Structures)

Bond Nalla Bunds are developed crosswise over streams having delicate incline and are plausible both in hard shake just as alluvial arrangement. The site chose for Nalla Bund ought to have adequate thickness of porous bed or endured development to encourage energize of put away water inside limited ability to focus time. The water put away in these structures is generally bound to stream course and the tallness is ordinarily about 2 m. These are planned dependent on stream width and overabundance water is permitted to stream over the divider.

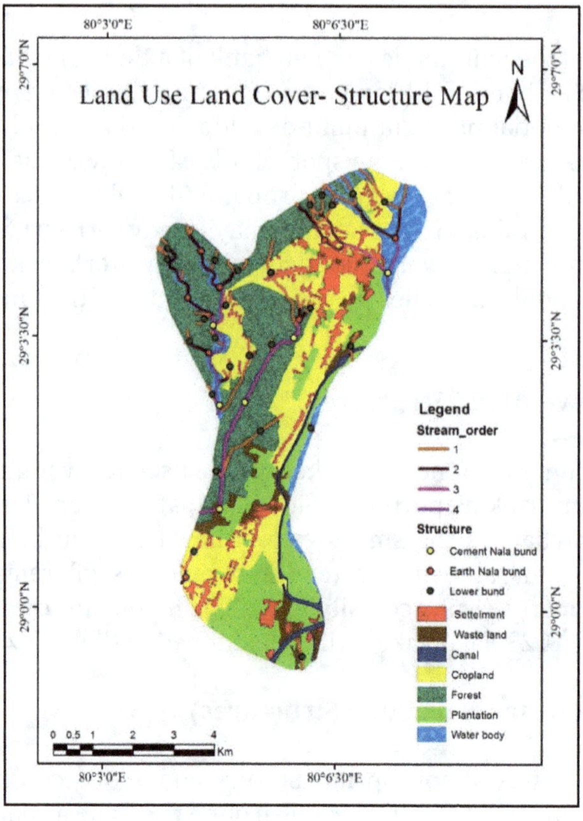

Figure 7. Land use land cover-structure map.

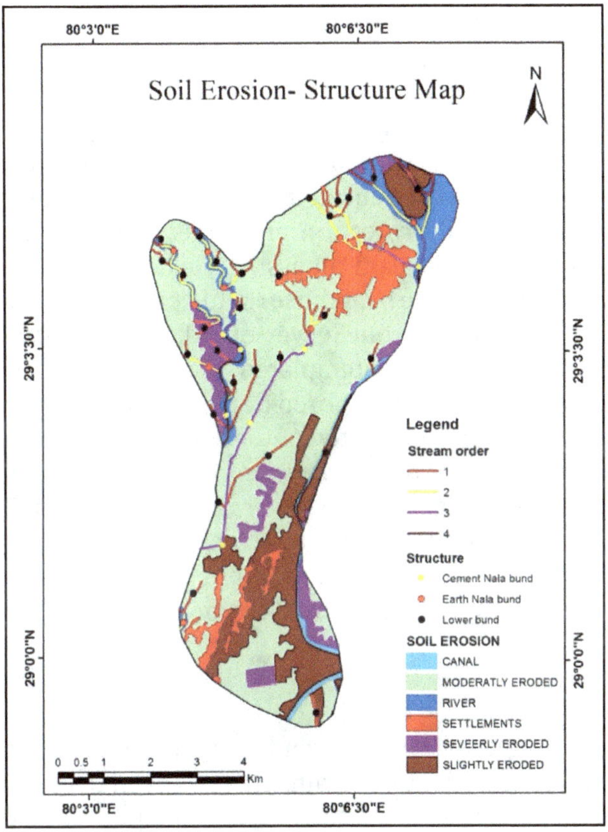

Figure 8. Soil erosion-structure map.

Some Other Methods for Soil and Water Conservation

The accompanying techniques might be embraced in regions with sandy loamy soil and low precipitation.

- Field bunds with waste weirs
- Land leveling and profound stopping
- Surplus water through field to field by stone waste weirs

Mechanical measures to avert water disintegration

- Graded bunds
- Bench patios

- Grassed conduits
- Diversion channels

Recommendation of various soil and water collecting structures assumes an extremely critical job which requires an all-around qualified and all around experienced labor with intensive learning in different water and soil protection programs. For this reason incline, stream heading and stream amassing yield raster maps alongside stream request (seepage), land use/land spread, and soil inclusions which have been set up from Survey of India toposheets and satellite symbolism ought to be contemplated. It is felt that field data is required to help structures proposed dependent on hydrology displaying. For example ground water potential can be assessed uniquely from serious documented perceptions of water table in any event during two seasons i.e., pre and post rainstorm periods. Correspondingly, for proposing structures the accompanying parameters like soil profundity, surface, landscape conditions and slant up and down the chasms at miniaturized scale level (on enormous scale) are required.

Conclusions

The present examination is an endeavor for creating characteristic asset based area explicit maintainable activity plan in watershed advancement exercises the executives of Tanakpur watershed, Champawat locale, Uttarakhand. For accomplishing this point, activity Plans for Watershed Development for the Tanakpur are proposed. The development of water protection structures like homestead lakes, check dams along the streams during the stormy seasons is fundamental. With the goal that it would revive the ground water zone and raises the degree of water table. Therefore the neglected grounds can be changed over to yield terrains and ranches. At fitting spots by doling out the weightages dependent on the needs utilizing ArcGIS programming. This model is exceptionally valuable for water gathering structure in watershed advancement ventures. Watershed the board arranging is a procedure that outcomes in an arrangement how to best ensure and improve the water quality and other common assets in a watershed. All the time, watershed limits stretch out over political limits into contiguous regions as well as states. That is the reason a far reaching arranging process that includes every single influenced district situated in the watershed is basic to effective watershed the executives. Water reaping structures are critical to moderate valuable common assets like, soil and water, which is exhausting step by step at disturbing rate. The Tanakpur zone of Champawat, lying in the lower regions of Shiwalik run, has soil and water the board issues, he great cultivable zones are being are changed over into waste and corrupted land due to over abuse of the land and water assets.

References

Agarwal, C.S. (1998). Study of drainage pattern through aerial data in Naugarh area of Varanasi district, U.P. *J Indian Soc Remote Sens* 26:169–175.

Anil, K., Tangirala Ramesh, S.V., Teegavarapu, Lindell. O. (2003). Modeling adaptive water quality management strategies using system dynamics simulation. *Environ Inform Arch* 1:245–253.

Chandra, S. (1990). Hydrology in ancient India. National Institute of Hydrology, Roorkee, p 106.

Choi, J.Y., Engel, B.A., Farnsworth, R.L. (2005). Web-based GIS and spatial decision support system for watershed management. *J Hydroinform* 7(3):165–174.

Chow, V.T., Maidment, D. R. and Mays L.W. (1988). Applied hydrology, McGraw-Hill Book Company, 1988.

Gajbhiye, S., Mishra, S.K., and Pandey, A. (2014). Prioritizing erosion prone area through morphometric analysis: an RS and GIS perspective. *Appl Water Sci* Springer 4:51–61.

Kanga, S., and Singh, S. K., and Sudhanshu (2017). Role of GIS in Creation of Spatial Socio Economic Indicators of Bilaspur District, H.P. (India), *Journal of Arts, Science and Commerce* International Refereed Research Journal, 2(10): 48-55.

Kanga, S., and Singh, S. K., and Sudhanshu (2018). Assessment of Geospatial Approaches Used for Classification of Crops. 2018. *International Journal of Mathematical, Engineering and Management Sciences*, 3 (3): 271-279.

Nag, S.K. and Chakraborty, S. (2003). Influence of rock types and structures in the development of drainage network in hard rock area. *J Indian Soc Remote Sens* 31(1):25–35.

Pakhmode, V., Himanshu, K. and Deolankar, S.B. (2003). Hydrological drainage analysis in watershed programme planning: a case from the Deccan basalt, *India. Hydrogeol J* 11:595–604.

Pandey, A., Chowdary, V.M., Mal, B.C. (2007). Identification of critical erosion prone areas in the small agricultural watershed using USLE, GIS and remote sensing. *Water Resour Manag* 21(4):729–746.

Santhanam, M.L., Yogananda, S. C. and Vijayakumar, S. (1982). Human and social factors in people's participation. *Journal of Rural Development*, 1(5):770-831.

Singh, S. (1995). Quantitative analysis of watershed geomorphology using remote sensing techniques. *Ann Arid Zone* 34(4): 243–25.

Singh, S. K., and Kanga, S. (2017). Mapping of salt affected and water-logged areas using geospatial technique, *International Journal of Recent and Innovation Trends in Computing and Communication*, 5 (5):1298-1305.

Singh, S. K., and Kanga, S. (2017). Role of Geoinformatics in Site Suitability Analysis of Infrastructures using PRA Approach, *American International Journal of Research in Science, Technology, Engineering & Mathematics*, 1(18):81-85.

Singh, V.P., Frevert, D.K. (2006). Watershed models. Taylor and Francis Group, Boca Raton, p 680.

Sreedevi, P.D, Subrahmanyam, K., and Ahmed, S. (2005). The significance of morphometric analysis for obtaining groundwater potential zones in a structurally controlled terrain. *Environ Geol* 47:412–420.

Strahler, A.N. (1964). Quantitative geomorphology of drainage basins and channel networks. In: Chow VT (ed) Handbook of applied hydrology. McGraw-Hill, New York, pp: 439–476.

Thomas, J., Joseph, S. and Thrivikramji, K.P. (2010). Morphometric aspects of a small tropical mountain river system, the southern Western Ghats, India. *Int J Digit Earth* 3(2):135–156.

Zhang, Z., Chen, X., Xu, C., Hong, Y., Hardy, J., Sun, Z. (2015). Examining the influence of river-lake interaction on the drought and water resources in the Poyang Lake basin. *J Hydrol* 522:510–521.

15

Managing Aquifers and Watershed in Urban Area through Participatory Approach to Sustain Urban Water Supply - A Case Study of Bhuj City (Gujarat)

Jadeja Yogendrasinh J.

Introduction

Bhuj City, the capital of the Kachchh district in western India, has a rich history deeply intertwined with-and dependent upon-water. The biggest and most successful city in an arid region, Bhuj owes much to its strategic water reserves, which have allowed the city to thrive and grow for more than 500 years despite minimal rainfall and frequent drought. The total urban area spreads over about 56 km^2. The city has total about 11 administrative wards. (Figure 1) The city's traditional water management system has been abused by rapid population growth and urbanization through three phases. Arid Communities and Technologies (ACT) - not for profit organization has evaluated Bhuj City's traditional water management system geo-hydrologically to assess feasibility to provide domestic water supply even after such changes.

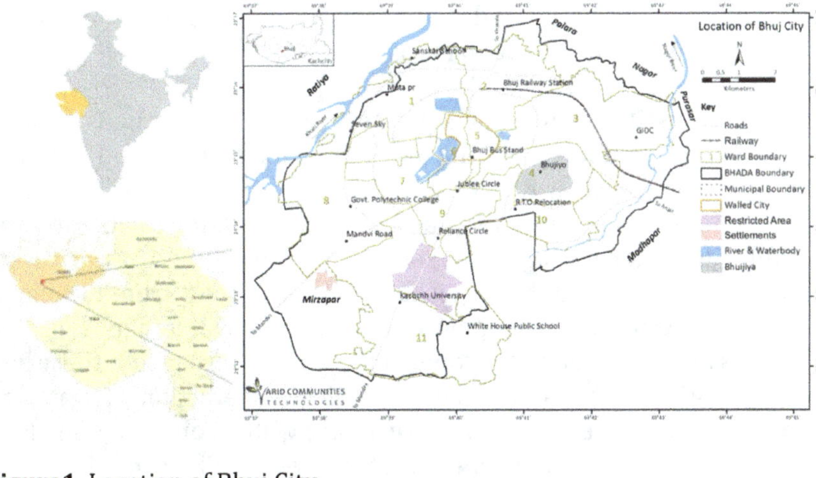

Figure1. Location of Bhuj City.

Approach

An integrated approach has been adopted and accordingly techno-social-institutional devices have been worked out for inclusion of participatory groundwater management considering its urban-peri-urban linkages. As far as technological aspects are concerned, the geohydrological assessment of entire project area has held through mapping of aquifer, study of water levels and groundwater quality, estimation of overland flow, aquifer properties and water budget. To quantify groundwater related problems groundwater monitoring network has established since year 2008 that successively periodically revised to study the impact. The findings of this investigation have been utilized for urban–peri–urban water management planning and to design demonstrations of specific activities those can help community and administration to allocate resources for execution.

In case of social aspects, intensive and series of social mobilization activities have organized to sensitize, educate and aware the community regarding importance of local resources from their protection, potential and their interlinked urban-peri-urban dynamics. As community involvement was most crucial part each activity has planned with specific objective and specific output. Social processes includes activities such as exposures, workshops and seminars, rallies, slideshows, competitions, drama, traditional celebrations on specific religious days. Another way of social mobilization and advocacy processes that adopted was demonstration of appropriate techniques for groundwater recharge and water supply management for various units of urban and per urban areas.

As far as institutional aspects are concerned, the methods adopted were capacity building, setting up of small institutions and group for focused objectives. As the aim of this institute is to maintain momentum of knowledge generation and to transfer to next generation an aquifer level forum has been set up by identifying a sensitive and community accepted distinguished representatives from urban and surrounding rural areas. Under the forum tank watershed level committees are being set up for some of the tanks of area. Both these kind of institutions have aim to keep watch on protection of water resources to manage supply part. Drinking water committee in slums and campus water management committees RRWHs committee have been constituted by the aquifer wide forum. Besides setting up of committees and forum, a youth team of 07 persons, known as Bhujal Jankars (BJs), has been trained for regular assessment of groundwater by collecting and analyzing data from observation network, to facilitate technical designs and planning of water supply, recharge, rain water harvesting activities by stakeholders including gram panchayats (Village secretariat) and municipality.

Results

The results of the study is discussed under various sub heads such as historical expansion and change in water management practices, terrain and geo-hydrological characteristics, groundwater potential of the city and action research activities carried out to develop participatory groundwater management approach.

Historical Expansion and Change in Water Management

Expansion process of Bhuj city has passed through three important events i.e. first, after independence during 1947 (nation-wide change in ruling system) second, when Municipal system came in existence and third post-earthquake where urban development authority has established. All these three development stages reveal how Bhuj city's local geo-hydrological characteristics based traditional water management system has ignored and fetching water from peri urban areas have affected land and water resources of both the areas and finally influenced life. Changes in land scape and land use has decrease land resources results into scarcity of agriculture and grazing lands for traditional livelihoods of peri urban areas. The same has also increased flood threat and over land flow. Water supply from groundwater resources from adjoining villages have resulted into depletion in water levels that again increased water scarcity for agriculture productivity as well drinking water supply for rural areas. Water level depletion led deterioration in groundwater quality has resulted into health issues. In addition, infrastructures like sewage disposal again polluted existing water bodies, groundwater and land resources again urban peri urban areas. The development and expansion city has placed several challenges and changes as mentioned in Table 1.

Table 1. Phase wise Urban Developmental Process and Emerging Trends in Bhuj City

Phase	Pre Independence Before 1947	Pre Earth quake of 2001 (1947 - 2001)	Post-Earthquake After 2001
Process	Establishment of Bhuj City as Capital of Kachchh	Expansion of Bhuj city in some part, Change in land use in surrounding of city, Development of Semi -Centralized water supply	New Development and expansion plans, New Infrastructures like road network, drainage network, water supply and distribution plans, Change in Land uses
Water Sources	Inter-linked Water Bodies	Local open wells and some bore wells from adjoining rural areas	Tube well in peripheral villages and Narmada Supply, Private bore well, Bottled supply, Tanker Supply
Management by	King and community	Municipality	Municipality GWSSB, GWIL, Private enterprises
Major Issues	Drought	Irregular supply, Encroachments in water ways and water bodies, Increase dependency on Municipality. Breakdown of traditional management system	Complete dependency on outside sources, Irresponsible citizens, Disappearing of water bodies and water channels, Floods and water logging, Decrease in Recharge, Local GW quality issue, Leakage and poor management of water and sewage infrastructure and discharge in downstream rural areas
Copping Mechanism	Inter-linked Catchments and Water body, Wells	New Bore well exploration, Bring water from distance source, Tanker supply	Disaster response Draining out of Strom water, Lack of long term copping mechanism

Physiogrpahy - Drainage and Watersheds

Physiographically the Bhuj city area has two important divisions i.e. southern dissected ridges resulted due to upliftment due to Katrol fault rocky and gravelly pediment zones. (Biswas, 1982) Both are regional nature division and have various local level landforms. Important landforms of the area are southern dissected hills, Bhujiya hill within city, Khari river gorge at western boarder of Bhuj city. Besides these there are various rivers and ridges of local scale define micro watersheds. It is important to mention that majority of landform in the project area are results of tectonic processes like faulting and ingenious activities.

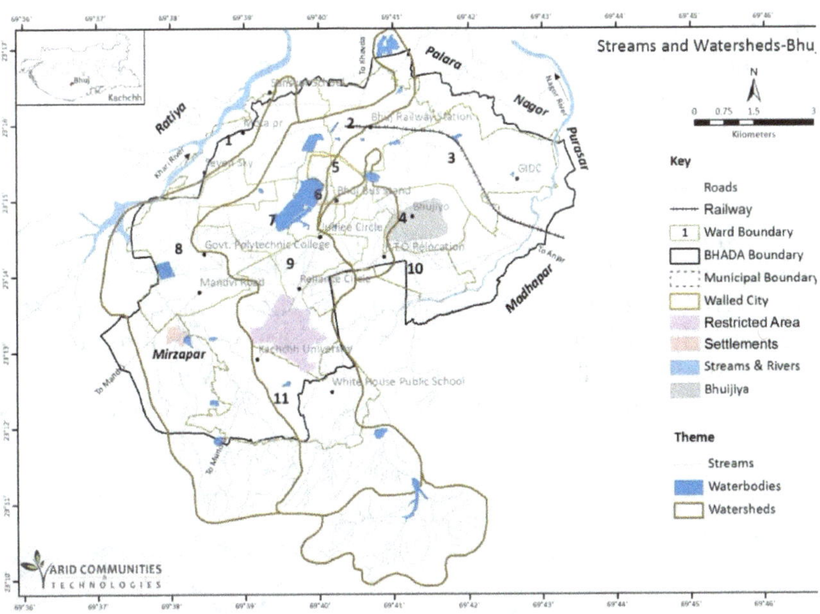

Figure 2. Drainage Pattern and Watershed of the Bhuj City.

Drainage and watershed wise the Bhuj city represents part of Rudramata basin that originates from Kachchh mainland ridge portion in south to merges into Banni in North. To manage city water requirement historically people have developed three lakes and their relevant watershed management. These lakes systems are (01) Hamirsar Lake System (02) Umasar lake system and (03) Deshalsar lake system. All three lake system and their watershed are interlinked with each other by developing canals and tunnels with an area of 40 km^2.

Geo-hydrological Characteristics

Geology: The rock formations of the area show their age ranging from Middle Jurassic to the recent age. The main rock formations of the area are sandstone and shale with a few basaltic intrusions. The oldest rock formations in the area are shale and fine grain sandstone of the Chhari and Katrol series of the Middle and Upper Jurassic. (Biswas, 1082 and Merh, 1995) There is frequent intercalation between the shale and sandstone, as well as with limited limestone deposits in the region. On average, the formation shows gentle dip towards south, ranging from 5° to 13°; at its shallowest the sandstone is only 18 m deep, and deepens to 123 m at its southern tip.

Aquifer Complexity: The belt of cretaceous sandstone is the region's only good aquifer, but has a risk of high salinity due to the naturally saline nature of the sandstone and proximity to iron and shale deposits. Due to Kachchh region's peculiar tectonic framework had divided Bhuj sandstone into various pieces and each piece has its own dynamic geo-hydrological characteristics. Figure 3. Shows various types of lineaments mapped within the Bhuj city (Biswas and Khattri, 2002; Jadeja, 2005). Due to dyke network the Bhuj formation as aquifer can divide into two zones north zone and south zone while stratigraphically sequence the same formation can divide into upper and lower Bhuj or shallow and deep aquifer respectively. Hence considering litho-tectonic set up the aquifer system of Bhuj can further cat-

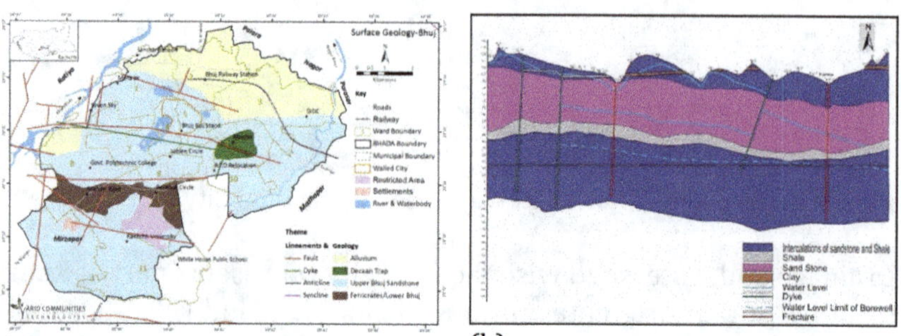

(a) (b)

Figure 3. (a) Surface Geology and Lineament Fabrics and (b) how these lineament are functioning as subsurface barriers and fracture systems within Bhuj City and Peri Urban areas, figure also portraying control of lineaments on groundwater levels.

egorize into four as north south shallow aquifers and north south deeper aquifers.

Groundwater Levels and Quality

It is well known that Kachchh region has very peculiar tectonic framework represented by various major fault, folds, fractures and volcanic structures. Bhuj city area also reveals lineament fabric that divides Bhuj sandstone into various compartments as modifies its hydraulic properties in term of secondary properties. Fig. 3 shows various types of lineaments mapped within the Bhuj city. Due to dyke network the Bhuj formation as aquifer can divide into two zones north zone and south zone while considering stratigraphic sequence the same formation can divide into upper and lower Bhuj or shallow and deep aquifer respectively divided by red colored lateritic zone exposed in middle of the city area. Hence considering litho-tectonic set up the aquifer system of Bhuj can further categorize into four as north south shallow aquifers and north south deeper aquifers (Arid Communities and Technologies, 2004, 2017).

There is a high concentration of pollutants in the groundwater that make it non-potable in many areas. The Total Dissolved Solids (TDS) concentration ranges from 750 mg/l in shallower areas to over 7000 mg/l in the deeper bores. This is an indication that, as the water table lowers, the concentration of dangerous pollutants increases. Additionally, and wells are dug deeper (beyond 125 m), they draw water from intercalation of sandstone with shale, yielding a very high saline content. It is crucial to improve surface water collection and percolation across the catchment to raise the water table to facilitate access to potable water across the region.

Water Balance

The quantitative estimation of total water resource of the Bhuj city has been attempted on annually replenishable components of recharge or discharge. There exists two such approaches viz. (01) Water Balance Approach and (02) Water Table Fluctuation/ Specific Yield Approach.

Estimation of water balance is useful tool to determine the groundwater recharge. It is accepted concept that a balance exists between the quantity of water supplied to the aquifer and the quantity of water stored in an aquifer over a specific period. In general, for balance all in-flowing and out-flowing parameters are considered. Based on this fact equation for groundwater recharge can be expressed as

Change in Storage = (Inflow into basin - Outflow from basin).

In a normal hydrologic cycle distribution of rainfall takes place into various parts when it reaches to the land surface, quantum of rainwater reaches to groundwater regime through percolation; water return to atmosphere through evapotranspiration; traps into soil as soil moisture; some part is stored into water impounding reservoirs; whereas remaining part flow out from the catchment as overland as well as sub-surface flow. During this cycle aquifer gains the water and this balance can be expressed as water balance (Schicht and Walton, 1996) and the equation to this effect can be derived as –

$$Rr + \text{Re} + Ri + I + Si = Se + O + Et + Dp + \Delta S$$

where,

Rr = Recharge due to rainfall

Re = Recharge due to canal seepage

Ri = Recharge due to return flow of applied irrigation waters

Se= Groundwater flow due to effluent seepage

I = Inflow from areas outside the basin

Si = Recharge due to inflow seepage from rivers, streams, reservoirs, lakes and ponds etc.

O = Outflow to the areas outside basin

Et = Evapotranspiration losses

Dp= Groundwater draft (pumpage)

ΔS = Change in groundwater storage.

Based on input and output parameters water resources have budgeted for Bhuj city shown in Table 2. This clearly shows that during year pre monsoon season of year 2014 to pre monsoon season of year 2015 there is net available water resources are 10.15 MCM that is more than the city's estimated annual water requirement i.e. 9.85 MCM considering about 2 Lakh population and 135 lit/day.

Table 2. Overall Water Resource Budget of Bhuj City

Net Watershed Area (A) = 72.6 sq. Km and Average Annual Rainfall (R) = 0.430 m			
Sr. No	Parameter	Inflow (MCM)	Outflow (MCM)
1	Rainfall (R×A)	31.218	
2	Recharge from rainfall (Rr)	4.005	
3	Recharge from Canal Seepage (Re)	0.000	
4	Recharge from return irrigation seepage (Ri)	0.460	
5	Recharge from surface storage (Si)	0.843	
6	Surface runoff (O)		15.448
7	Evapotranspiration (Et)		8.849
8	Ground water draft (Dp)		8.400
9	Change in ground water storage (ΔS)		-6.318
	Total	36.526	26.379
Balance : (Inflow – Outflow)			10.147

Discussion

As groundwater management is not only technical solution it also requires aware society and administration. Keeping this in mind the Bhuj city groundwater management plan has been prepared with strong consideration of local geo-hydrological conditions and with systematic social awareness process aimed with competent citizen's institution. The management plan includes both the aspects i.e. resource strengthening with efficient supply management and competent citizen's institution. As far as resource strengthening is concerned, recharge plan has been proposed considering aquifer potential and its hydraulic characteristics for Bhuj city (Figure 4) and to maintain recharge in to aquifer water ways and water bodies requires special attentions. Therefore, for water channels specific width of each channel has calculated and recommended to urban development authority to carry sufficient and guided rain water volume towards recharge structure.

During the project each aquifer has considered for specific recharge plan. The techniques suggested for recharge area (01) Providing recharge bed and percolation well in common plot in societal areas; (02) each campus should have rainwater harvesting and recharge bore well; (03) each rivulet should have percolation structure at regular 300 m interval; (04) each water channels should have specific width to transfer storm water load based on run off from respective micro catchment. (06) all non-registered water body should be registered as municipally or gram panchayat's property; (07) identified water body must have recharge bore well. With all these techniques Bhuj city recharge plan has developed as shown in Fig. 4. It is important to mention that all techniques suggested for this planning have properly demonstrated through community participation and proven their successes based on regular monitoring of groundwater levels and quality with respective rainfall condition.

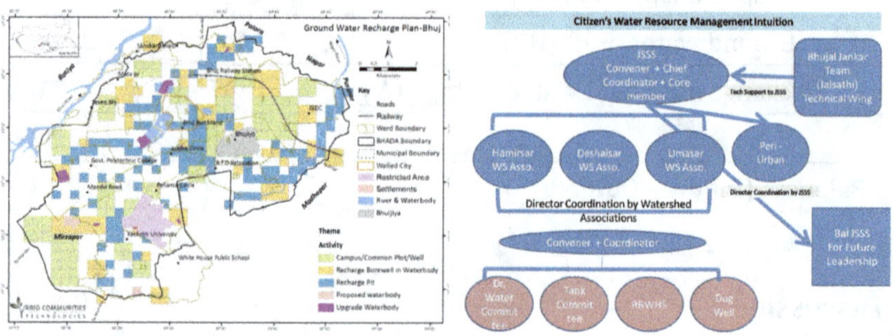

(a) (b)

Figure 4. (a) Groundwater Recharge Planning for Bhuj City and (b) Aimed Institutional Structure for Urban-Peri-Urban Water Resource Management (JSSS).

It is proven that without participation of stakeholder resource management in any area may not provide sustainable solution therefore, ACT has begun to educate citizens of Bhuj city to fulfil the aim of establishment of citizen's empowered institution. Figure 4 explains the institutional structures that evolved after citizen's awareness actions and emerging institution in the name of *"Jal Srot Sneh Samvardhan Samiti"* (JSSS).

Conclusions

The Participatory Groundwater Management (PGWM) experiences of

Bhuj City have provided the directions how one can though about management of water resources for rural as well as urban areas during rapid urbanization phase of development. To prevent local water resources especially in adjoining rural areas and to sustain urban water resources in decentralized but people centric manner following conclusions can be made.

Detailed geo-hydrological characterization of each city is an important decision making tools that need to be consider as one of parameter for developing urban area development plan. Involvement of all level of stakeholders with proper education through awareness is non-negotiable condition especially for increasing ownership of citizen's on local water resources. Municipality and NOGs should jointly made efforts to establish empowered and technically strengthened citizen's forum along with Bhujal Jankar as bridge between community and government. While supplying urban water priority must be given to local sources and therefore, while designing centralized supply form distance sources its allocation should be based on availability of local water. Experiences like Bhuj city need to be given attention at state level.

Acknowledgments: Authors are highly thankful to Bhuj Municipality, Bhuj Urban Development Authority, citizens of Bhuj, and various partners such as Sahjeevan, Bhuj Setu, KMVS, CEPT, GWSSB, Prof. K. C. Tiwari, ACT members and Arghyam Trust for support. Authors also acknowledge organizers of Peri Urban Conference, 2017 for providing opportunity to present their experiences in the seminar.

Abbreviations

The following abbreviations are used in this manuscript:
ACT: Arid Communities and Technologies
BJs: Bhujal Jankars
CEPT: Center for Environment Planning and Technology
GWIL: Gujarat Water Infrastructure Limited
GWSSB: Gujarat Water Supply and Sewage Board
JSSS: Jal Srot Sneh Samvardhan Samiti
KMVS: Kachchh Mahila Vikas Sangathan
PGWM: Participatory Groundwater Management

References

Arid Communities and Technologies (2004). Proposal Revival of Hamirsar Lake and Its Catchment – Wetland and Urban Rural Relationships.

Arid Communities and Technologies (2017). Understanding Investments, Economics and Pricing Principles of Drinking Water in Bhuj City and Prospects of Local Water Resource Development.

Biswas, S.K. and Khattri, K.N. (2002). A geological study of earthquakes in Kutch, Gujarat, India. Jour. Geological Soc. India, v.60 (8), pp. 131-142.

Government of Gujarat (2011). Census Handbook of Kachchh
Central Ground Water Board (2011). State Profile – Ground water Scenario in Gujarat.

Central Public Health and Environmental Engineering Organization (1999). Manual on Water Supply and Treatment.

Gujarat Ecology Commission (1997). Eco Regions of Gujarat.

Jadeja, Y. J. (2005). Studies on Water Resource Development and Planning of Pachchham Area Kachchh District-Gujarat, Ph.D. Thesis, M.S. University of Baroda, Vadodara.

Ministry of Urban Development Government of India (2011). Status Report, Service Levels in Urban Water and Sanitation Sector.

16

Impact of Climate Change and Monsoon Variability on Groundwater Level Fluctuation: A Case Study of Semi-Arid Region of Rajasthan

Amit Daiman, Narendra Gupta, Gaurav Tiwari and Saurabh Shiva

Introduction

We are living in a period of climate change brought about by increasing atmospheric concentrations of greenhouse gases. Global mean temperatures have risen 0.3–0.6^{0}C since the late 19th century. It is also reported that the expected global rise in temperature over the next century would probably be greater than observed in the last 10,000 years. As a direct consequence of warmer temperatures, theater cycle will undergo significant impact with accompanying changes in the rates of precipitation and evaporation (Intergovernmental Panel on Climate Change (IPCC, 1995).

The All-India area-weighted mean summer monsoon rainfall, based on a homogeneous rainfall data set of 306 rain gauges in India, developed by the Indian Institute of Tropical Meteorology, is widely considered as a reliable index of summer monsoon activity over the Indian region. It is interesting to note that there have been alternating periods extending to 3-4 decades with less and more frequent weak monsoons over India (IITM, 2016).

The Indian summer monsoon (June to September) rainfall is very crucial for the economic development, disaster management, hydrological planning for the country. Earlier, Mooley and Parthasarathy (1984), Parthasarathy *et al.,* (1993), Parthasarathy *et al.,* (1994). Rapid industrial development, urbanization and population growth have led to climate change. Water is indispensable for life, but its availability at a sustainable quality and quantity is threatened by many factors, of which climate plays a leading role. The Intergovernmental Panel on Climate Change (IPCC) defines climate as "the average weather in terms of the mean and its variability over a certain time-span and a certain area" and a statistically significant variation of the mean state of the climate or of its variability lasting for decades or longer, is referred to as climate change. Evidence is mounting that we are in a period of climate change brought about by increasing atmospheric concentrations of greenhouse gases. Atmospheric carbon dioxide levels have continually increased since the 1950s. The continuation of this phenomenon may significantly alter global and local climate characteristics, including temperature and precipitation.

Climate change can have profound effects on the water cycle through precipitation, evapotranspiration, and soil moisture with increasing temperatures. Regional impacts of climate change on hydrological processes and water resources are becoming more important. Water resource management plans increasingly need to incorporate the effects of global climate change in order to accurately predict future supplies. Groundwater is the major source of water across much of the world, particularly in rural areas in arid and semi-arid regions. Aquifers generally are replenished by effective rainfall, rivers. A change in the amount of effective rainfall will alter recharge, but so will a change in the duration of the recharge season and rainy days. Increased winter rainfall, as projected under most scenarios for mid-latitudes, generally is likely to result in increased groundwater recharge. It is to be pointed out that groundwater is the principle source of drinking water in the rural areas. About 85% of the rural water supply in India is depending on groundwater. The most optimistic assumption suggests that an average drop in groundwater level by one meter would increase India's total carbon emissions by over 1%, because the time of withdrawal of the same amount of water will increase fuel consumption. Climate change is likely to affect ground water due to changes in precipitation and evapotranspiration.

Description of Study Area

Jaipur is one of the first planned city of India, located in the semi-desert land (semi-arid region) of Rajasthan. Jaipur district, covering geographical area of 11,061.44 sq. km and extending between north latitudes 26°25' and 27° 51' and east longitudes 74°55' and 76°15'forms east-central part of the Rajasthan State. According to census 2011, total population of Jaipur district was 66,26,178 with rural population of 31,54,331 and urban population of 34,71,847 and decennial growth of 26.91% (period 2001-2011).

A distant place from Arabian Sea as well as Bay of Bengal gives rise of continental climate. During monsoon period from July to September and occasionally during rest period of the year in the wake of western disturbances humidity, cloudiness and rainfall activities increase. The summer in Jaipur is very hot while winters are extremely cold. The maximum temperatures hover at 40°C to 47°C in May. The winter minimum temperatures remain about 4°C to 9°C and fall below zero degree. Rainfall increases from the month of June when thundery activities start and July and August are the rainiest months. Monsoon generally sets around 25th June, and last up to middle of September. Rainfall decreases sharply in October and November.

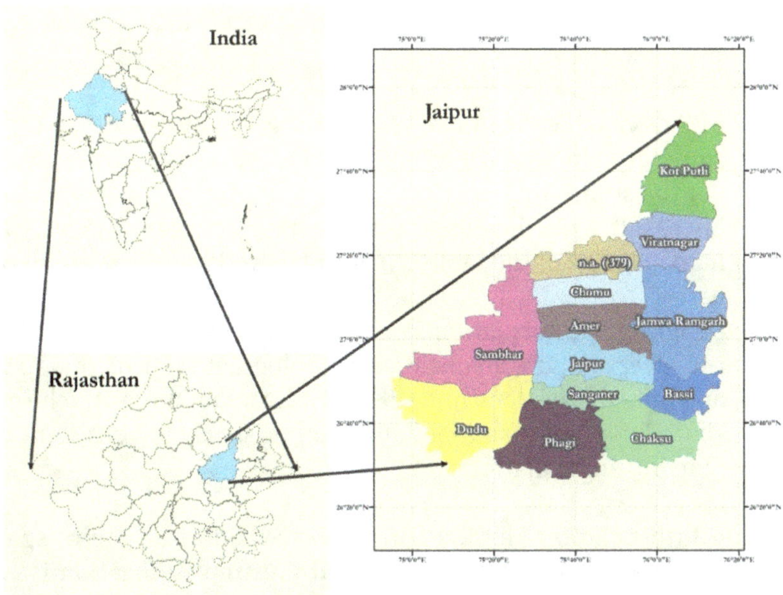

Figure 1. Location Map of Jaipur District.

Objectives, Data and Methods

The potential impacts of climate change on groundwater resources have long been recognized, although there has been comparatively little research relating to groundwater. The principle focus of climate change research with regard to groundwater has been on quantifying the likely direct impacts of changing precipitation and temperature patterns. The research objectives are:

- To develop a conceptual model of the Impact of Climate Change on Groundwater fluctuation of the study region;
- Undertake a statistical analysis to separate climate into regional and local events and determine the role of each in contributing to groundwater recharge.
- Identifying the Indian Monsoon Pattern.
- To investigate how regional and local weather events affect groundwater.
- To determine potential impacts of climate change on groundwater for the study area.
- Recommendations for groundwater management.

The following methods have been used for exploring the groundwater level fluctuations:

- GPS device has-been used for finding out the locations of piezometer and dug wells, which have used for groundwater depth estimation.
- Microsoft Excel has used for database creation, Analysis and management of the monsoon, rainfall and groundwater depth data.
- Geographic Information System (GIS) is very good platform to retrieve, analyze, manage and implementation of the spatial and non-spatial data. And GIS tool has capability to attaché non-spatial data with spatial data, then it is easy to represent in a map on single platform.

Kriging Interpolation Method can be described as a type of spatial moving average, where the value at an unsampled location is estimated as a weighted average of the known observations. The weights assigned to the known values are based on spatial trends and correlations that may be present. The Kriging method has been used for the estimation of the potential of groundwater depth by help of ArcGIS software.

Kriging weights the surrounding measured values to derive a prediction for an unmeasured location. The general formula for Kriging interpolators is formed as a weighted sum of the data:

$$\hat{Z}(s_0) = \sum_{i=1}^{N} \lambda_i Z(s_i)$$

where:
Z(si) = the measured value at the ith location
λi = an unknown weight for the measured value at the ith location
s0 = the prediction location
N = the number of measured values

Analysis & Results

1. Climate Change and Indian Monsoon

The all-India area-weighted mean summer monsoon rainfall, based on a homogeneous rainfall data set of 306 rain gauges in India, developed by the Indian Institute of Tropical Meteorology, is widely considered as a reliable index of summer monsoon activity over the Indian region. Long time series of this index since 1871 have revealed several interesting aspects of inter-annual and decadal-scale variations in the monsoon and its regional and global tele-connections.

During the period 1995-2015 (2002, 2004, 2009, 2014, 2015), there were five major drought years, defined as years with AISMR. Jaipur district also impacted by flux of Indian monsoon.

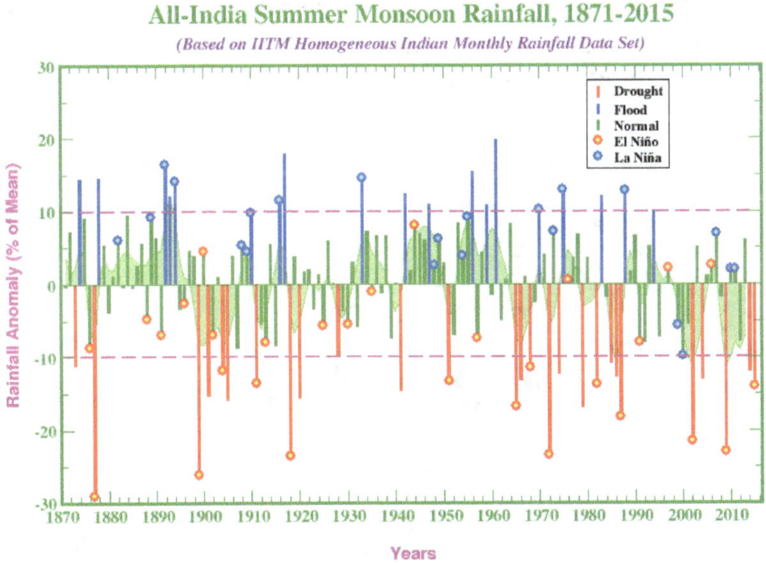

Figure 2. All-India Summer Monsoon (June-September) Rainfall (AISMR) Anomalies during 1871-2014 (Source: IITM).

2. Rainfall Pattern

The Jaipur district receives normal annual rainfall of 527mm (1901-71) while average rainfall during the period 1995 to 2015 has been 615.11 mm. Over 90% of total annual rainfall is received during monsoon. Total annual potential evapotranspiration is 1744.7mm. Annual total rainfall variation of Jaipur District during the period 1995-2015 is shown in Figure 3, which shows that the average annual rainfall is 615.11 mm with standard deviation (SD) 237.66 mm .The highest annual rainfall 1283 mm and the lowest 238 mm have been observed during the years 1995 and 2015 respectively. However, Jaipur city has experienced floods during 1995 & 2010. The district is prone to drought spells as witnessed during 1999 to 2002.

3. Impact of Rainfall Pattern on Groundwater Level

Systematic Hydrogeological Surveys in the district were first carried out by Geological Survey of India during 1965 - 1968. This survey has continued by Central Ground Water Board, after its establishment in 1972. Analysis of decadal pre-monsoon and post-monsoon water level data (May, 1995-2015) shows that almost the entire district has experienced falling trends of water level except a few isolated patches of rising trend mostly in the blocks of Dudu, Phagi and Chaksu (Figure 3f, 3i, 3j). Rising trends in the district range from 0.01 m/year to 0.04 m/year and falling trend range from 0.05 m/year to 0.34 m/year.

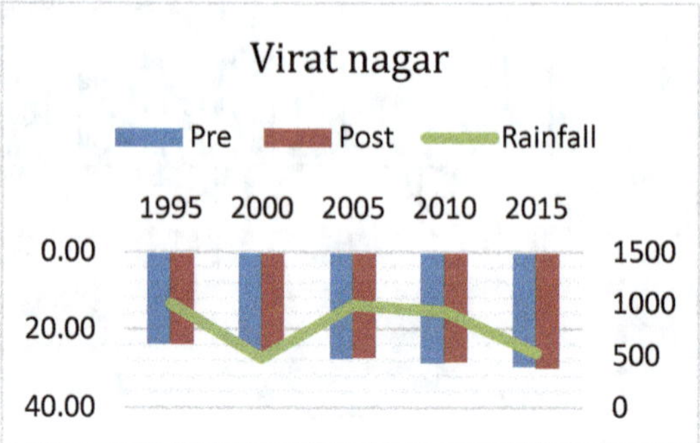

Figure 3(a)

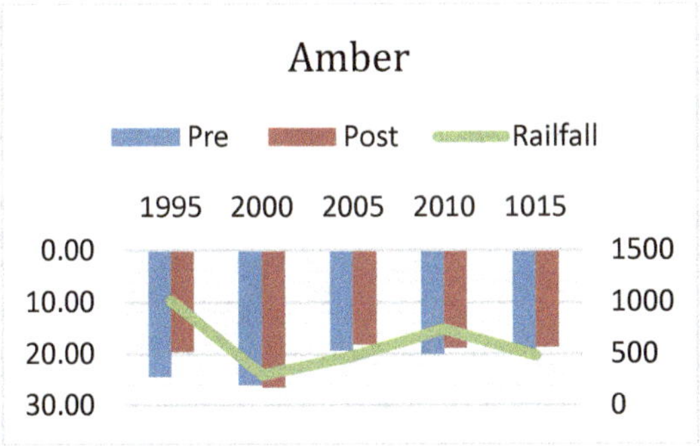

Figure 3(b)

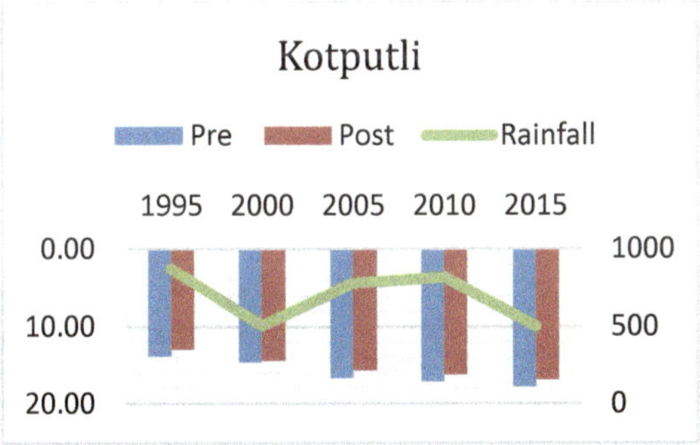

Figure 3(c)

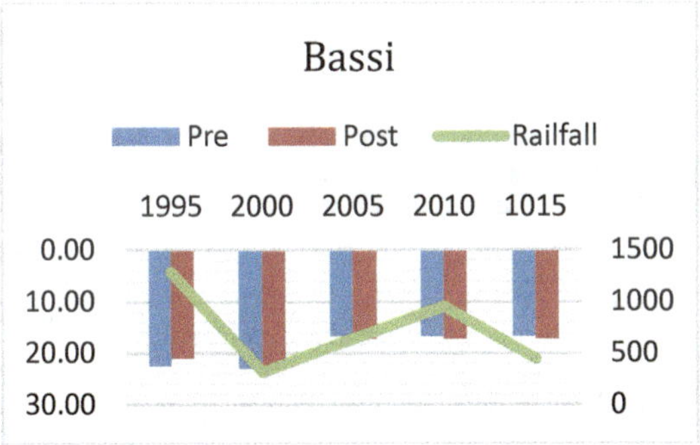

Figure 3(d)

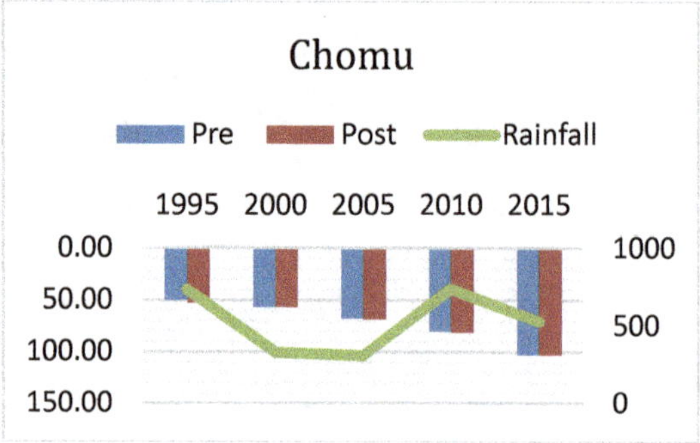

Figure 3(e)

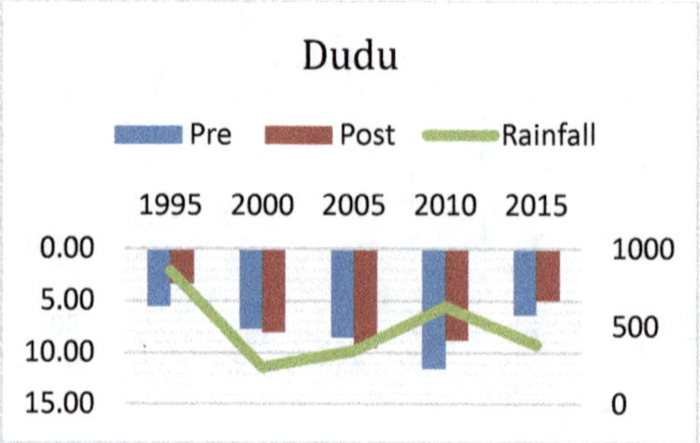

Figure 3(f)

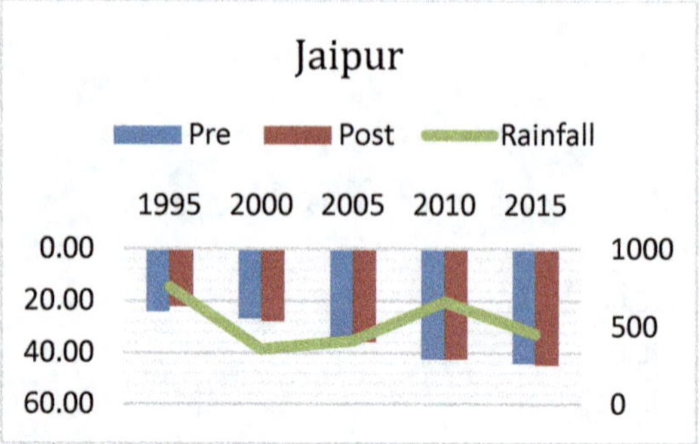

Figure 3(g)

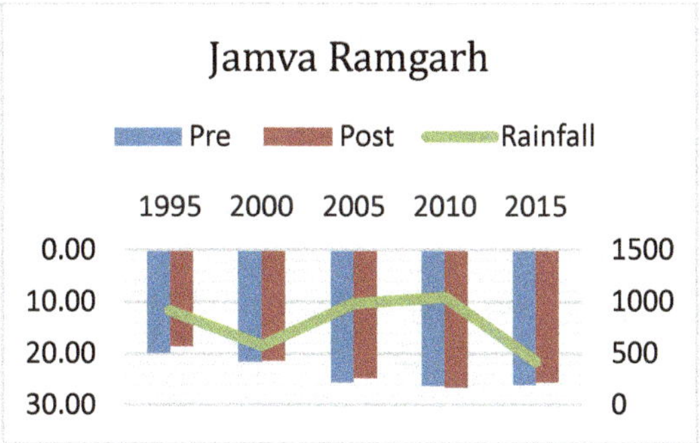

Figure 3(h)

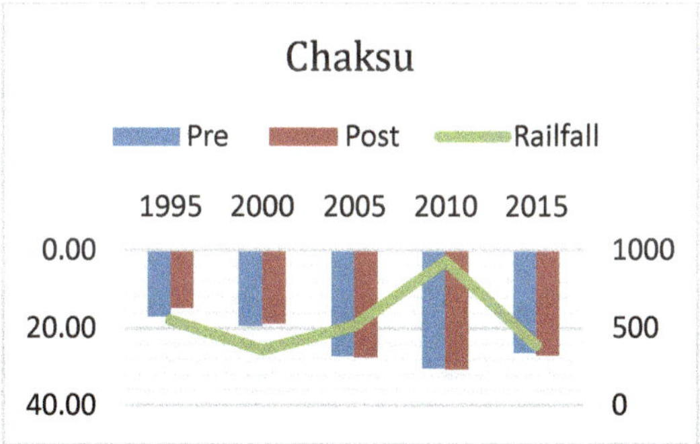

Figure 3(i)

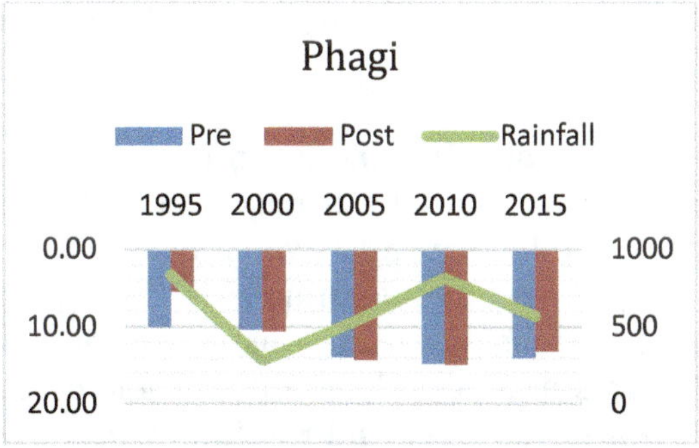

Figure 3(j)

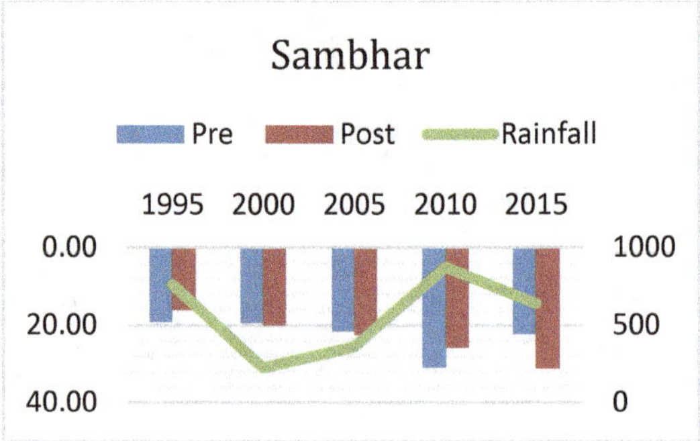

Figure 3(k)

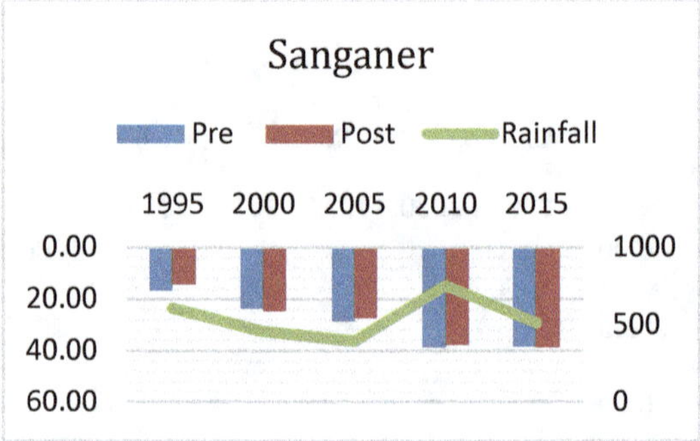

Figure 3(l)

Figure 3. Tehsil wise graphical representation of the groundwater depth and rainfall.

Long-term water level data (pre monsoon 1995-2015) have indicated declining water level trend in the district. All blocks except one block fall under over exploited category, which is necessitates regulation and control of ground water withdrawals through notification of blocks and further imposing ban on construction of ground water abstraction structures except under indispensable cases. Dudu, Sambhar, Chaksu, Jamwa Ramgarh and Kotputli blocks have affected by salinity problem. Further many areas in the district have problem of fluoride contamination in ground water because those areas facing the problem of declining water level. The maps generated for average depth of water level fluctuation, rainfall and increase/decrease in rainfall have been analyzed. The spatial variation maps of average depth of water level and fluctuation are shown in Figure 4.

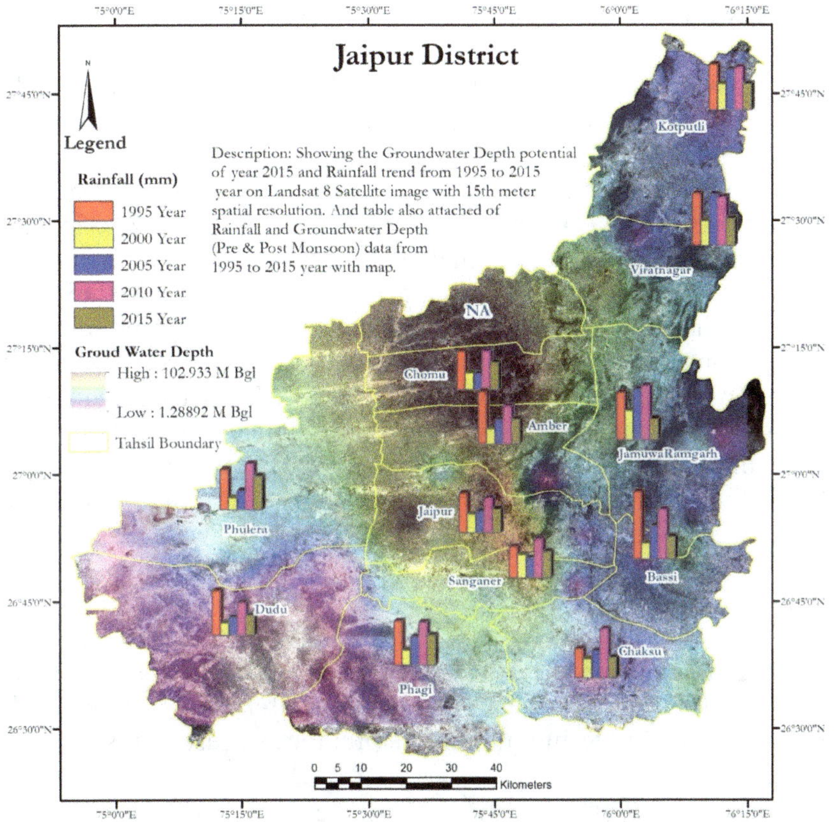

Tehsils	Rainfall in mm					Ground Water Depth									
	Year					Pre-Monsoon					Post-Monsoon				
	1995	2000	2005	2010	2015	1995	2000	2005	2010	2015	1995	2000	2005	2010	2015
Amber	1008	293	470	731	483	24.37	26.08	19.48	20.12	19.35	19.59	26.57	18.16	18.88	18.81
Bassi	1283	312	634	945	435	22.48	23.02	16.70	16.70	16.70	21.09	23.29	17.15	17.15	17.15
Chaksu	551	356	513	928	388	16.86	19.49	27.21	30.27	26.28	14.78	18.77	27.44	30.44	26.96
Chomu	733	325	306	737	523	50.66	56.85	67.60	80.00	102.50	52.92	57.24	68.60	81.30	103.00
Dudu	857	238	344	627	383.5	5.48	7.72	8.56	11.54	6.36	3.33	8.01	9.49	8.76	4.98
Jaipur	755	354	409	659	456.5	23.92	26.82	35.20	42.28	43.96	22.19	27.84	35.67	42.56	44.59
Jamuwa Ramgar	916	569	980	1039	420	19.97	21.52	25.65	26.30	26.07	18.60	21.41	24.74	26.62	25.59
Kotputli	871	502	776	816	500	13.81	14.55	16.70	17.20	17.83	12.94	14.42	15.72	16.22	16.92
Phagi	834	281	535	810	567	10.01	10.30	13.92	14.84	14.06	5.49	10.62	14.19	14.89	13.08
Phulera	760	224	361	872	639	19.23	19.43	21.57	31.00	22.32	16.08	20.15	22.57	25.90	31.25
Sanganer	604	455	392	750	512	16.70	23.62	28.74	38.86	38.39	14.25	24.69	27.41	37.68	38.63
Viratnagar	995	472	976	923	519	23.67	25.57	27.55	28.58	29.58	23.62	25.66	27.29	28.33	29.83

Figure 4. Groundwater depth potential and rainfall pattern on Landsat-8 satellite image.

Conclusion

The present study showed that although the average rainfall in the Jaipur District is above 615.11 mm, yet the average depth to water level in this region is very deep particularly during pre-monsoon in May, where the water level lies below 103 mbgl at various places. In August although the average depth to water level lies at 25.7 mbgl in Jaipur district. It is observed that by the pre-monsoon and post-monsoon. The study shows the potential of employing GIS based interpolation techniques for geospatial analysis of ground water trend for deciphering areas for groundwater development in accordance with the temporal rainfall pattern over the District. Thus, it could inferred that these regions are more prone to drought like conditions due to a constant decline in the water level particularly the places, which are facing the minimum rainfall condition where small water harvesting structures should be made to augment the ground water.

Recommendations

After detailed scientific studies, massive programme needs to be taken up for artificial recharge to groundwater using suitable techniques, Following techniques are helpful for increase the groundwater level,

- Ground water draft is very high in the District. Stage of ground water development in the district has reached 207% due to indiscriminate use. It has to be controlled by preventing further development.
- Revival of traditional rainwater storage system i.e. Baori, open wells, tanka etc. for rainwater conservation for use in day to day life will reduce groundwater draft.
- Awareness programme on rainwater harvesting will be beneficial to check decline in water level and justified use.
- Taking advantage of uneven topography of the hard rock area small water harvesting system or earthen dams, up streams of irrigation commands at suitable sites may be constructed to store rainwater. This will increase recharge of ground water which ultimately results in increase yield of wells.
- Modern agriculture management techniques have to be adopted and optimum utilization of the water resources n and High water requirement crops should be discouraged. Proper

agricultural extension services should be provided to the farmers so that they can go for alternate low water requirement economical crops.

- Pricing of ground water for irrigation use may also be considered to reduce the stress on ground water.
- Reuse of domestic wastewater for gardening, recharge etc.
- Promoting economic use of water in bathing, cleaning, cooking etc.
- Leakage from domestic taps, pipelines for water supply to urban/ rural areas be checked effectively.
- Treatment of industrial effluents so as to check pollution of fresh groundwater resources.
- Use of treated industrial wastewater for irrigation, horticulture and recharge to Groundwater.

References

Central Ground Water Board (2006). Dynamic Ground Water Resources of India. Ministry of Water Resources, Government of India, Faridabad.

Daiman A., Gupta N., (2015). Identification of groundwater recharge zone and rainwater harvesting sites using geoinformatics: A case study of karwan watershed in Sagar district of Madhya Pradesh" in ISG Journal of Geomatics, Vol. 9, ISSN: 0972-642X, pp: 198-202.

IPCC Second Assessment, Climate Change (1995). A Report of the Intergovernmental Panel on Climate Change.

Jat J.P., Gupta N., Daiman, A. (2015). "Hydrogeomorphological Studies for Exploring Ground Water Potential of Bamel Watershed in Sagar, Madhya Pradesh" in Annals of The Rajasthan Geographical Association, Volume No. XXXIl, ISSN: 0975-4652.

Mooley, D. A., and B. Parthasarathy, (1984). Fluctuations in all-India summer monsoon rainfall during 1871-1978. Climatic Change, 6, 287-301.

Parthasarathy, B., K. Rupa Kumar, and A. A. Munot (1991). Evidence of secular variations in Indian monsoon rainfall-circulation relationships. J. Climate, 4, 927-938.

Postel, S. (1999). "Pillar of Sand: Can the Irrigation Miracle last?" Norton: New York.

Singh R.P., Daiman A., Kumar A. (2014). "Spatiotemporal Characteristics of Ground Water Level Fluctuation in Jaipur Urban Area Rajasthan India" in Global Journal of Research & Review, ISSN: 2393-8854, Vol. 1, No 3.

Tirkey, A.S., Pandey A.C., and Nathawat, M.S. (2012). "Groundwater Level and Rainfall Variability Trend Analysis using GIS in parts of Jharkhand state (India) for Sustainable Management of Water Resources" International Research Journal of Environment Sciences Vol. 1(4), ISSN 2319–1414, pp: 24-31.

17

Geospatial Applications in Land Degradation and Watershed Management in Semi-Arid Areas

S. Rama Subramoniam, C. Mathuvanthi Nachiyar, H.M. Ravishankar, K.S. Ramesh and K. Ganesha Raj

Introduction

Land being a natural component of ecosystem not only refers to soil but also includes other elements such as water, flora and fauna. Land can only be considered degraded with respect to some use to which it may be put. Land degradation and desertification is a complex process which deals with both geographical and socio-economic aspects. Land degradation starts once the balance between man's use of land and the land's ability to renew its quality is disturbed. Ecosystem on Earth is affected by some form of land degradation. When land is degraded, the ecology is damaged, which may lead to serious effects in terms of soil erosion, loss of soil fertility and thus reduced plant growth or crop productivity, clogging up of rivers and drainage systems, extensive floods and water shortages.The deforestation and other agricultural practices have led to the changes in the hydrologic environment and caused widespread land degradation problems.

Most of the ecosystem services depend on water, and water management is crucial to maintaining and increasing ecosystem services. For an effective land management we should understand that the water cycle and land management are inextricably linked. In other words, it can be stated that every land use decision is a water use decision. Gains in agricultural water productivity, therefore, will only be obtained alongside improvements in land management. A global survey suggests that 40% of agricultural land is already degraded to the point that yields are greatly reduced, and a further 9% is degraded to the point that it cannot be reclaimed for productive use by farm level measures. Soil erosion, nutrient depletion and other forms of land degradation reduce water productivity and affect water availability, quality, and storage. Reversing these trends entails tackling the underlying social, economic, political and institutional drivers of unsustainable land use (Deborah Bossio, 2010). Watershed plays a major role in maintaining the fragile balance between productivity functions and conservation practices through monitoring and identification of critical areas, existing agricultural practices, crop rotation, energy efficient farming methods and reclamation of under-utilized lands. Watershed management is fast becoming the blue print for dry land agriculture development in our country. It aims at conserving soil and moisture, as well as to put the lands to use according to their capabilities to improve the overall productivity of the catchment. The major emphasis is on increase of food production, reduction in regional disparity between irrigated and dry land/rain-fed areas, increasing employment opportunities and restoration of ecological balance within watershed through integrated approach. Integrated watershed management, which involves soil and water conservation coupled with suitable crop management, is another excellent strategy for mitigating soil erosion. The goal of watershed development is to sustain the agriculture production without land degradation. Under watershed development, suitable measures are recommended for stabilization and improvement of drainage system, construction of soil and water harvesting structures, wasteland reclamation to establish vegetation cover, agro-forestry, and agro-horticulture development etc.

Land Degradation

Land degradation in arid, semiarid and dry sub-humid areas resulting from various factors, including climatic variations and human activities are termed as Desertification (UNCCD, 1994). Land degradation is caused by change in frequency and amount of rainfall, reduction in vegetal cover, wrong agricultural management practices, cultivation on marginal lands, over-exploitation of the natural resources, excessive grazing, etc. Land degradation may be referred as loss in certain intrinsic qualities or a decline in the land's capability to perform vital functions. In terms of agriculture, Land degradation can be defined as a loss in fertility of the soil which leads to decrease in production both in terms of quantity and kind. In agriculture, systematic crop residue removal without fertilization, poor cultivation practices and extensive soil salinization are examples of mismanagement that may lead to land degradation. Numerous definitions of land degradation have been suggested by various authorities to express the degree of deterioration in the potential of the land in terms of the physical, chemical and biological aspects. It has both environmental as well as economic consequences. The information on land degradation is needed for a variety of purposes such as to improve productivity levels in degraded lands through reclamation programmes, rational land use planning and also to bring more areas under cultivation. Conventional methods for assessing land degradation are either not accurate or costly. In contrast, geospatial technologies which use satellite remote sensing data and various GIS technique has shown potential to assess land degradation in addition to efficient vegetation cover mapping. A reliable, cost effective and timely method is needed to determine the extent of these problems. Hence, an appropriate action plan is required for assessing and managing land degradation which requires information on the area and the spatial distribution of the land undergoing different processes of degradation. Remote sensing data acquired in the visible and near infrared regions have shown encouraging results in providing information on spatial extent of degraded lands. By virtue of its synoptic viewing capability, repetitive coverage, and spectral sensitivity over large areas, it is a valuable tool for obtaining relevant data on degraded areas rapidly. Backed up by adequate ground truth and extra information such as topographic data, soils and geology maps as well as imageries from different dates, the accuracy of mapping can be improved to a great extent.

Land Degradation Process

Land is the most valuable natural resource, which needs to be harnessed according to its potential. Geist and Lambin (2004) carried out a worldwide review of the causes of desertification, and from 132 case studies they identified four major type of basic factors that drive Land degradation: (1) increased aridity; (2) agricultural impacts, including livestock production and crop production (3) wood extraction, and other economic plant removal; and (4) infrastructure extension, which could be separated into irrigation, roads, settlements, and extractive industry (e.g., mining, oil, gas). They concluded that only about 10% of the case studies were driven by a single cause (with about 5% due to increased aridity and 5% to agricultural impacts). About 30% of the case studies were attributable to a combination of two causes (primarily increased aridity and agricultural impacts), while the remaining cases were combinations of three or all four basic factors. From the results they identified general global patterns that lead to degradation of land. The study identified that factors causing degradation derive from underlying forces associated with particular combinations of socio-economic and biophysical factors. For example, two underlying forces, climate and technological factors were the key drivers of degradation in the majority (54%) of the case studies in southern Europe.

Due to over exploitation and mismanagement of natural resources coupled with socio-economic factors (like land shortage, inappropriate land use, severe economic pressures on farmers, poverty, population growth etc.). The problems of land degradation are on the rise. Therefore, management of land resources is essential for both continued agricultural productivity and protection for the environment. This requires inventory of degraded lands to initiate appropriate reclamation / treatment plans. The information on the extent and spatial distribution of various kinds of degradation/ degraded lands is thus essential for strategic planning of development of degraded lands. The land degradation process cycle is described in Figure 1.

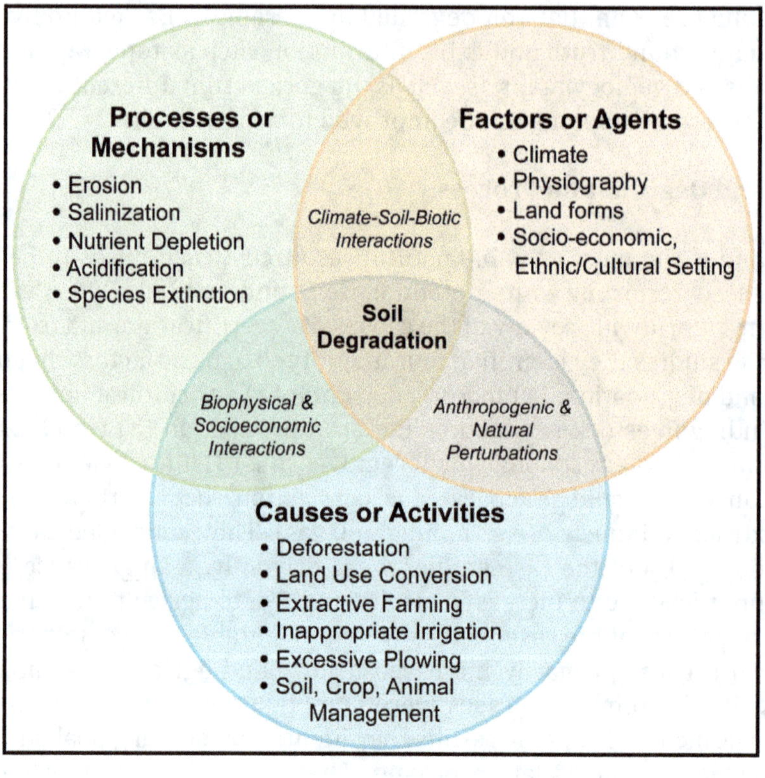

Figure 1. Land degradation process cycle.

Stages of Land Degradation

There are various stages of land degradation. Land degradation leading to loss in total productivity is not a sudden phenomenon. Rather, it tends to occur along a retrogressive pathway. In the initial stages, the reduction in potential productivity is low. 'Slight degradation' represents a persistent productivity loss of 10% (Dregne and Chou, 1992) to 15% (Sehgal and Abrol, 1994). It can be overcome by adopting appropriate management practices. If land degradation is allowed to proceed, productivity declines further, a stage is reached when conventional agronomic practices are found inadequate to regain this lost capacity. This stage of degradation is called 'moderate degradation' and corresponds to a potential productivity loss of 10 to 25% (Dregne and Chou, 1992) or 33% (Sehgal and Abrol, 1994). Only ameliorative management can restore these degraded lands. Land degradation is considered severe when the productivity loss reaches 50% to 66% (Dregne and Chou, 1992; Oldeman, 1988; and Sehgal and Abrol, 1994). Upto this point, land degradation is reversible because restoration of land is still possible, although at high costs and with major engineering works. Further degradation makes a land un-reclaimable economically. This stage of land degradation is practically irreversible. Reversible degradation process should be monitored constantly and serviced timely. Land might be degraded for one purpose and not necessarily for another. Apparently, it is use specific, management-sensitive, and thus not necessarily permanent (Johnston and Lewis, 1995).

Indicators of Land Degradation

Land degradation is not a sudden phenomenon but a gradual process. The costs of preventing land degradation are not high when action is taken early. Once severe, the land must be abandoned because its reclamation becomes economically non-viable. Indicators need to be developed that can predict the onset of land degradation. Potential indicators should meet the requirements such as quantitative, sensitive enough to give early warning of the impending change, widely applicable, able to assess the present status and trend, able to distinguish change due to natural cycles as opposed to anthropogenic interventions, and relevant to ecologically significant phenomena.

Various Estimations on Land Degradation in India

Out of 329 million ha geographical area of our country, nearly 175 ha land is subjected to some kind of land degradation (National Commission on Agriculture, 1976). Department of Agriculture and Co-operation reported 107 million hectares of area under various types of degraded lands (DAC, 1994). Another study published degraded lands increased from 130 million hectares in 1987 to 188 million in 1993 (Annon, 2002). There is no reliable data available on status of land degradation in India until recent time. The most dominant category is water erosion which covers more than 80% of degraded land. The other important categories include waterlogging, salinization/alkalization and wind erosion. NRSA / DOS reported salt affected soils in the country covers 6.73 million hectares assed based on satellite remote sensing and adequate ground truth during the period 1986-1987 (outcome of the project entitled 'Mapping of Salt Affected Soils of India'). Direct observations, parametric methods as well as remote sensing techniques have been used for assessment of land degradation on a global level (FAO, UNDP and UNEP, 1994).Recent studies indicate that the areas affected by water and wind erosion are about 150 m ha, salinity and alkalinity is about 60 m ha and shifting cultivation constitutes about 4.0 m ha respectively. Vast retches of lands suitable to be put under non-agricultural use and for reclamation for better productive use are available in the country. Not less than 37.5 m ha of such lands are expected for viable treatment. The findings of these studies highlight that remote sensing tool provides a rapid, relatively inexpensive means to gather land information where no other sources are available or where the quality of available information is low. It permits an exact delineation or clear physiographic boundaries. It gives the opportunity for repetitive, annually or seasonally multi-spectral examination.

Remote Sensing Applications in Land Degradation Studies

Multi-temporal satellite imageries from satellites like Landsat, SPOT and IRS have been operationally used in India to derive information on degraded lands, and monitor them periodically in time and space domain (Sharma and Bharghava, 1987; NRSA, 1981; Venkataratnam, 1984; Karale *et al.,* 1988; Rao *et al.,* 1991; Venkataratnam and Ravisankar, 1992). There exists a variation in the methodology adopted for the land degradation assessment by various individual/

agencies/organizations. On-screen visual interpretation of IRS P6 LISS III FCC (23.5 m spatial resolution), and subsequent field verification have led to the identification of different types of land degradation.

Land Degradation in India

According to National Commission on Agriculture (1976), about 175 million hectares of land constituting 53.3 percent of the total geographical area of 329 Mha is subject to various kinds of degradation (Ravisankar and Sreenivas, 2010). According to the latest reports of Department of Agriculture and Co-operation (DAC, 1994) 107 million hectares of area was found under various types of degraded lands (Tables 1 & 2).

Table 1. Land Degradation Status in India (Area in lakh ha.)

S.N.	Type of degradation	MOA, 1985 Assessment	MOA, 1994) Assessment
1.	Water Erosion	1071.96	571.55
2.	Wind Erosion	177.96	104.60
3.	Ravines	39.75	26.78
4.	Salt Affected	76.16	63.23
5.	Waterlogging	85.26	31.97
6.	Mines & Quarry Wastes	--	2.53
7.	Shifting Cultivation	49.12	23.78
8.	Degraded Forests	194.94	248.97
9.	Special Problems: Coastal Sandy Areas *Included in Other Types	27.30 14.65	0.89 --
	TOTAL	1736.40	1074.30

(Draft Report on status of Land Degradation in India, Department of Agriculture and Co-operation, 1994)

Table 2. Statistics on Degraded Lands in India

Agencies/Organisations	Mha
1. National Commission on Agriculture, Govt. of India (1976)	175.0
2. National Bureau of Soil Survey and Land Use Planning (1994)	187.7
3. Department of Land Resources (2000)	63.8

Table 3. Methodologies used for land degradation assessment

Year/ organisa-tions	Input data used Source of Information	Criteria selected	Scale and output
1977, 1983, 1986	Global maps and statistics	State of vegetation, state of soil erosion, state of salinization decrease of crop yields	Global/Continental, statistical tables small scale maps
1986	NOAA (AVHRR)	Vegetation density & productivity	Regional Small scale maps
1986	Satellite images/ ground surveys	Degree of soil salinity, soil salinity	Statistical tables; Regional scale
1977, 1978	Small scale soil maps statistics, soil descriptions	Soil salinity, degree of soil aridity	Small scale maps/global
1981, 1982	Statistical infor-mation	Erosion, Salinization	Statistical tables at Global/regional level
1984	Satellite images and aerial photo-graphs, ground surveys	Status of vegetation, erosion, salinization	Medium scale maps statistics, regional scale
NRSA, 1990	Satellite imagery and ground truth studies	Degree of Soil erosion	Small scale Maps/regional level
NRSA, 1995	Satellite imagery and Ground truth studies	Degree/extent of salin-ity and/or sodicity	Small scale maps/National level

NRSA, 1996	Satellite imagery and Ground truth studies	Salinity/Sodicity	Medium Scale Regional level
NRSC,2010	Satellite imagery and Ground truth studies	Erosion/salinity/Alkalinity	Medium scale/National Level

Mapping Land Degradation

The procedure involved in mapping land degradation using satellite data which consists of input data, preparatory work and methodology. The methodology normally adopted for mapping at any scale consists of preparation of base map, on-line visual interpretation of satellite data, development of legend, ground truth collection, analysis of soil samples, classification of degradation classes and finalization of maps in light of field information and analytical data. For delineation and mapping of land degradation classes, multi-temporal geo-rectified data acquired during major cropping seasons like kharif, rabi and zaid seasons are used. Temporal data set helps to address the seasonal variability in intensity of problem in degraded lands. The satellite data is geometrically corrected for further processing, like visual interpretation, ground truth collection and land degradation map preparation. The general methodology adopted for national level degradation mapping at 50K Scale using LISS 3 data is depicted in Figure 2 and details are presented in Table 3.

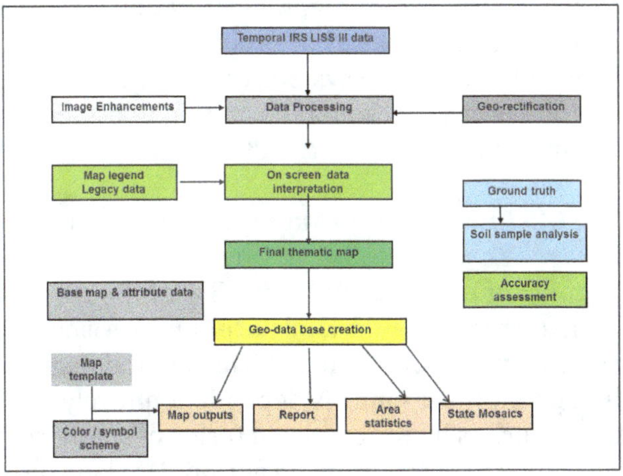

Figure 2. Flow chart showing methodology of Land degradation mapping.

Watershed Approach to Reclaim the Land Degradation

There are salient mitigation techniques for reversing land degradation in India one of which includes watershed approach. Integrated watershed management, which involves soil and water conservation coupled with suitable crop management, is another excellent strategy for mitigating soil erosion. Development and management of watershed resources to achieve optimum production without causing deterioration in the resources base is integrated watershed management.

Watershed as a unit for developmental planning is the natural choice of present time. The sustainable development of natural resources in a watershed is based on maintaining the fragile balance between productivity functions and conservation practices through monitoring and identification of critical areas, existing agricultural practices, crop rotation, energy efficient farming methods and reclamation of underutilized lands. Watershed management is fast becoming the blue print for dry land agriculture development in our country. This program aims at conserving soil and moisture, as well as to put the lands to use according to their capabilities to improve the overall productivity of the catchment. The major emphasis is on increase of food production, reduction in regional disparity between irrigated and dry land/rain-fed areas, increasing employment opportunities and restoration of ecological balance within watershed through integrated approach. The goal of watershed development is to sustain the agriculture production without land degradation. Under watershed development, suitable measures are recommended for stabilization and improvement of drainage system, construction of soil and water harvesting structures, wasteland reclamation to establish vegetation cover, agro-forestry, and agro-horticulture development etc.

Role of RS & GIS in Watershed Management in Semi-Arid Areas

Dry land farming in India accounts for 63% of the cultivated land of 144 million hectare. These drylands are impoverished, deficient in plant nutrients; suffer from various forms of land degradation. The crop production on these lands is dependent entirely on the natural precipitation, which is highly erratic in terms of spatial and temporal distribution during the crop-growing period. Dry land farming is also affected by socio-economic conditions of smallholdings, expensive credit, low fertilizer use, poor infrastructure, occurrence of frequent droughts, non-adoption of improved cultivation practices and crop varieties etc. Realizing the importance of dryland development both Government of India and Government of Karnataka have accorded high priority for development of rain-fed areas through adoption of various watershed development projects such as DPAP, NWDPRA, JRY, Hill Area development Program, Sujala - a World Bank Assisted concurrent Monitoring Programme etc.

As the watershed development project involves the efforts of various departments and considerable budget, we advocate use of remote sensing and GIS as useful tools in prioritizing the watersheds, developing natural resources database and generating action plan maps in GIS environment and finally as a monitoring and evaluating tool. For effective planning and implementation of watershed development, information on nature, extent, magnitude and spatial distribution of land, including its potential and limitations and temporal behavior is paramount important. Remote sensing data hold great potential for deriving timely and reliable information.

With the launch of Indigenous Indian Remote Sensing (IRS) satellites, satellite data is available from 1986 onwards. Currently, IRS 1A/1B, IRS P2, IRS-P3, IRS-1C/1D, Resourcesat, Cartosat, Landsat, spot satellite data are available. The spatial resolution has significantly improved and the spectral coverage increased to cover middle infrared thermal and microwave wavelength regions and receptivity increased with multiple satellites. Microwave remote sensing is helping to see through clouds enabling monitoring through monsoon clouds. The availability of high resolution data with the capability to provide unbiased, holistic and synoptic view has facilitated the generation of basic inputs required for taking up prioritization of watersheds, generation of natural resources database in GIS environment and action plans for watershed development and impact assessment and evaluation of implementation of watershed programmes. As demonstrated under many national level projects and other user funded projects, the satellite data provides required/essential information on current land use/land cover, hydro-geomorphic conditions, soil characteristics, status of drainage and surface water bodies etc., which are vital for watershed management. The IRS 1C/1D provides multispectral LISS III data with 23 m and PAN data with 5.8 m resolution. The composite hybrid data of (LISS III + PAN) not only improves identification of features but also helps in mapping at cadastral level providing detailed information on 1:12,500 scales. Thus, the high-resolution data provides ample scope to monitor the activities at farm level with frequent intervals for tracking implementation, apply mid-course corrections and assessing effectiveness of implementation.

Over exploitation of natural resources for meeting requirements of ever growing population for food, fuel and fiber has led to serious environmental degradation. Sustainable development calls for optimal utilization of available natural resources based on their potential and limitations.

Evaluation of Watershed Management Projects

A. Prioritization of watersheds

The availability of high resolution, multi-sensor and multi-frequency satellite data with the capability to provide un-biased, holistic and synoptic view has facilitated the generation of basic inputs required for taking up developmental works on watershed basis as well as it has provided a means for monitoring the implementation works in the watersheds. As demonstrated in many studies the satellite data provides required/essential information on current landuse/land cover, hydro-geo morphological conditions, soil characteristics, status of drainage and surface waterbodies, slope and aspect, etc., which are vital for the prioritization of watershed in a given study area.

Watershed approach of conserving and developing the natural resources of an area or a region is accepted as the most appropriate method. State/District administration are carrying out developmental works on watershed basis under several schemes like JRY, DPAP, NWDPRA, RVP" etc. In the changed scenario of equity, it is mandatory to identify at least 2 to 3 watershed s of size 500-1000 ha in each taluk every year and treat them over a period of two to three years rather than selecting a large single 20,000 ha sub watershed and treat over a period of 5-6 years. The district administration finds it difficult to identify such critical watersheds in each taluk and hence desired to use remote sensing and GIS techniques to get a quicker and reliable solution of prioritizing the watersheds on scientific basis. It is in this connection, the present study was initiated by the then Director, ADP, RD & for Dakshina Kannada district of Karnataka envisaging the use of satellite technology for prioritizing the micro watersheds. The project was sponsored by the Planning Commission, Govt. of India.

In the present study, IRS 1C/1D LISS III satellite data has been used along with collateral data available for generating information on current land use/land cover, hydro-geo-morphological conditions, soil characteristics, status of drainage and surface water bodies, slope & aspect etc., which are vital for the prioritization watershed in a given area.

The methodology adopted consists of two steps:

- delineation of watershed s from basin level to micro-watershed level (500 ha) and coding each micro-watershed using alpha-numeric coding system and
- prioritization of micro-watersheds using an elimination technique & considering the factors like current vegetation cover, wastelands, soil types, erosion status, slope, scope for development etc.

Each of these factors are given weightages in spatial domain using GIS to arrive at five categories like viz., very high, high, medium, low and very low priority micro-watersheds. The Management Information System (MIS) for entire district is developed using Arc-Info GIS which helps in keeping track of the status of implementation programmes. Fifty four watersheds with at least 2-3 in each taluk have been identified as critical watersheds in Dakshina Kannada district which require immediate attention. Currently about 19 such watersheds have been identified for treatment during the current year.

Criteria Used

An elimination technique was adopted:

1. Micro-watersheds/Sub-watersheds with forest cover > 80% are demarcated as low or very low priority watersheds.
2. The satellite data in conjunction with the SOI topo-maps are used to estimate the aerial extent of wasteland in the watersheds.
3. The watersheds with vegetation (forest/agriculture) > 50% were identified and checked for erosion status, scope for devel-

opment etc., and classified as low, medium or high priority watersheds.

4. Then the watershed with open/degraded/uplands with an area of > 20% to 60% was further studied with respect to soils of the area, geology, slope, present landuse/landcover ground water potential, scope for future development, erosion status qualitatively and relatively (at Micro-watershed and Sub watershed level) and classified them as high or very high priority watersheds.

The maps were generated taluk-wise for all the 8 taluks. 7 to 8 critical priority watersheds are identified for immediate treatment in each of the taluk.

The outputs provided/generated include:

- The digital mosaic of entire district generated using IRS 1C LISS III (3 scenes) on 1:250,000 scale,
- Taluk-wise maps showing drainage, watershed boundaries (from basin level to micro-watershed levels) and priority levels indicated as very low, low, medium, high and very high levels,
- An MIS of entire district showing taluk-wise details and
- All the layers are digitized and are available in Arc-info environment at NRDMS Centre, Mangalore.

All the maps were displayed in a one-day workshop conducted at Mangalore ZP's office and discussions were held amongst scientists of RRSSC, NRDMS and officers of line departments of Dakshina Kannada. All the suggestions emerged out were incorporated in the maps and maps were finalized.

A Management Information System (MIS) for entire district at taluk level was generated. It provides for each micro watershed the information on to which taluk it belongs, to which region/basin /sub-basin/ sub-catchment/ major watershed/sub-watershed it belongs, its geographical area etc. The MIS helps in keeping track of the treated watersheds over a period of time.

The taluk wise maps have been provided to CEO, ZP, and Dakshina Kannada district. Now ZP has taken up 19 watersheds (i.e. 2 watersheds per taluk) and are implementing watershed development plans using the fund received by them. There is a plan to implement watershed development programmes in all the 54 critically identified watersheds under this project over a period of 5 years. After this ZP Kodagu have come forward with a request to generate priority map to the entire district and the work is in final stages. Similarly there are request coming from ZP's of Mysore, Mandya, Davanagere, Raichur etc., for doing the prioritization of watersheds in their respective districts.

B. Natural resources data base preparation and action plans generation

The information on natural resources with regard to their nature, extent, spatial distribution and potential and limitation is a pre-requisite for optimal landuse planning and for initiating any other developmental activities at district/taluk/watershed level. The optimal land utilization of available land and water resources based on their potential and limitation is a key to sustainable development in a watershed. Space borne multispectral data is available since Landsat – 1 in 1972, for generating information on natural resources and subsequently for preparing optimal landuse. By virtue of fairly large area coverage at a frequent interval, the satellite data have immense potential for providing timely, reliable and cost effective information on various natural resources and environment.

The methodology adopted involves generation of thematic maps showing current land use/land cover, types of wastelands, forest cover/types, surface water resources, drainage system/pattern, potential ground water zones, land forms (geomorphology), geology (rock types, structural features, mineral occurrence), soil resources maps (soil type, irritability, land capability, crop suitability maps) etc., using data from Indian Remote Sensing Satellite data (IRS). In addition maps like slope, village location, boundary and transport network maps were also generated using SOI and other collateral information. Socio-cultural, socio-economic, demographic and meteorological data were collected from available existing databases. By integrating the information in a GIS environment, derivative maps called composite mapping units showing resource availability of the individual land parcels, land characteristics, status of land erosion/limitations etc., are prepared. These maps are validated through the adequate field checks and by using the authentic existing information. Specific developmental plans in the form of action plans for water resources development and alternate land use development were generated on watershed basis in consultation and close coordination between space scientists, state line department/watershed officials, agricultural universities/research centers and local farmers so as to ensure the technical feasibility and cultural acceptability.

Under the National project "Integrated Mission for Sustainable Development" coordinated by National Remote Sensing Agency, Hyderabad in collaboration with State Remote Sensing Service Centers, Regional Remote Sensing Service Centers, Private Agencies and other Central/State Organizations, the natural resources data base creation and action plan generation for taking up developmental work was successfully completed for 19 districts completely and one block in each of the 80 districts identified by Ministry of Rural development. The GIS data creation for identified districts in Arc-Info environment for most of these districts and blocks is undertaken under the National project "Natural Resources Information System" through the participation of concerned State Remote Sensing Centers and Regional Remote Sensing Service Centers of ISRO.

The hard copy outputs on all natural resources and action plan maps have been handed over to the district administration for further utilization. As desired by many district administration in many districts/taluk, watershed wise maps on natural resources and action plans were generated and handed over to them for taking up watershed development programmes.

The implementation of such action plans have been reported from many taluk/districts from the states of Karnataka, Andhra Pradesh, Gujarat, Rajasthan, Uttar Pradesh, Maharashtra, Madhya Pradesh, Orissa etc., on watershed basis. In Bijapur district, Karnataka at least 8 watersheds have been identified and implementation of watershed programmes were entrusted to NGO's by Zila Panchayat, Bijapur using the natural resources data base created under IMSD, action plans were generated in close interaction with the Remote Sensing scientists, department officials, scientists from Research centers, field functionaries, etc. The same action plans for water resource development and alternate agriculture resources development generated through RS & GIS were successfully implemented and monitored.

Watershed Monitoring

The repetitive coverage of the satellite provides us an excellent opportunity to monitor the land resources and evaluate the land cover changes and its impact on environment through the comparison of images acquired for the same area before and after the treatment period. Changes that could be derived through satellite data include

- Cropped area (change in areas extent of agricultural crops, cropping pattern, extent of wet land/irrigated croplands)
- Plantations (increase/decrease under horticulture and forest plantations)
- Wastelands (change in areas extent),
- Alternate land use (switchover from marginal lands to agro-horticulture/agro forestry),
- Water body (change in no. and areal spread) and
- Biomass (overall change in productivity).

However, the socio-economic indicators like land value, employment opportunities, changes in ground water levels, agro based and allied industries, migration of labor and cattle etc., are collected from field survey. This monitoring and evaluation study would provide insight in to the impact of developmental activities in the watershed and also provides feedback for making necessary modifications, plan revisions and preparation of fresh watershed developmental projects.

The digital analysis of satellite data involved acquisition of post and pre-treatment satellite data, loading of satellite data, registration with respect to SOI map, digitization of watershed boundary and cultural boundaries, extraction of study area, ground truth data collection and classification of post and pre-treatment satellite data, comparison of land use and land cover and NDVI (Normalized difference vegetation index images) depicting biomass changes, generation of statistics and output products. Monitoring one watershed (Cauvery Watershed) using remote sensing depicted in Figure 3 (LULC Change) and Figure 4 (Biomass Change) respectively.

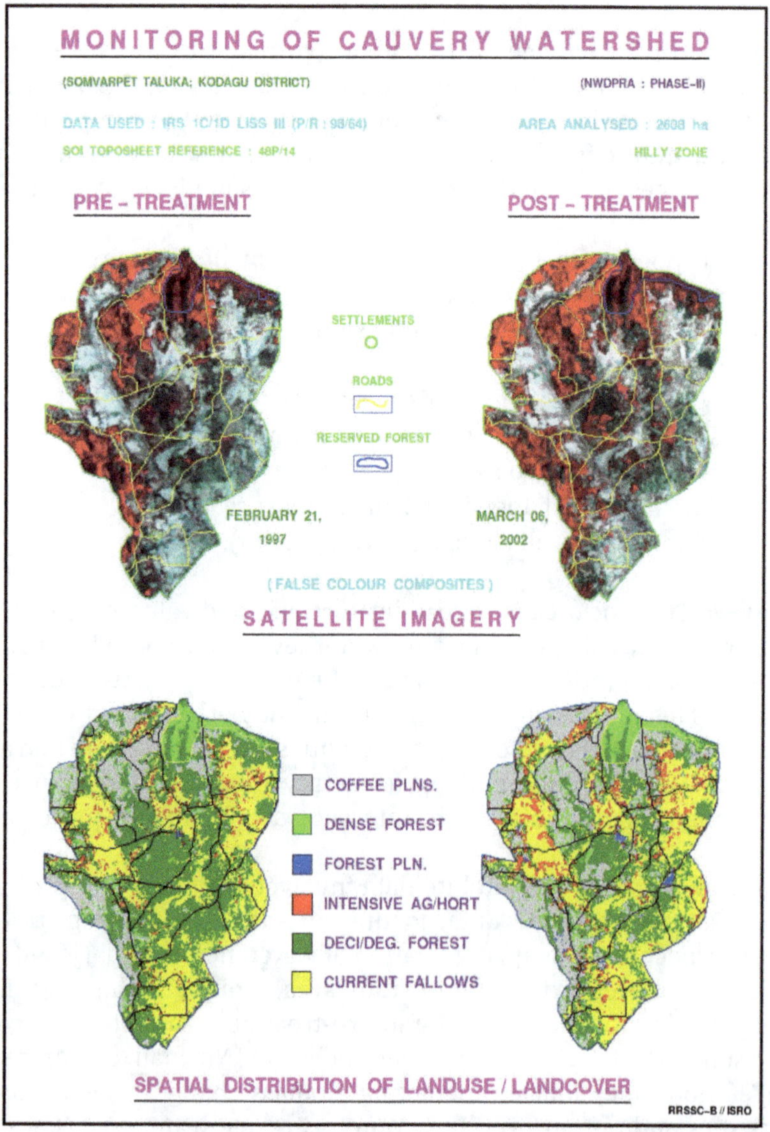

Figure 3. Watershed Change analysis with Land use/Land cover.

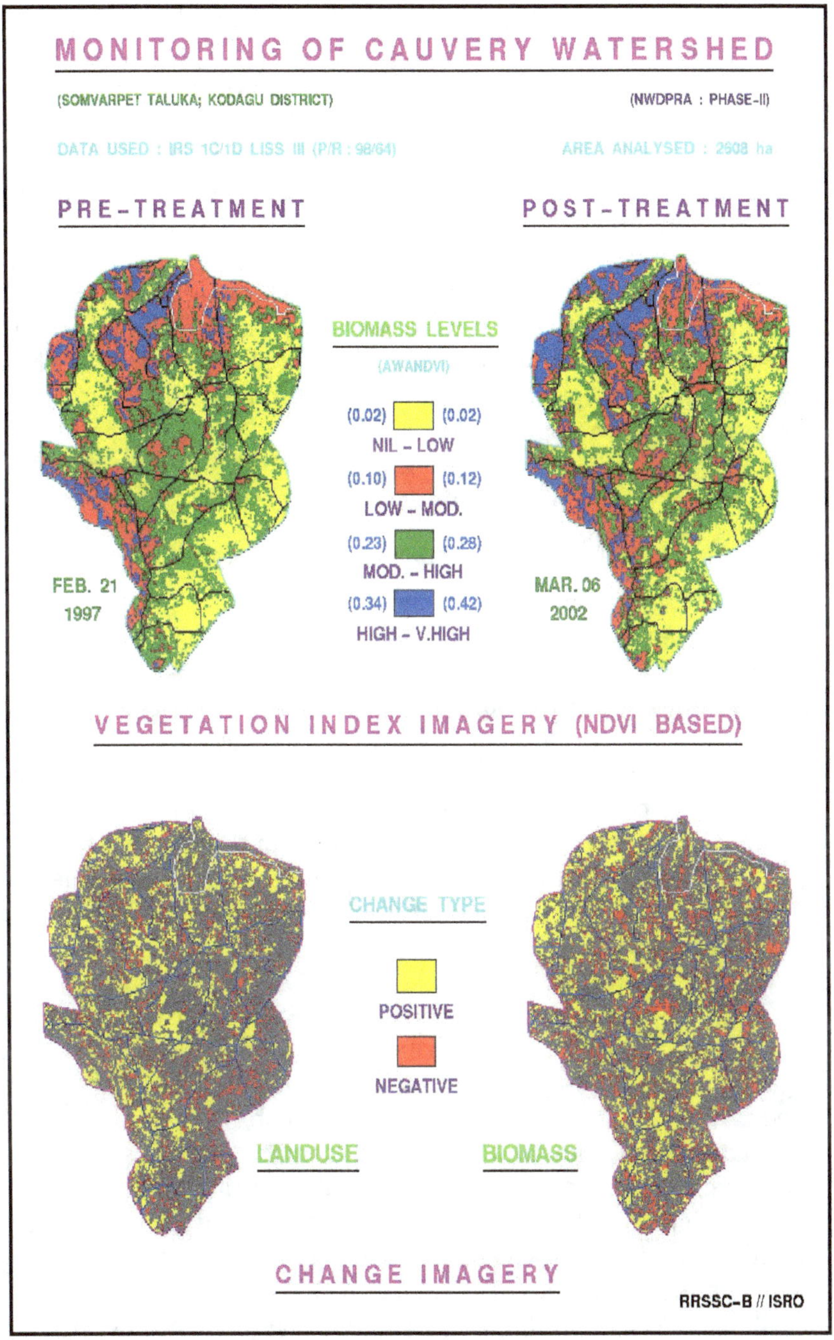

Figure 4. Watershed Change analysis with Bio mass.

The study conducted over 19 watersheds indicated increased area under forest/horticulture plantation, expansion of area under cultivation, change in cropping pattern, increase in the irrigated area, alternate landuse practices (i.e. increase in area under agro-forestry/agro-horticulture), more number of water bodies, change in water spread area, reclamation of wastelands for productive use, more area under fruit crops and forest.

The experience indicates that the remote sensing technology is quite cost effective to assess the status of watershed development. It is observed that the cost of monitoring and status assessment is about one-fifth the cost of such assessment using conventional techniques. The watershed development authorities find RS & GIS techniques as the appropriate tool to monitor and evaluate the impact of watershed development programmes, as provides more accurate, timely reliable and objective information.

The overall improvement in the land and water resources development has been observed varying from 7 to 27% in the 20 watersheds of Karnataka. The success is evident at places, perhaps in favorable climatic zones or due to additional inputs, commitments and conservation measures. Because of other causes like lack of co-ordination, integrated approach, community participation and improper appreciation of the concept, the visible changes are not proportionate to the inputs in cumulating. The performances of these watersheds are found to be better in dryland rain-fed farming areas. Watersheds of the medium rainfall zones had given good results than high or low rainfall regions.

The Rain-fed Farming Systems Division (RFSD), Ministry of Agriculture and Cooperation, New Delhi entrusted a work of monitoring and impact evaluation of 60 treated NWDPRA watersheds using Remote Sensing and GIS in the states of Madhya Pradesh, Maharashtra, Orissa, Rajasthan, Tamil Nadu and Uttar Pradesh. The study also involved assessment of implementation of NWDPRA developmental activities and their impact on both arable and non-arable lands as well as agro-climatic zone-wise analysis of the performance of developmental activities in the watersheds. The entire work was completed by involving all five RRSSCs and the state wise reports have been submitted to the RFSD, Ministry of Agriculture, Government of India, and New Delhi. Subsequently, RFSD, Ministry of Agriculture, New Delhi have accepted and have requested RRSSC to further conduct monitoring and impact evaluation of another 60 watersheds in the states of Andhra Pradesh, Gujarat, Haryana, Rajasthan, Orissa, Uttaranchal, Uttar Pradesh, West Bengal and Tamil Nadu using RS and GIS in the II phase.

The impact evaluation carried out for 11 watersheds in Tamil Nadu using RS & GIS in Phase I by RRSSC-B indicated an overall change from as low as 2.37% to 27.38%. The satellite based monitoring and evaluation of selected treated NWDPRA watersheds of Tamil Nadu indicates increased area under cultivation both dry and irrigated, horticulture agro-horticulture/agro-forestry, more number of water bodies, and reclamation of wastelands for productive use. Fewer efforts are made in forestry sector. Though the area of implementation selected is relatively big, the implementation is limited to small areas and very much scattered in the watershed area. The concept of watershed is lacking and most of the implementation works are carried out based on sectored approach. Similarly, the monitoring and evaluation work for atleast 10-12 watershed have been completed in the states of Madhya Pradesh, Maharashtra, Orissa, Uttar Pradesh, Rajasthan and state wise reports have been submitted indicating the overall performance of individual watersheds.

The studies so far carried out indicates that there is a greater need to stream line the implementation process by increasing manpower of the department, better coordination between identified interdisciplinary team from various participating departments, utilization of better maps for preparation of implementation plans and record keeping and greater need for protection, maintenance and conservation of developmental works. In Karnataka a sound beginning has been made in many watershed programmes like World Bank aided 'SUJALA' watershed development programme. In this project NGO's and local committees are involved in actual plan preparation and implementation making use of Remote Sensing derived maps as well as GIS for integrating all the information and monitoring the entire process of implementation at cadastral level. This approach makes the watershed management more meaningful and productive.

Conclusion

Land is a precious resources on which all life depends. In such cases when land is degraded it affects the whole ecosystem. We should fight the land degradation menace by the use of available conservation technologies. Satellite remote sensing has the capability of providing adequate information for assessment, monitoring, management and control of land degradation processes which are dynamic and temporal in nature. People's participation and co-operation is must in organizing soil and water conservation, especially in dealing with fragmented and scattered land holdings. There is a great need to educate the farmers on the benefits of soil and water. In spite of the fact that erosion control procedures developed at experimental stations in different countries have proved to be quite effective, their wide application is hampered because of a number of financial and socio-economic reasons. However, the role of prevailing climatic variations and influence of global warming in intensifying land degradation need to be considered when designing strategies for combating it. Results shows that with proper land use planning and use of integrated soil, water and crop management option the land degradation can be controlled. Promising watershed management technologies developed at the project sites provide a good framework for increasing productivity and income of farmers on sustained basis, while improving the soil and water resources. Finally, it can be concluded that the availability of high resolution satellite data at frequent intervals provides sufficient timely reliable information in generation of natural resources data base and action plans in GIS environment and finally for monitoring and impact evaluation of land degradation and watershed monitoring. The technology allows planner to evaluate the development programmes continuously and apply mid-course corrections if required.

References

Annon, (2002). Agriculture inputs and Soil Degradation. http://www.teriin.org/reports/rep02/repo2o6.thml

Barrow C. J. (1991). Land Degradation: Development and Breakdown of Terrestrial Environments, *Cambridge University Press*, Cambridge, UK.

DAC, (1994). Draft Report on status of Land Degradation in India, Department of Agriculture and Co-operation.

Deborah Bossio, Kim Gehe, William Critchley (2010). Managing water by managing land: Addressing land degradation to improve water productivity and rural livelihoods, *Agricultural Water Management*, Volume 97, Issue 4, P: 536-542. https://doi.org/10.1016/j.agwat.2008.12.001

Desertification and Land Degradation Atlas of India (Based on IRS AWiFS data of 2011-13 and 2003-05), Space Applications Centre, ISRO, Ahmedabad, India, P:1-219.

Dregne H. E. and Chou Nan-Ting (1992). Global desertification dimensions In: Degradation and Restoration of Arid Lands, *International Centre for Arid and Semiarid Studies*, Texas Tech University, Lubbock, Texas, USA, P: 249-282.

Dregne H. E. (1978). Desertification: man's abuse of the land, *Journal of Soil and Water Conservation,* Volume 33, P: 11-14.

FAO (undated on-line), The problem of land degradation, Available on-line *www.fao.org/docrep/v9909e/v9909e02.htm*.

FAO (1976). A Framework for Land Evaluation, *FAO Soils Bulletin 32*, Rome, Italy.

Helmut Geist J, Eric Lambin F. (2004). Dynamic causal patterns of Desertification, *Bio-Science*, Volume 54, Issue 9, P: 817-829. https://doi.org/10.1641/0006-3568(2004)054[0817:DCPOD]2.0.CO;2

Johnson Douglas L & Lewis Laurence A. (1995). *Land degradation: creation and destruction*, Oxford, UK, Cambridge, USA, Blackwell.

Karale, R.L., Saini, K.M., & Narula, K.K. (1988). Mapping and monitoring ravines using remotely sensed data, *Journal Soil Water Conservation India,* Volume 31, issue (1, 2), P: 76.

National Commission on Agriculture (1976). *National Commission on Agriculture*, Govt. of India, Ministry of Agriculture and Irrigation, Dept. of Agriculture in New Delhi.

NRSA (1981). Satellite remote sensing survey for soil and land use in part of Uttar Pradesh, Project Report, *National Remote Sensing Agen-*

cy, Hyderabad.

Oldeman L. R. (1988). Global Assessment of Soil Degradation (GLASOD), Guidelines for General Assessment of Status of Human Induced Soil Degradation, *International Soil Reference and Information Center*, Wageningen, The Netherlands.

Rao, B.R.M, Dwivedi, R.S, Venkataratnam L, Ravi Sankar T, Thammappa S.S, Bhargawa G.P, and Singh A.N. (1991). Mapping the magnitude of sodicity in part of Indo-Gangetic Plains of Uttar Pradesh, using Landsat-TM data, *International Journal Remote Sensing* Volume 12, Issue 3, P: 419-425.

Ravisankar, T. and Sreenivas, K. (2010). Soil and Land degradation. *Remote sensing Applications.* NRSC, Hyderabad.

Sehgal, J. L. and Abrol, I. P. (1994). Soil Degradation in India: Status and Impact. *Oxford and IBH Publishing Co. Pvt. Ltd*, New Delhi, India.

Sharma, R.C. and Bhargava, G.P. (1987). Operational use of SPOT-1 image in mapping and management of alkali soils, *Proc. Nat. Sympo. Remote Sensing in Land Transformation and Management*, Hyderabad.

UNCCD (1994). Elaboration of an international convention to combat desertification in countries experiencing serious drought and /or desertification, particularly in Africa, Paris: General Assembly of the United Nations, P: 58.

UNDP, FAO and UNEP (1993). *Land degradation in South Asia: its severity, causes, and effects upon the people [World soils research report no. 78].* Rome: FAO, P: 100.

UNEP, (1992), World Atlas of Desertification, Edward Arnold, Sevenoaks, UK.

Venkataratnam, L. (1984). Monitoring and managing soil and land resources in India using remotely sensed data, *Proc. 3rd Asian Agricultural Symp.* Chiang Mai, Thailand.

Venkataratnam, L., and Ravisankar, T. (1992). Digital analysis of Landsat TM data for assessing degraded lands, *Remote sensing applications and Geographical Information Systems – Recent Trends, Tata McGraw Hill publishing Company Ltd.*, New Delhi, P: 87-190.

18

Utilization of GIS Integrated with Asset Management System for Public Water Utilities

Ir. Arpat Suparyatna

Introduction

Water is the main requirement for every human being, for bathing, washing, cooking and other important needs. Clean water needs are increasing along with population and environmental growth. To maintain the quality of water and the availability of clean water, this means the importance of sustainable management of clean water and uniform supply for all customers. Clean Water Company provides clean water and distributes it to its customers. Clean water companies must be able to do water management well by maintaining quality, quantity and continuity. To maintain the continuity of water availability, a good management system is needed. One of them is doing a good asset management system so that the utility assets owned can be controlled and monitored both in the process and development.

Clean Water Supply Asset Management System

Why do we need Asset Management? Some considerations that cause us to need asset management are:

- Demand for drinking water supply needs has increased steadily as population increases
- Limited resources
- Age of infrastructure
- The decreasing efficiency of production and distribution
- Target coverage of services that are not affordable
- Continuity of water management business

To fulfill that, all new investments that are very expensive and not optimal are needed. Therefore, a more reliable asset management system is needed to monitor the movement of assets and to meet the needs of providing clean water.

GIS Technology for Public Water Utilities

At present GIS technology has been widely used to assist inventory in managing asset information in the clean water management utility business. What are the functions of GIS in the clean water management utility business?

1. Monitor changes in pipeline network assets either from changes in length, diameter, or because of changes due to the age of the pipeline network.
2. Monitor the addition and movement of customer meters.
3. Mapping all asset locations owned, making it easier in planning asset development for either infrastructure or customers.
4. As a basis in planning the design and needs of the master plan
5. Providing information to customers regarding affected areas due to network leaks, information related to consumption of customer usage, etc.
6. Monitoring the location of pipeline network leaks and quick repairs.
7. Monitoring the location of water quality in each network (can be used as monitoring the water content of each network point in all locations, making it easier to carry out the analysis).
8. To model the pipeline network plan by utilizing GIS as a data model (Hydraulic Model).
9. Monitoring elevation due to infrastructure development in a management area.

There are many more that can be utilized from GIS for clean water management needs. In terms of utilizing GIS for Asset Management needs, what can be mapped in GIS? The following are asset data that can be mapped in GIS for Asset management inventory requirements:

- Customer location data (Meter Number, Customer Number, address, etc.)
- WTP location data including asset information
- Reservoir location data including asset information
- Pipe network data installed
- Data location of pipe network accessories installed
- Elevation and topographic data
- Road map based map

Utilization of GIS integrated for Public Water Utilities

For the record, GIS cannot manage assets, but GIS provides an inventory of what and where asset locations are different. With current technological advances, all systems can be integrated in one system. GIS can be integrated with Inventory Systems, Asset Systems, SCADA systems, Billing systems, Meter Reading systems, data ware houses, resource planning, etc. By integrating the GIS system with the asset management system in the company, management will be easier by looking directly at the historical geographical and current location of an asset or just representing asset location for monitoring or control purpose.

GIS Integrated Asset Monitoring System

With the advancement of internet network technology, which has now entered the Internet of Things (IOT) era and also the industrial revolution 4.0, GIS technology is currently not only accessible in the internal environment, but can be accessed everywhere using the internet network. With current technological advancements it is possible to integrate GIS with other systems such as the Asset management system. The advantage of using GIS integrated asset management is to become more powerful, affordable and easier to manage assets. In this system GIS is very helpful in providing geographical information on assets and historical asset conditions. Some of the benefits of using GIS Asset management are;

1. Can track all assets, add or provide detailed information
2. Make analysis or report related to asset conditions
3. Real time tracking assets
4. Optimization and efficiency of asset management

Within the scope of the water management business, monitored assets can be in the form of pipelines, pipeline accessories, customers, infrastructure and movable assets such as operational vehicles. For example, a pipeline installed in location A can be traced to its location and can be analyzed via GIS whether it is necessary to replace the pipe or repair it. With GIS Maps, pipelines will be easily traced to their location and can be grouped based on the desired classification according to the data.

The planning department will find it easier to identify which networks need to be upgraded or which areas need to be added to assets and infrastructure. With GIS maps, the planning department can also make management planning easier and structured. The process of interaction between parts can be done by making work orders, making it easier to handle quickly and efficiently.

The warehouse or warehouse section can identify the availability of material in each branch. Besides that, it can also serve material needs for several projects according to Work orders made between sections. Systemically, GIS integrated asset management is very helpful in better and easier asset efficiency and control, from the Planning section it will be easy to identify the asset condition directly, in the field it will be easy to accelerate to the location of asset handling, in the warehouse section it will be easy to identify material needs.
An example of a GIS Integrated Asset Management System can be described as shown in Figure 1.

Figure 1. GIS asset management system.

GIS Integrated Operation System

The management of the water utility business cannot be separated from the handling of the operational and asset management system so that the continuity of the water business runs well and sustainably.

With the help of GIS technology, managing the water utility business will be easier because we will be able to plan, monitor, manage and run operations more controlled.

In the operation system GIS plays an important role in helping to get field information related to asset conditions and help identify asset conditions. For handling GIS operations on a Flow diagram can be formed in a system as shown on Figure 2.

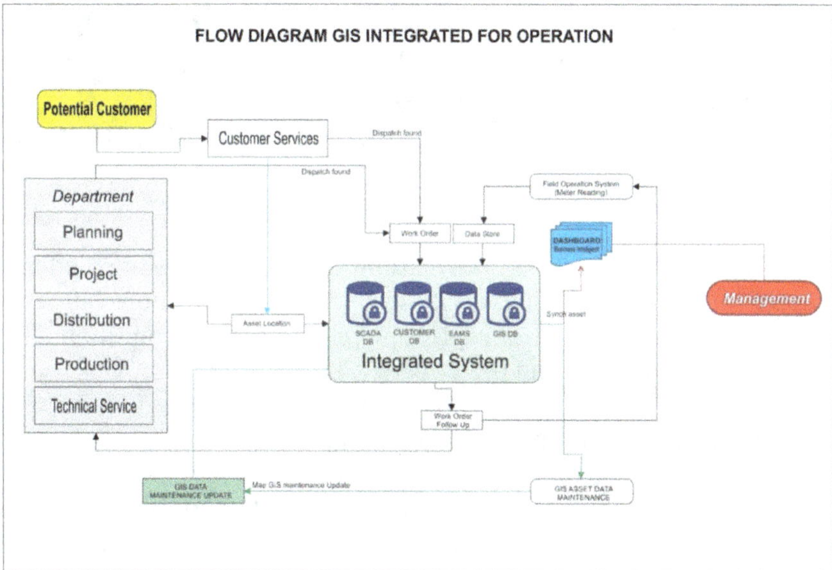

Figure 2. Flow diagram GIS Integrated for operation.

The figure above is an example of a schematic flow of GIS integration with other systems in handling operations both in terms of service or handling reports from customers regarding assets owned. In the flow diagram in the schematic is not only the scope of the department involved, but the control and monitoring process can be done by Management in this case BOD, Director or Shareholder. Process monitoring and progress can be done in the form of a dashboard display or in the form of reports that can be seen by Management. Example simple display of Dashboard representation as a control and monitoring for Management and monitoring of each department are as follows Figure 3.

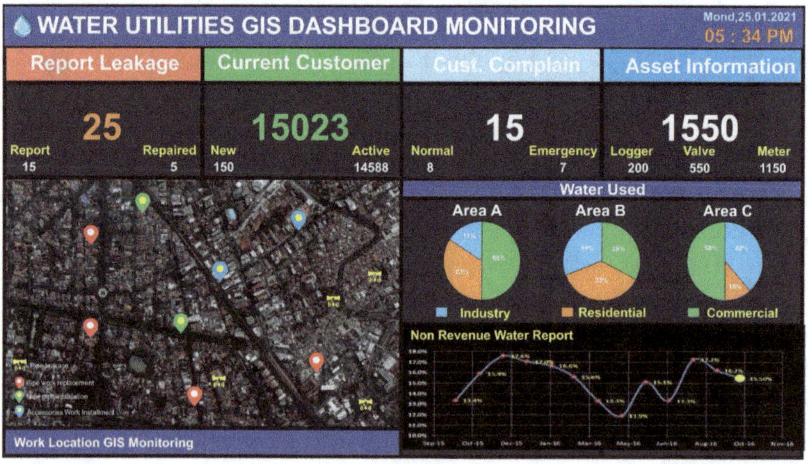

Figure 3. GIS Dashboard.

GIS helps display location information that needs to be handled and represents asset conditions. With a system of integration, GIS technology can also be used to represent reservoir level conditions live by integrating SCADA and GIS.

GIS Integrated Leakage Monitoring System

Another function of GIS that can be utilized in the operation process is in the compatibility of the pipeline network backup. GIS is integrated with various systems such as asset inventory, network repair systems and call center systems. GIS will help in displaying the location of the leak in real time and requesting a specification of the pipeline that fixes the leak.

Leak detection can be done with two events namely manual detection or automatic detection. For automatic detection, it can be done by installing a leak detection device in each pipeline that will be monitored. The detection device will send data to the system server in the event of a leak, from the data the system will automatically make a work order to be followed up. Data will then be processed in the system and approved in GIS for approval of leak locations. Based on information on work orders, this operating team will repair the leaky pipelines according to the location in the GIS. With GIS the repair team will easily find location information, supported by mobile applications such as Android.

For complete information, sources of information can be obtained from customer reports or discoveries directly in the field based on Visual. This information is sent to the call center and from the call center sends data to the System and an automated work order is made to be followed up with corrective steps by Tim's operation. Besides the team's operations can also check the material permitted in the warehouse through the warehouse inventory system. Repair results are released to the system and progress can be monitored in the GIS Leakage System. In the GIS Scheme, an integrated leak control and monitoring system can be illustrated in Figure 4.

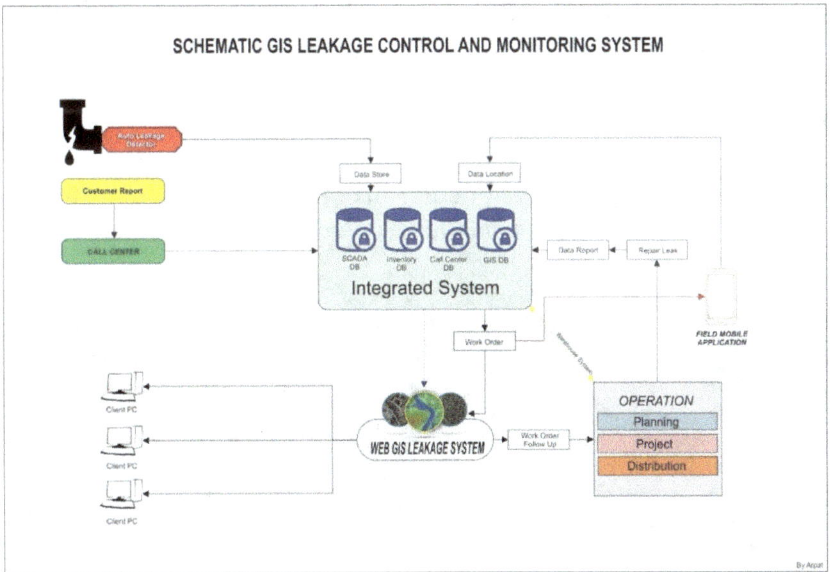

Figure 4. Flow Schematic GIS Integrated Leakage Control and Monitoring System.

Conclusion

Entering the industry 4.0 era of integration technology is growing rapidly, GIS is not only used as a display, but is a function that supports the industry. SCADA technology integrated with GIS has developed in several companies and includes several clean water utility management companies, where GIS is used as a function of monitoring an area of water pressure to control and monitor pressure in an area of DMA or District Meter Area. With the integration of GIS systems with other systems, the GIS function can be utilized in planning, developing and monitoring a work in the management of clean water.

GIS as a system is very helpful in providing geographical information in managing management information assets within the scope of clean water management. With the advancement of technology, GIS Feature has now been developed in the form of a 3-Dimensional GIS Map, and does not rule out the possibility of an asset image being mapped in 3D so that the use of GIS will be more maximal. Besides that, combining GIS with Real Time Camera makes it easy to control assets in the field live. Thus, the use of GIS integrated with other systems will further develop in the future.

References

Wikipedia, the free encyclopedia, *Geographic information system,* *https://en.wikipedia.org/ wiki/Geographic_information_system*

19

Conservation of Wetlands - Issues and Potentials: A Case Study of the Sambhar Salt Lake (Rajasthan)

Kartar Singh, Mili Ghosh, Shubha R. Sharma and Retinder Kour

Introduction

Wetlands are unique, distinctly productive, and vital resources because they provide habitat for migratory water birds, groundwater recharge, carbon storage, and healthy environment in many ways (Silvius *et al.,* 2000; Berkes and Jolly, 2001; Prasad *et al.,* 2002). The anthropogenic pressure is considered as the most influential factor for the wetland degradation worldwide (Finlayson and Rea, 1999; Davis and Froend, 1999; Li *et al.,* 2014). A constant monitoring system is required for these fragile aquatic systems to protect from the changes, their causes, and consequences. Nowadays, the various forms of degradation threaten these wetlands and such detrimental effects are more prominent in salt lakes. These lacustrine systems are economically viable and considered as an integral part of the natural ecosystem, which play a vital role in regional ecology, but the existence of these systems is at risk (Golubkov *et al.,* 2007; Williamson *et al.,* 2009). Generally, these lakes appear in endorheic drainage basins, and known as 'Saline lakes', 'Salt pans', 'Playas', 'Salinas', and 'Sabkhas' (Yechieli and Wood, 2002).

These lakes act as sinks for the surface and subsurface drainage effluents (Martínez-López *et al.,* 2014). The property of shallowness makes these lakes distinctive habitats for colonial water birds. The local geomorphology, evaporation, precipitation, geochemistry, and the unique watersheds control these systems, because the lake and its watershed are often deemed as a single ecosystem. The lakes provide insights about the sedimentary antiquities that develop knowledge attributed to the palaeo-climatic studies. The specific biological attributes and extreme environments create the curiosity about salt lakes. These lakes support valuable ecological services by sustaining complex microorganisms, which cannot be replaced by other ecosystems. Unfortunately, the salt lakes are one of the most threatened habitats of the world, which are facing several anthropogenic pressures and only a few are subjected to active conservation (Williams, 1993). The disproportionate use of water in the salt industry caused a rapid decline of lake resources, which creates a hydro-ecological disturbance leading to the loss of entire ecological integrity. The salt lakes are very sensitive to the climatic fluctuations, which negatively influences their water regime leading to degraded carrying capacity (Cui *et al.,* 2013). The degraded condition of the shallow lakes severely affects the migratory water birds, and has a critical influence over the wetland conservation strategy. A comprehensive hydro-geomorphological assessment with dissemination of datasets on the changes of socio-ecological structure is therefore a prerequisite to monitor these lakes. The scientific inventories accompanied by a viable management plan are necessary to maintain distinctions of the Ramsar convention (1970).

Site Description and Previous Studies

The Sambhar Lake (referred as 'SSL' hereafter) is situated at 26°58′N, 75°05′E, having a nearly elliptical shape with its major axis in a northeast to southwest direction (Figure 1). The lake catchment covers the Jaipur, Nagaur, Sikar, and Ajmer districts of the Rajasthan with total area of 7560 km^2. The lake bed height varies from 360-365 m (msl) with a water surface area extends from few km^2 to ~230 km^2 subjected to the amount of water received from direct rainfall and the lake inlets. In summers, the maximum temperature ranges from 40-48°C whereas in winters it lies between 2-11°C. An inconsistent temporal precipitation is the key characteristic of this region. In its geographical extent, this lake is having ~22.5 km length, 3.5-11 km width, with 0.6 m and 3 m as mean and maximum water depth respectively. In its northwest and west basin area, the lake is surrounded by the 'Aravali' hills. This lake is a playa depression and has two major lifelines named the 'Rupangarh' in the southwest and the 'Mendha' river in the northeast. Significantly, the salinity of this lake is a matter of geoscientific discussion, but it has been considered that the amount of soluble salts due to chemical weathering of the surrounding hills is the major source of the salinity in the lake.

The regional population is dependent on the rainfed agriculture, animal husbandry and salt mining activities. This lake is one the major commercial salt production sites in India with ~2 lakh tons of raw salt, and 40,000 MT processed salt production every year, which has been planned to increases up to 10 lakh tons per annum (TPA) by 2019-2020. This lake has a great potential for production of other chemical compounds (i.e., sulphate of potash, derivatives of value added bromine and magnesium chloride, β-carotene, and have opportunities for solution mining, production of packaged drinking water through industrial reverse osmosis). It also has the possibilities for different tourism activities.

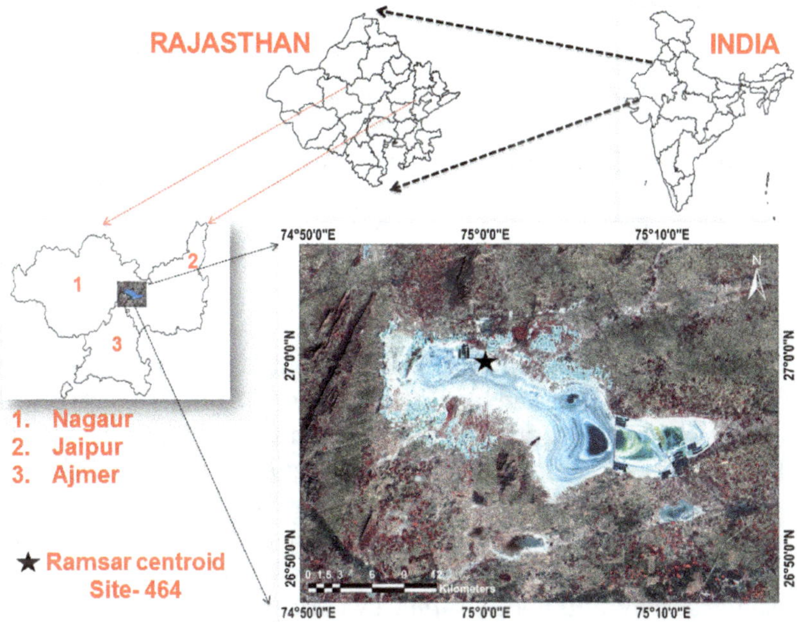

Figure 1. Geographic location of the Sambhar Salt Lake (SSL), India.

This hypersaline-alkaline shallow (HSAS) lake has very rich biological diversity with a variety of zoo benthos, primary producers, and waterfowls that show the seasonal fluctuation with changes in percentage salinity. The SSL appears dark green due to the rapid growth of the blue-green algae and diatom species after the rainfall. During the low salinity conditions, the zooplanktons generally occur in the lake and the presence of typical arid and semiarid vegetation can be easily observed. A comparative analysis with some of the important salt lake systems of the world shows that the SSL has significant ecological characteristics. Unfortunately, the wetland research groups interested in such lakes have overlooked the SSL ecosystem in spite of its universal importance, presence of idiosyncratic organisms, and aesthetic values. In winters, the SL provides a vital breeding ground for thousands of migratory flamingos, but enormous anthropogenic activities are affecting the lake ecosystem. Currently, this lake is facing industrial pressure, leading to its complete desiccation. The monitoring records about this lake are scarce that limits the comprehensive ecological assessment, which necessitates a systematic research about the environmental condition of this lake.

The SSL has not been much explored as compare to the other saline lakes. The palaeo-environment and geomorphology of this lake have been studied regularly, but research related to its watershed management, hydrometeorology, and water bird census is scarce. The inclusive notes on the genesis and conservation values of this lake have been discussed (Mathur and Mathur, 2007; Rahaman, 2007). The scientific literature about this lake is of multidisciplinary interest, but it is infrequent and dispersed in a variety of magazines and journals. This lake has been designated as an international bird area (IBA) with A1, A4i, and A4iii criteria, and also identified as a Ramsar wetland; irrespective of these global alliances, the national level biodiversity projects are not designed for this lake. The 'IBAs in Danger'- priority sites for immediate action initiative 2014 has completely ignored this lake. The Indian media reports provide significant information about conservation and management issues related to this lake. This study aims to assess the lake deterioration factors, condition, conservation issues, and their consequences using landsat satellite data, GPS based baseline survey, and photographic evidences.

Baseline Survey

At first, an initial study was conducted for the review of current condition of the lake based on the Google Earth images and reports published in the national newspapers and magazines. Afterward, in December 2012, the first field survey was conducted for the collection of baseline information about the lake. During the years 2013-2015 and 2018-2019, a number of GPS survey were conducted to verify the threats and issues associated to the lake with collection of real time photographic evidences. The presence of illegal bore wells, salt waste dumping, trespassing, and industrial development were the main threatening issues found in the main lake area.

Estimation of Water Surface Extent

Procurement of satellite data

Landsat MSS, TM, ETM, and OLI images were acquired from United States Geological Survey (USGS) GLOVIS portal. All Landsat images used are of L1T product type and geo-referenced with precision better than 0.4 pixels. The imageries were all free of clouds. Descriptions of the Landsat images are presented in Table 1.

Table 1. LANDSAT data used in this study

S.N.	Imagery date	Path/Row	Scene ID	Sensor ID
1	10/11/1975	159/41	LM21590411975314AAA05	Landsat MSS
2	10/03/1977	159/41	LM21590411977069XXX03	Landsat MSS
3	14/06/1981	159/41	LM31590411981165AAA03	Landsat MSS
4	18/10/1981	159/41	LM31590411981291AAA03	Landsat MSS
5	02/05/1998	148/41	LT51480411998122XXX02	Landsat TM
6	10/11/1998	148/41	LT51480411998314XXX01	Landsat TM
7	03/04/1999	148/41	LT51480411999093AAA01	Landsat TM
8	13/11/1999	148/41	LT51480411999317XXX02	Landsat TM
9	09/12/2000	148/41	LE71480412000344SGS00	Landsat ETM
10	28/12/2001	148/41	LE71480412001362SGS00	Landsat ETM

11	13/11/2002	148/41	LE71480412002317SGS00	Landsat ETM
12	22/04/2003	148/41	LE71480412003112ASN00	Landsat ETM
13	08/11/2009	148/41	LT51480412009312KHC00	Landsat TM
14	03/05/2010	148/41	LT51480412010123KHC00	Landsat TM
15	26/10/2010	148/41	LT51480412010299KHC00	Landsat TM
16	06/05/2011	148/41	LT51480412011126KHC00	Landsat TM
17	14/11/2011	148/41	0101203079484_00001 (Request ID)	Landsat TM
18	12/05/2013	148/41	LC81480412013339LGN00	L8 OLI-TIRS
19	28/04/2014	148/41	LC81480412014118LGN00	L8 OLI-TIRS
20	27/12/2015	148/41	LC81480412015361LGN00	L8 OLI-TIRS
21	29/06/2019	148/41	LC81480412019180LGN00	L8 OLI-TIRS

The L1T data provides geometric accuracy (i.e., root mean square error < 0.5 pixels), by incorporating geodetic accuracy and ground control points. Since the L1T Landsat products are geometrically corrected, no such corrections were applied in the pre-processing. The Landsat images, acquired in the form of digital numbers, were calibrated to planetary surface reflectance using Landsat Data Users Handbook. In case of the shallow waters, the implementation of the atmospheric correction method is always quite complicated. In this study, to perform the atmospheric correction, the improved dark object subtraction (IDOS) method was employed using ERDAS Imagine modeler 2014. The IDOS corrected the small inconsistencies of atmospheric conditions for the area under consideration. The conversion of apparent to true planetary reflectance (ρλ) was performed as follows.

$$\rho_\lambda = \pi[(L_{sat} - L_{haze})/E_o \cos TZ] \quad \text{(Eq.1)}$$

where, π = 3.14, L_{sat} stands for sensor radiance, L_{haze} represents the atmospheric path radiance, E_O is the mean solar exo-atmospheric irradiance, and TZ stands for solar zenith angle.

Generation of Water Indices

Using the modeler function in ERDAS Imagine 2014 Software, the Normalized Difference Water Indices (NDWI) were generated from 1975 to 2019. The shoreline areas were carefully examined to decide the accuracy of each indexed image. The total water surface area of the lake for respective years was calculated using ERDAS measurement tools. The water surface dynamics and development of saltpans around the lake shoreline were analyzed using change detection maps (Figure 4). The GPS points of bore well units were superimposed on the satellite imagery, and a danger zone map was prepared to assist the lake management plan (Figure-9). Significantly, the threats and conservation issues scrutinized during the baseline survey were analyzed and verified using geoinformatics techniques. The active conservation plan has been proposed for this lake with potential management solutions.

Results and Discussion

Threats and challenges

In the last decade, the SSL environment has been rapidly degraded, which is creating a risk for the regional livelihoods because it contributes valuable socioeconomic benefits to the country. The present lake scenario with altered eco-hydrology is destructive for migratory water birds. The potential threats and conservation issues related to the SSL are as follows.

1. The rapid development of the private salt manufacturing units (PSMUs) around the lake.

2. Water diversion/restriction due to the establishment of PSMUs in the riverbed and dried channels.
3. Industrial salt waste dumping (ISWD) in the lake area that pollutes the water and degrading the soil quality.
4. The incessant cutting/flattening of the hills for the establishment of new PSMUs near the lake shoreline.
5. The installation of illegal borewells and setting up the electric power lines, circuit panels, and open wiring in the lake area.
6. The agriculture land transformation by means of converting crop fields into the saltpans.

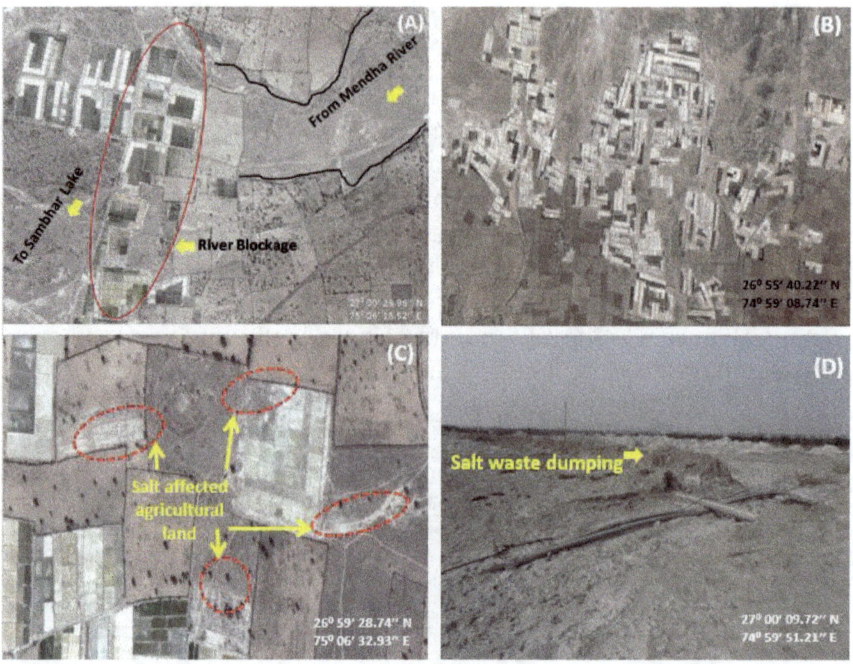

Figure 2. showing encroachment in lake catchment and near the river inlets: (A) the diversion or restricted water flow of the 'Mendha' river (B) development of PSMUs in the 'Rupangarh' dry riverbed; (C) agricultural land transformation practices; and (D) evidences of industrial salt waste dumping. (Source: *real time photographs captured during the years 2015, 2017 and 2019*)

Evidently, the associated environmental threats and lake issues are the subsequent results of the establishment of these PSMUs (Figure 2A and B). In this area, the salt manufacturing plays a major role in the local economic growth, but instead of a great potential for eco-tourism, the state administration is lacking in strategies the lake action plans.

These PSMUs are continuously using the lake water as well as ground water for their saltpans; they are not considering the lake ecological values, which are disrupting the entire ecosystem. Therefore, "the PSMUs are contravening the international treaties and designations associated with this lake". The central ground water board (CGWB) India has notified the SSL region for the control and regulation of ground water because the adjacent district named 'Nagaur' is facing scarcity of drinking water. Despite of such threatening facts the PSMUs are drilling bore wells in this area, and constructing saltpans near the lake boundary to reduce the cost and time, which deteriorates the lake shoreline (Figure 3 and 7).

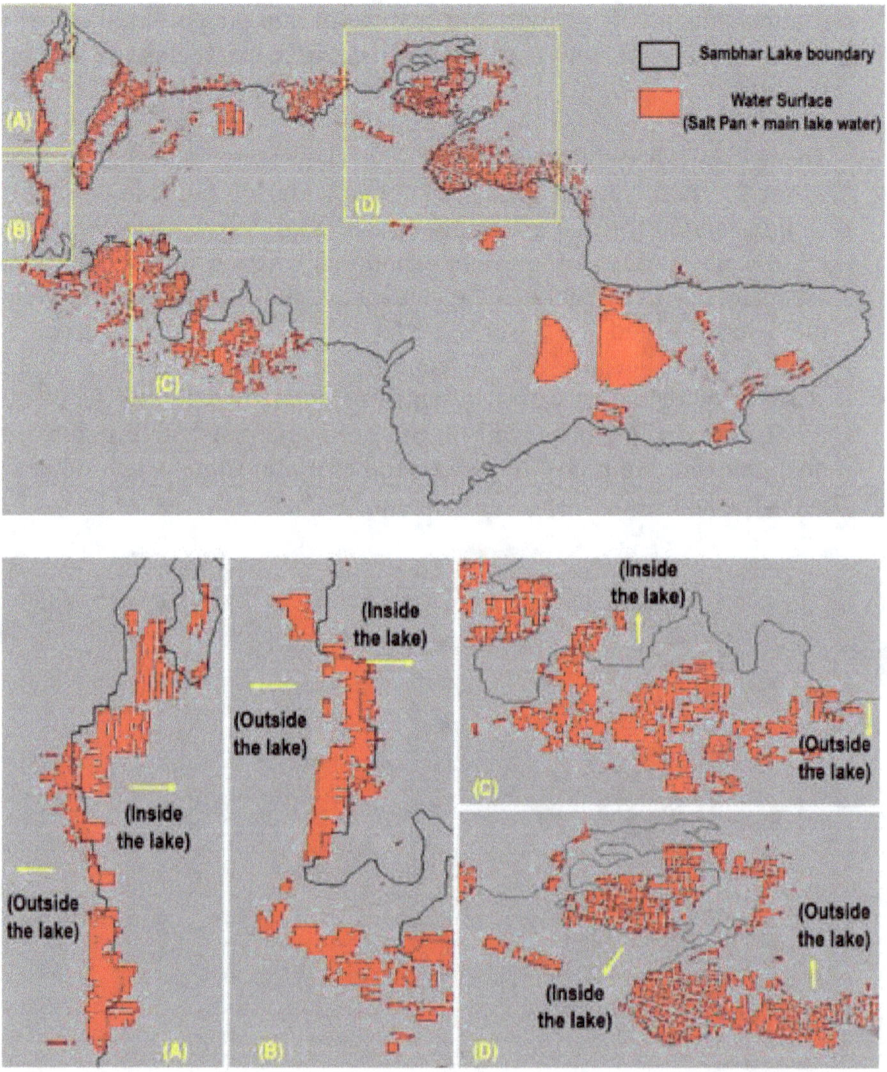

Figure 3. The spectral index image shows the rapid development of saltpans in and outside of the lake boundary. This dearth can convert a prey-rich habitat into the desert; therefore, the consequences and management aspects related to the lake environment have been discussed in this work.

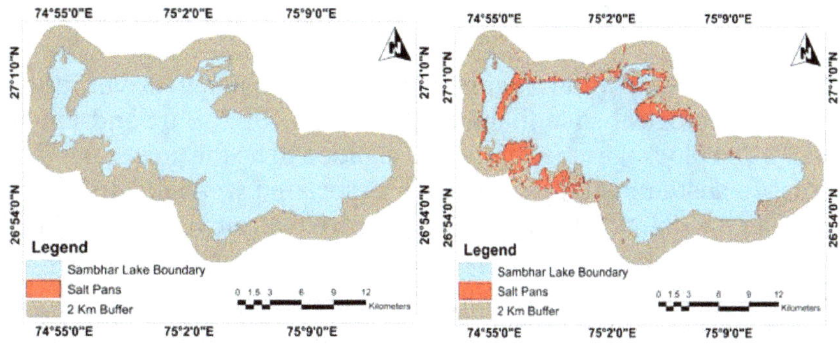

Figure 4. The change detection map developed using Landsat images of the 1981 and 2014, showing the development of private saltpans with in the 2 km buffer zone of the lake boundary.

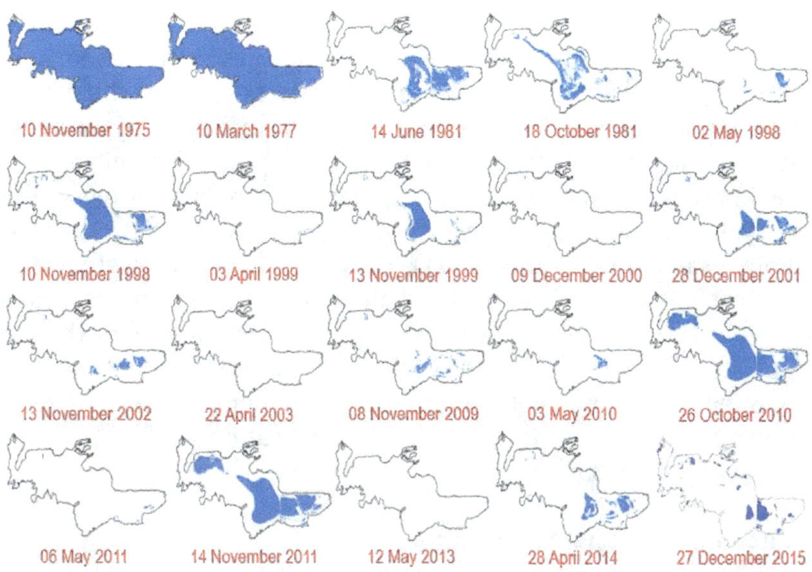

Figure 5. Showing the spatio-temporal dynamics of water surface area in the SSL.

Water Diversion/Restriction

The unrestrained water regime is identified as a key 'driver' of the wetland ecological sustainability (Poff *et al.,*1997; Hart and Finelli, 1999). The blockage in water flow is claimed to be the most relentless threat to sustaining rivers and their associated wetlands. In case of this lake, the 'Mendha' and the 'Rupangarh' were two main lifelines, which nourish the lake, but currently these ephemeral rivers have been encroached (Figure 2A and B).

In the northeast of the SSL, a rapid development of the PSMUs is observed in the main course of the 'Mendha' river that introduced a pressure to halt the river flow (Figure 2A). Similarly, in the southwest the 'Rupangarh' river has been encroached, which lead to the complete desiccation of the lake (Figure 2B). The primary impact of river encroachment can be of two fold: (1) water scarcity in the lake, (2) increase in the salinity, and these two can initiate several consequential effects on the lake ecosystem. The decreased lake volume can expose the entire lake bed that creates significant rise to the emission of dust particles and phytotoxins, which affects the waterbird population in the lake area. The increased salinity can be one of the subsequent results of less water availability in this lake. A significant decline in the wetland biodiversity and change of halosensitive species with halotolerant one, signify the biological responses to increased salinity gradient (Williams, 2001). The altered salinity gradient leads to the various physio-chemical and biological changes, and decreased lake volume can affect the thermoregulatory operations of the lake organisms. The encroachment in dry riverbed can lead to affect the catchment hydrology. Note that the encroachment activities affecting the water regime and intensifying desertification lead to the rapid conversion of the SSL into a barren salt scape.

Industrial Salt Waste Dumping

The ISWD is one of the major threats, which is polluting the entire lake area (Figure 2D). These practices are common in this area and thus affecting the biochemistry of this lake. Apparently, for industries, dumping of the salt waste into the lake can be the easiest way to manage the industrial effluents, but it is disastrous for the lake ecosystem. These effluents not only contribute to the rising height of the lake bed, but also altering the lake's micro-climatic condition. In the shallow lakes, several trophic interactions exist (i.e., desorption and sorption, sediment resuspension), which are helpful in the release of nutrients by diffusion exchange process in the water column. The waste sediments and excessive presence of Phosphorous (P) and Nitrate (N) can initiate the eutrophication or algal bloom in the lake. A moderate growth rate of algal species is an indicator of the trophic state of the lake. Therefore, it is necessary to assure the development pattern of phytoplankton biomass in the lake. An increase in phytoplankton biomass, turbidity, sedimentation, anoxic condition, and decrease in species diversity are the major consequences of the ISWD. The industrial effluents carry a variety of sediments, which limits the availability of sunlight for benthic organisms of the lake. The ISWD can lead to the increase in organic matters, which results in reduction of the dissolved oxygen in the lake water. The shoreline areas are sensitive to the external sediments, and these areas are the first potential sites of the lake to get affected by the waste effluents. Therefore, the increased external sediment load can alter the wind driven movements in the pelagic zone, which negatively affects the lake resuspension process and enables the expansion of undesired vegetation growth. Specifically, the siltation process can influence the littoral vegetation composition and initiate drastic changes in the biological structure of the lake. The ISWD can contribute as heavy metal contamination, which is harmful for waterbirds, and have a possibility to accumulate in the aquatic organisms. Inclusively, the ISWD causes radical changes in primary production and mineralogy of the lake, therefore, such type of unlawful practices should be prohibited in the lake area.

Incessant Cutting/Flattening of the Hills

The 'Mohanpura hill', part of the 'Aravali' hills (26^{0}59'59.02"N, 74^{0}55'54.88"E), has been flattened almost 70% for the establishment of new PSMUs (Figure 6). Possibly, the Mohanpura hills are hydrologically active, and the sedimentary attributes of this area can be used to scrutinize the hydrogeological history or genesis of the SSL. The Mohanpura hills can have major impact on the hydro-chemistry of this lake, if the hill or rock matrix contributes the considerable solutes to this lake. The rock weathering and wind erosion are mainly responsible for the removal of solutes from the sedimentary structures. The rock components eroded form the Mohanpura hills make this lake hydro-geologically interesting. The minerals interflow with rainfall from this hilly terrain causes the salinity fluctuation in the lake water. Significantly, this hilly terrain regulates the wind circulation effects and contributes to the microclimate of this lake. In shallow waters, the sub-surface wind circulations are significant and decisive in evaporation processes leading to the changes in water level fluctuations. Possibly, these wind effects regulate the thermal structure and salinity, which can influence the occurrence of migratory water birds.

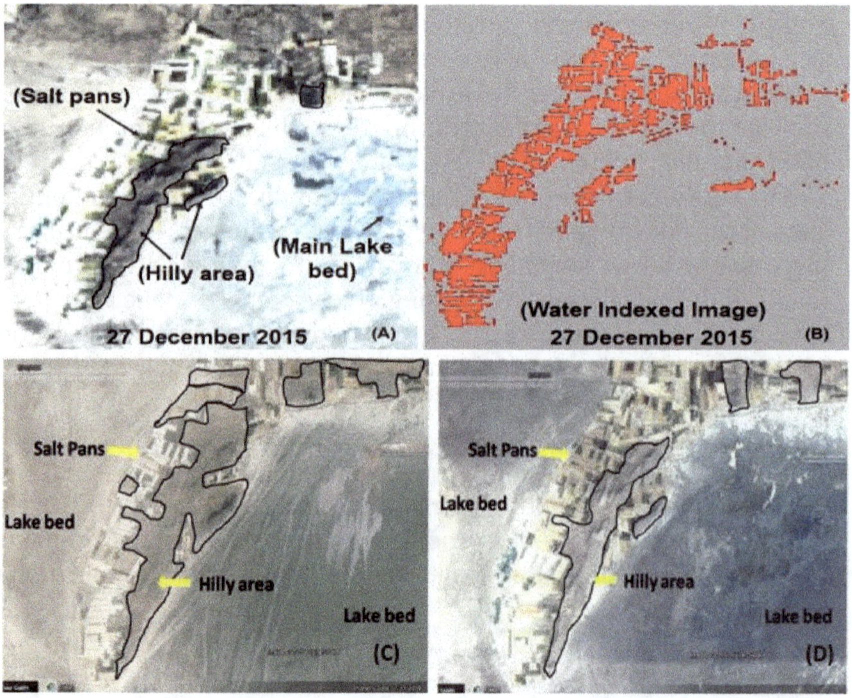

Figure 6. The condition of the Mohanpura hills after the flattening activities: (A) the FCC developed using L8 OLI-TIRS imagery; (B) the water index image, showing the development of PSMUs, and reduced area of the hilly terrain; (C), (D) the degraded condition of the Mohanpura hills [Google earth images of 20th March 2008 and 27th February 2014, respectively].

The incessant flattening of the Mohanpura hills can have severe consequences on the SSL ecosystem; therefore, to sustain the ecological equilibrium, all the related hydro-geomorphological phenomena need to be accounted for.

Installation of Illegal Borewells and setting up the Power Lines

Nowadays the over extraction of ground water using illegal borewells in the SSL region is one of the common practices and no regulatory action has been initiated against these activities. Consequently, the water table has decreased up to 60 m in this area. These borewells are installed by the salt industries, and several kilometres of power lines have been laid down to connect these borewells to the electric supply (Figure 7D). At several locations in the lake, these power lines have been found open, which is not safe for living beings (Figure 7B and C). As per the investigation committee reports, ~15-20 borewells have been drilled out ~per 0.16055 ha in the lake.

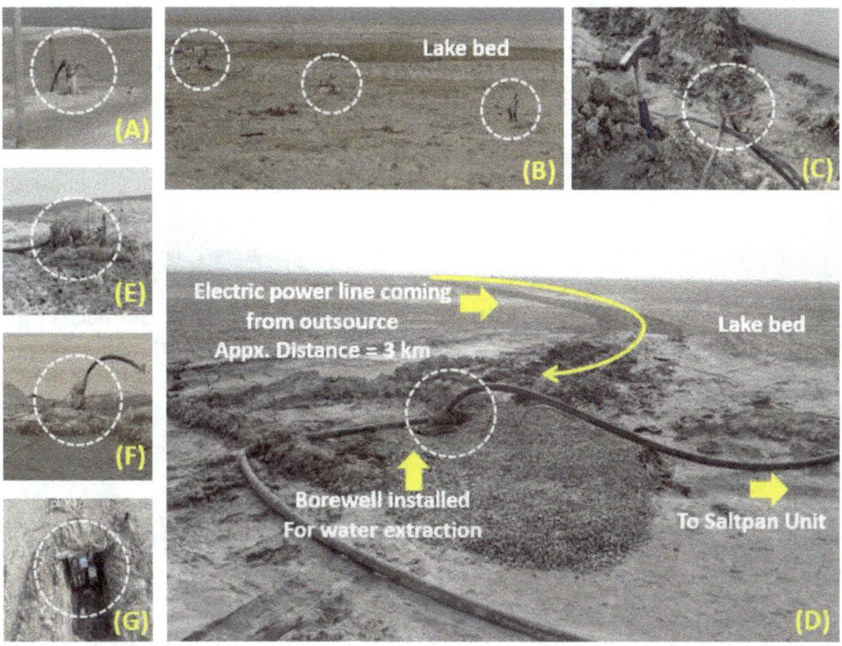

Figure 7. The real time photographic evidences of the unauthorized borewell units and open electric wiring in the main lake area.

In summers, due to the dry events, excessive trespassing causes removal of the top soil that drastically changes the overall soil profile of the lake. The altered soil quality influences the salinity gradient, which can adversely affect the aquatic biodiversity. This research work strongly advocates the urgency to build rapid action plan to control the excessive anthropogenic pressure on the lake resources.

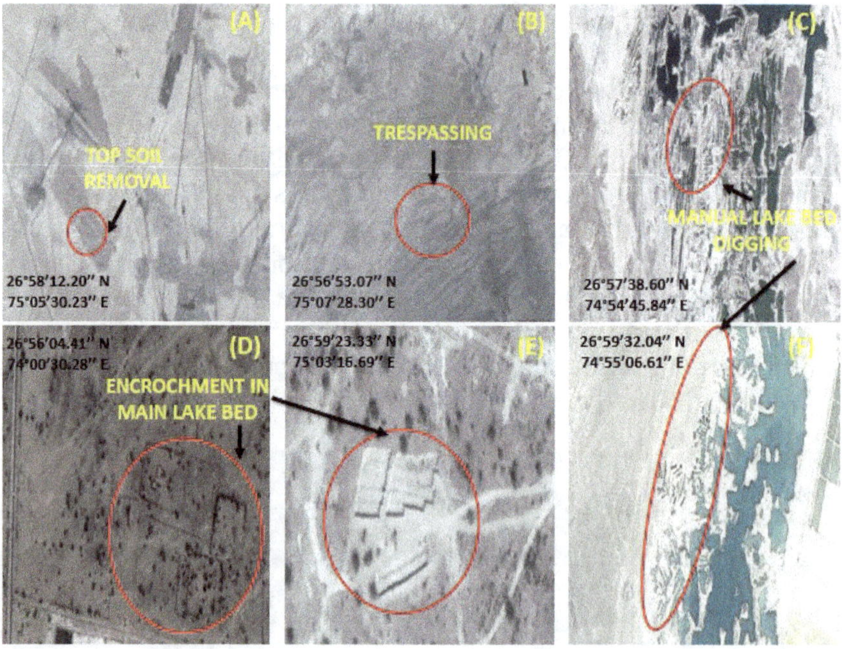

Figure 8. (A) Evidences of the lake's top soil removal; (B) trespassing in the lakebed during the dry events; (C), (F) showing the illegal activities related to the digging of lakebed for rainwater storage; and (D), (E) evidences of the encroachment in the lake area. (Source: Google earth images)'

Agricultural Land Transformation

In this region, the increased salinity can adversely affect the overall agricultural productivity. The net fertile cropland is decreasing due to the rapid conversion of these agricultural fields into the saltpans, thus it is provoking desertification in this region. These crop fields are degraded due to the accumulation of salt residues, which causes drastic land transformation and limits the goals of sustainable agriculture development in the arid and semi-arid areas of the country (Figure 2C). On the other hand, this lake is facing severe environmental problems (i.e., top soil removal and intensive trespassing (Figure 8A and B), manual lake digging to store rainwater for saltpan units (Figure 8C and F), and unapproved colonization (Figure 8D and E). Certainly, for this lake the environmental impact assessment (EIA) is needed to determine whether the socioeconomic benefits of commercial salt production justify the healthy ecosystem.

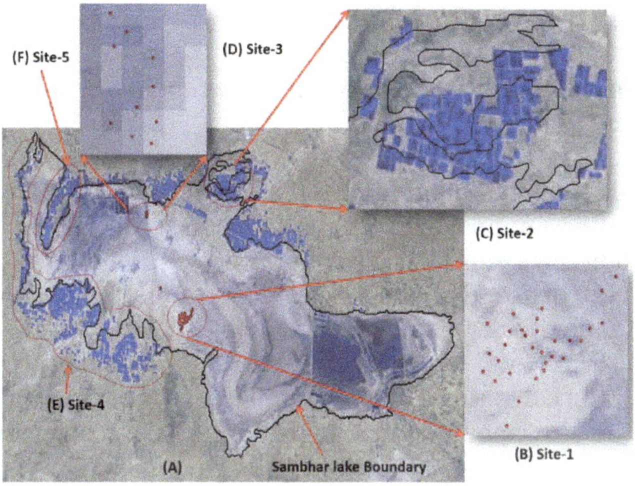

Figure 9. (A) Synoptic view of the SSL; (B) site-1 with GPS locations of the illegal borewells (C) site-2 the PSMUs development; (D) site-3 with the illegal borewells location near the road causeway; (E) site-4 the uncontrolled growth of the PSMUs; and (F) site-5 with cutting of the Mohanpura hills.

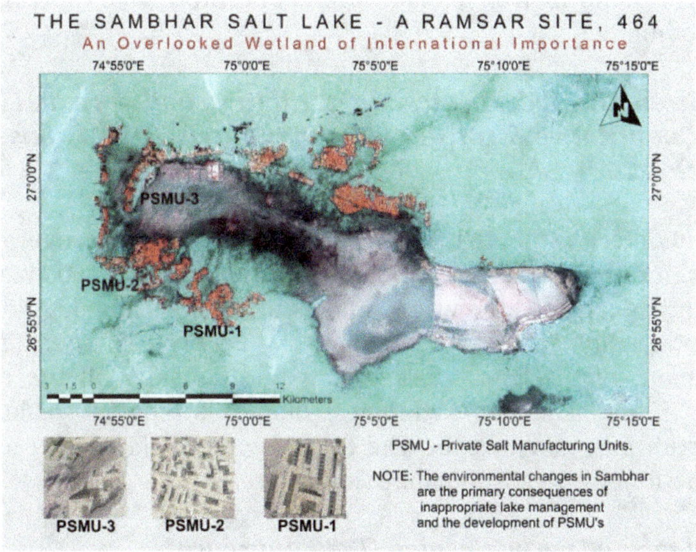

Figure 10. The PSMUs development around the SSL. Salt pans (Red colored) in and around the lake boundary. The eastern area of the lake is mostly used by the SSL authority for salt production and refineries. (*NOTE : These three PSMU areas are only for indication, unfortunately a lot of area is covered by illegal salt miners*). [*Map prepared using L8 OLI/TIRS Image, June, 2019*]

Potential Management Solutions

A sustainable adaptive governance management (SAGM) based approach

The biophysical degradation of lakes is a global phenomenon and, thus, is not exceptional to the SSL in India. In spite of performing important functions, this lake is becoming ecologically dead. There is a need to restructure the land-use system around the lake related to the salt industries to ensure sustainable utilization of the lake resources. The SAGM based mechanism is required for realistic assessment to ensure that the PSMUs understand the lake issues; therefore, it increases the community participation in lake management. The management fails due to a lack of good governance in cooperation between the economic and lake ecological benefits. The mechanism must involve the grassroots-level institutions that monitor and identify the responsible indicators across the spatio-temporal scales (Figure 5). This study suggests the implementation of site-specific SAGM method developed by Alipour *et al.,*(2015), to explore the lake administrative policy as well as public perception to revive this prey-rich environment.

The Strengths, Weaknesses, Opportunities, and Threats (SWOT) and Quantitative Strategic Planning Matrix (QSPM) based approach

A rapid deterioration in the SSL condition has caused a severe water stress in this area. High water consumption by the PSMUs have significantly decreased the water inflows to the lake, and degraded the habitat quality, especially for migratory waterbirds. There is a requirement of the framework that would consider all the driving forces in the lake basin, including ecological strengths, weaknesses, opportunities, and threats. The SWOT with QSPM can be used to achieve long-term sustainability in this area.

An Integrated Early Warning (IEW) approach

Rapid development of the PSMUs has increased water consumption in the SSL area, leading to water supply shortages and quality degradation, which consequently increase the water-stress risks and threats to the environmental security. Currently, methods that widely consider the resource-based water stress is lacking, and development of such methods is needed to manage the water demands of economic development in the shadow of ecological integrity of this lake. This study recommends implementation of the IEW method to estimate the water stress and multiple relationships among driving forces and the lake ecosystem status. This method is helpful in forecast of water stress trends and delivers early warning signs (Wang *et al.,* 2013). The lake administration should consider the resources-induced as well as pollution-induced water stress prevention. Positively, the IEW method would ascertain potential water stress in this lake subject to intensive anthropogenic pressure.

The Aquatic Real time Management System (ARMS) and index of sustainable functionality (ISF) based approach

The transition toward a sustainable development of the shallow lakes is founded upon the complex relationships between socioeconomic and environmental issues. The ISF has been found very effective to quantify the sustainability of dynamic systems (Kristiana *et al.,* 2011). The ISF measures the functionality of the processes of particular ecosystem weighted by the stakeholders' priorities and provides an objective based assessment of the systems' ability to engross various influential factors. Possibly, the ISF based assessment with ARMS is required for the SSL to measure its sustainable functionality over the extended periods. The ISF formulation is provided below.

$$ISF = \sum_{j=1}^{J^i} \sum_{K=1}^{K^i} \left[\frac{W_{jk}^i}{L_{jk}^i} \left\{ \sum_{l=1}^{L_{jk}^i} \frac{1}{M_{jkl}^i} \sum_{m=1}^{M_{jkl}^i} I_{jklm}^i \right\} \right] \quad \text{(Eq.2)}$$

where
i = subdomain 1, 2, 3,. . .
j = perspective 1, 2, 3,. . .
k = system 1, 2 ,3,. . .
l = function 1, 2, 3,. . .
L = number of functions in the matrix element (j, k)
I = indicator 1, 2, 3,. . .
M = normalized indicator 1, 2, 3. . .
W = weightings

An approach based on paleolimnology

An operational management of lakes not only requires an integrated scheme that considers both catchment and lake environments, but also a cognizance of ecology as well as information of the environmental stressors. Therefore, the conditions before the disturbance of this lake need to be examined for convincing management objectives. The paleolimnological techniques provide this essential knowledge through lake sediments and samples from its catchment for quantification of the multiple lake stressors. Diatoms presented in the lake sediments are very useful to explore long-term history of trophic state dynamics because of these are ideal bio-indicators with defined ecological optima and tolerance to different lake conditions. Ultimately, the paleolimnological study with long-term spatio-temporal monitoring data (Figure 5) can contribute to the development of active lake management plans.

Shoreline physical disturbance index based management

Monitoring of the shoreline habitat condition is essential because in case of the SSL these areas are especially vulnerable to the PSMUs encroachment that deleteriously affecting the lake environment. This disturbance in structural complexity of physical habitat can affect lakes' aquatic organism. Thus, it is needed to assess the structural complexity of these areas using index-based methods (i.e., USPEA 2007a), to preserve biological integrity of the SSL ecosystem. In this view, the lakeshore anthropogenic disturbance index (RDis_IX) is appropriate to calculate the physical disturbance around the lake (Kaufmann *et al.,* 2014).

$$RDis\_IX = (Disurbance\ Intensity + Disturbance\ Extent) + \ /2 \quad \text{(Eq.3)}$$

where, disturbance intensity can be represented by separate sums of the mean proximity-weighted values of near shore non-agricultural and agricultural disturbance types, and extent is the area of the shore with the occurrence of any type of disturbance.

An institutional and conflict management based approach

The institutional schemes are decisive in an effective implementation of natural resource policy. Sometimes, the enactment of various policies is not realistic due to site-specific public interventions. The management of system resources is an institutional problem (Acheson 2000). In case of the SSL, institutional set up is scarce to apply and manage water resources at the ground level. In addition to the central and state government, the Sambhar salt limited has shared the accountability for the SSL management that shows multiple flaws. These flaws include inefficient monitoring systems at different administrative levels, weak enforcement directives, absence of distinguished wetland laboratories, lack of cooperation between ministries and research departments. The conflict management and traditional dispute resolution mechanisms provides an understanding of institutions' capabilities to establish balance between local legality and legalization (Jianping and Haizhou, 2015).

'The greatest challenge facing this lake is the conflict between biodiversity conservation programs and the economic benefits of using lake resources'

The clashes over enterprises between the Sambhar salt limited and the PSMUs have also emerged. The conflict resolution is complex because benefits are differing from conservation priorities, such as establishment of PSMUs without considering the existence of the lake.

Other possible lake management methods

Implementation of approaches like 'together for environmental assessment and management' (TEAM), volunteers lake monitoring program (VLMP), nested adaptive system using socio-ecological attributes, contingent valuation method (CVM), ecosystem service values (ESV) based quantification of environmental flow, and consensus based resource management can be performed for the SSL.

Conclusion

This study provides the state of knowledge and facts about the SSL and explores the anthropogenic disturbances, which are degrading the lake ecosystem. The environmental changes in the SSL's ecosystem are the primary consequences of inappropriate lake management, and the development of PSMUs. The GPS based baseline survey and satellite imagery (i.e., Landsat TM, ETM, and OLI), have been used to explore the SSL's current physical condition and reveal that 40-50% land of the lake surroundings has been covered by the PSMUs.

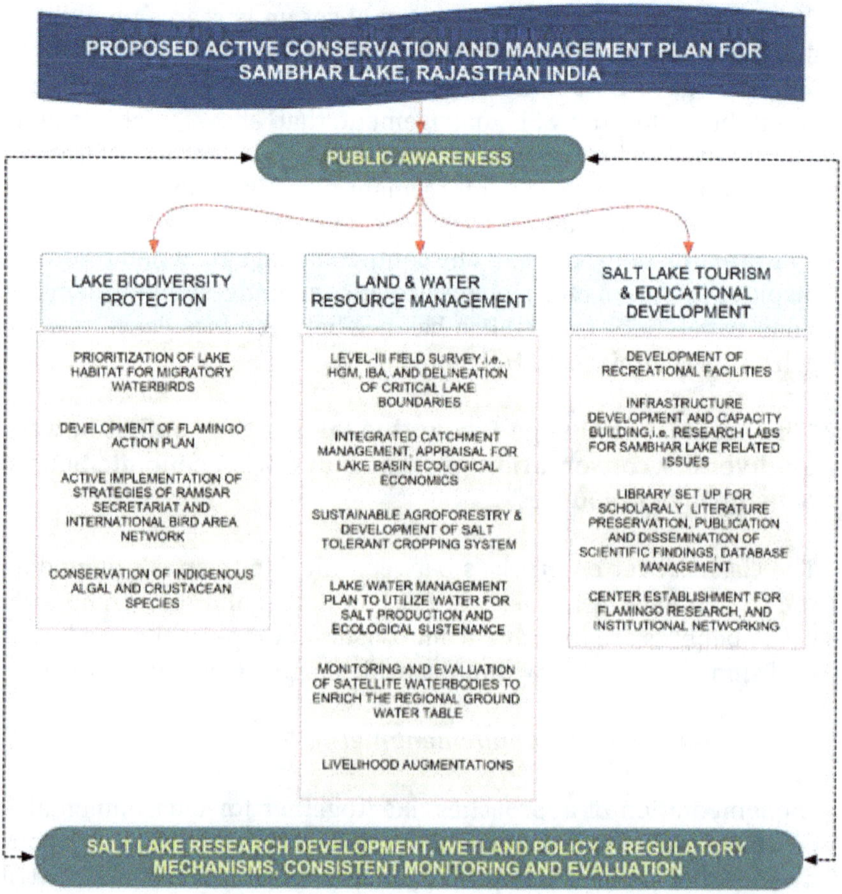

Figure 11. The proposed active conservation and management plan for the Sambhar Salt Lake.

The remote sensing technique with ground based survey helped in the analysis related to the land conversion, hill cutting, and identification of the ISWD areas in this region. The rivers named 'Rupangarh' and 'Mendha' have been impinged by the PSMUs. These conservation issues necessitate the in-depth awareness about the changes in the lake's ecological integrity due to the human induced activities. This lake has suffered from the prolonged anthropogenic disturbances, and there is an urgent need to quantify the anthropogenic as well as hydro-climatic effects. Thus, the geoinformatics based lake danger zone map has been prepared to guide the lake authority for the better management (Figure 9).

This study witnessed that thorough scientific research and political action is essential to protect this aquatic system. Our study suggests that a restricted zone of 5 km has to be declared around the lake where establishment of the PSMUs should be banned and all the bore-well units have to be detached from the lake area (Figure 10). Furthermore, the threatening scenario of this lake needs some international measures to control its grievous condition. The comprehensive ground based scientific inventories are necessary to accelerate the process of rejuvenation of this lake. A conservation and active management action plan have been suggested for this lake, including biodiversity protection measures, land-water resource development, and salt lake tourism with institutional networking (Figure 11). The participation of international wetland organizations can play a vital role to combat with the lake environmental crisis. This work is 'A Call for Action' to develop scientific awareness in the public about current condition of the SSL, which will help to protect this lake from a desolate environment.

References

Acheson, J. (2000). Clearcutting Maine: Implications for the theory of common property resources. *Human Ecology*, 28(2), 145-169.

Alipour, H., & Olya, H.G.T. (2015). Sustainable planning model toward reviving Lake Urmia. *International Journal of Water Resources Development*, 31(4), 519-539.

Berkes, F., & Jolly, D. (2001). Adapting to climate change: social-ecological resilience in a Canadian Western Arctic community. *Conservation Ecology*, 5, 18.

Cui, L., Gao, C., Zhao, X., Ma, Q., Zhang, M., Li, W., *et al.,*(2013). Dynamics of the lakes in the middle and lower reaches of the Yangtze River basin, China, since late nineteenth century. *Environmental Monitoring and Assessment*, 185(5), 4005-4018.

Davis, J. A., & Froend, R. (1999). Loss and degradation of wetlands in southwestern Australia: underlying causes, consequences and solutions. Wetlands Ecology and Management, 7(1), 13-23.

Finlayson, C. M., & Rea, N. (1999). Reasons for the loss and degradation of Australian wetlands. *Wetlands Ecology and Management,* 7(1), 1-11.

Golubkov, S., Kemp, R., Golubkov, M., Balushkina, E., Litvinchuk, L., & Gubelit, Y. (2007). Biodiversity and the functioning of hypersaline lake ecosystems from Crimea Peninsula (Black Sea). Fundamental and Applied Limnology, 169(1), 79-87.

Hart, D. D., & Finelli, C. M. (1999). Physical-biological coupling in streams: The pervasive effects of flow on benthic organisms. Annual Review of Ecology and Systematics, 30, 363-395.

Jianping, W., & Haizhou, Ma. (2015). The effectiveness of small to medium river basin organizations in China: a case study of the Erhai Lake Basin Management Commission. *International Journal of Water Resources Development,* 31(4), 589-603.

Kristiana, R., Vilhena, L. C., Begg, G., Antenucci, J. P., & Imberger, J. (2011). The management of Lake Burragorang in a changing climate: The application of the Index of Sustainable Functionality. *Lake and Reservoir Management,* 27(1), 70-86.

Kaufmann, P. R., Peck, D. V., Paulsen, S. G., Seeliger, C. W., Hughes, R. M., Whittier, T. R., *et al.,*(2014). Lakeshore and littoral physical habitat structure in a national lakes assessment. *Lake and Reservoir Management,* 30(2), 192-215.

Li, L., Li, J., Yao, X., Luo, J., Huang, Y., & Feng, Y. (2014). Changes of the three holy lakes in recent years and quantitative analysis of the influencing factors. Quaternary International, 349, 339-345.

Martínez-López, J., Carreño, M. F., Martínez-Fernández, J., & Esteve, M. A. (2014). Wetland and landscape indices for assessing the condition of semiarid Mediterranean saline wetlands under agricultural hydrological pressures. *Ecological Indicators,* 36, 400-408.

Mathur, R. P., & Mathur, L. N. (2007). Conservation of Sambhar wetland in Rajasthan. *Journal of Social Policy Research Institute,* 2, 122-129.

Poff, N. L. R., Allan, J. D., Bain, M. B., Karr, J. R., Prestegaard, K. L., Richter, B. D., *et al.,*(1997). The natural flow regime A Paradigm for river conservation and restoration. BioScience, 47(11), 769-784.

Prasad, S. N., Ramachandra, T. V., Ahalya, N., Sengupta, T., Kumar, A., Tiwari, A. K., *et al.,*(2002). Conservation of wetlands of India - a review. *Tropical Ecology,* 43(1), 173-186.

Rahaman, A. A. (2007). Sambhar Lake: Wetland an assessment. *Journal of Social Policy Research Institute,* 2, 130-135.

Silvius, M. J., Oneka, M., & Verhagen, A. (2000). Wetlands: lifeline for people at the edge. Physics and Chemistry of the Earth, Part B: Hydrology, Oceans and Atmosphere, 25(7-8), 645-652.

[USEPA] US Environmental Protection Agency. (2007a). Survey of the nation's lakes. Field operations manual. Washington (DC): EPA 841-B-07-004.

Williamson, C. E., Saros, J. E., Vincent, W. F., & Smol, J. P. (2009). Lakes and reservoirs as sentinels, integrators, and regulators of climate change. Limnology and Oceanography, 54(6, part 2), 2273-2282.

Williams, W. D. (1993). Conservation of salt lakes. *Hydrobiologia,* 267(1), 291-306.

Williams, W. D. (2001). Anthropogenic salinisation of inland waters. *Hydrobiologia,* 466(1), 329-337.

Wang, X., Su, J., Cai, Y., Tan, Y., Yang, Z., & Wang, F. (2013). An integrated approach for early warning of water stress in shallow lakes: a case study in Lake Baiyangdian, North China. Lake and Reservoir Management, 29(4), 285-302.

Yechieli, Y., & Wood, W. W. (2002). Hydrological processes in saline systems: playas, sabkhas, and saline lakes. *Earth Science Reviews,* 58(3), 343-365.

20

Restoration of Water Supply Potential: A Case Study of Semi-Arid District of Rajasthan

Milan Kumar Yadav and Himanshu Singh Mahawar

Introduction

Ajmer City has always been dependent upon surface water sources to meet its demand. Piped water supply in Ajmer started in 1884 from Anasagar reservoir to the then population of 50,000. In 1892, Foysagar Lake was constructed to cater the increased population of the town. Historically, the city is understood to have an efficient system of rain water collection, through interlinked lakes. The Foysagar Lake, with its extensive catchment area, was connected to the Anasagar Lake (By Bandi River), which in turn was connected via channel to Pal Beechla for carrying surplus and run off water. Eventually, with increase in population the adequacy of the source reduced.

Hence in 1909, Bhanwta well field was developed on Sagarmati River about 18 Km south west of Ajmer city to cater increased water demand for an increased population. In 1962-68, water supply to Ajmer city was augmented from ground water source near Banas River, which is about 110 km away. Subsequently, second phase of this was developed during the years 1978-83. Apart from these, some more measures for development of sources of water for Ajmer were taken to improve the water supply to the city.

The height of Foysagar dam was increased by 1.22 m to increase the holding capacity of the lake. In 1982 wells were drilled at Budha Pushkar to supply water to Ajmer and Pushkar. Tube wells were also drilled at Leela Seori during 1990-1995 to supply water to Ajmer as well as Pushkar town. The water supply scheme from Budha Pushkar and Leela Saori has since been abandoned. The present study analyzed various alternate water sources around Ajmer city to cope with increasing population demands. According to geological surveys by Britishers and recently by ISRO (Bhuvan) the groundwater is abundant in south of the city which is valley region of Sagarmati river flowing through Ajmer.

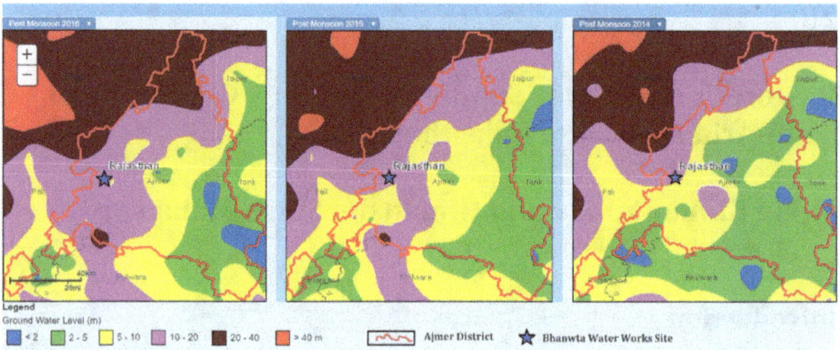

Figure 1. Map Showing Groundwater Levels of Ajmer District 2014-2016, INDIA-WRIS.

Hence, the Bhanwta Supply Works located on this valley could be developed as a major water supply source for the city. It served Ajmer city for 100 years till Bisalpur supply reached Ajmer. Also, the basic infrastructure is already present there and is in good condition therefore, with small funds it could be restored rather than finding a new supply source elsewhere.

Present Water Supply System

Bisalpur Phase-1: Presently Ajmer city is dependent on Bisalpur Dam for its water supply, which is located about 115 Km away from the city. The Bisalpur-Water Supply Project (BWSP) Phase-I was commissioned in year 1995 to supply water to Ajmer city, along with supply to other cities like Beawer, Kishangarh, Nasirabad, Kekri and Sarwar. Water in Bisalpur comes mostly from the catchment areas of Bhilwara, Chittorgrah and parts of Ajmer, but this year, Ajmer got 254 mm of rain, against 550 mm and Bhilwara got only 210 mm of rain in comparison to 540 mm. The full reservoir level of Bisalpur dam is 315.50 m and live capacity of the reservoir is 893.91 MCM. Presently, Bisalpur is supplying water to 3 Districts namely Ajmer, Jaipur and Tonk. The total supply output from the dam was 380 MLD for Ajmer District, of which 120 MLD of water was allocated for Ajmer City. But due to siltation the water level in the dam has been reduced as a result from October onwards, Ajmer District will get 160 MLD water daily out of which 80 MLD will be of Ajmer City.

Local Sources

Apart from the above, approximately 0.6 MLD of water is extracted from Foysagar Lake, and pumped into the local distribution network, on intermittent basis. There is one tube-well which is used for water extraction from Foysagar Lake. Tanker Supply to the newly developing area is maintained by means of Municipality Tankers or Private Vehicles, for which again water is extracted from Tube Wells.

System Adequacy

The water supply network coverage in the city is approximately 90%. Presently, the water supplied by Municipality is for 1-1.5 hours which is very short duration. The supply timings within the city are unpredictable and ranges from once in Two days to once in Three days, depending on the Season. The per capita water available in Ajmer is 135 LPCD, although the supply to end-user is estimated to be lower. This lower availability at the users end can be attributed to lack of proper water management systems, uneven distribution of water and high transmission and distribution losses. The total available storage capacity is presently 47 percent.

Table 1. Water Demand in Different Categories of Ajmer City

Category of Water User	Water Demand in MLD	Water Supply in MLD
Domestic	38	31.15
Industrial	0.267	1.04
Commercial	4.69	2.27
Institutional	2	-
Any Others	4	3
TOTAL	48.96	37.46

Source: Public Health Engineering Department

Due to the uneven topography of the city and the present system of distribution zones, it is noticed that some areas of the city do not receive adequate pressure and supply. Storage capacity augmentation and re-zoning of water distribution is needed to maintain the water supply according to growing urbanization.

Table 2. Status of water supply service levels

S. N.	Indicators	Present Status	MoUD Benchmark
1	Coverage of Water Supply Connections	85%	100%
2	Per Capita Supply of Water	135 LPCD	135 LPCD
3	Extent of Metering of Water Connections	70%	70%
4	Extent of Non-Revenue Water	30%	20%
5	Quality of Water Supplied	100%	100%
6	Cost Recovery in Water Supply Services	33%	-
7	Efficiency in collection of Water Supply related charges	80%	90%

*MoUD:- Ministry of Urban Development
*LPCD – Litres per capita per day
Source: Public Health Engineering Department (2017)

Water requirement for the City, based on present population figures and URDFPI norms of 135 LPCD for the City, is estimated at approximately 81 MLD. This implies that 4 MLD of surplus availability provided Bisalpur Phase I is operational to its full capacity. This situation points out that there should be no immediate need for fresh works for source augmentation. But discussions with PHED indicated that, currently the transmission mains are conveying only 68 MLD collectively to meet the water demand of Ajmer City, as against 85 MLD water allocation under Bisalpur Phase-1.

Therefore, besides proper supply system there is still a deficit of 13 MLD as per demand of present population. Hence, the authorities need to think of alternate sources of water supply for the city otherwise situation will get worse in the near future.

Strategies to Meet Demand

To meet out the above deficit other sources of water supply that can be harnessed are: Bhanwta, Foy Sagar and Budha Pushkar. Out of which, Bhanwta has maximum potential and can be developed economically because basic infrastructure such as wells, pipelines, pumping motor, power source, service points, etc. are already present and in working conditions.

Historical Overview of Bhanwta Water Works

Bhanwta village is located 18 km south east of Ajmer city. It lies on the Sagarmati valley. In the year 1907 there was an acute shortage of water supply in Ajmer city which lead for search of alternate water supply sources for meeting demands of people. This brought Bhanwta as an important water supply source for Ajmer city. It was first discovered by M. Vrendenburg (From Geological Survey of India) who mentioned it in his report on the possibilities of various sources of supply in Ajmer Region. Mr. A.E.Silk (Sanitary Engineer of Bengal) in 1906 vides his report on Water Supply in Ajmer – Merwara Appendix B suggested various suggestions and recommendations for the same.

By the 25th of July, 1907 the positions of affairs had become most critical as no water had been received in the Foysagar and Anasagar, and the total amount of water available for supply was reported by the Mr. S.O. Heinemann (Executive Engineer, Ajmer Provincial Division) to be not more than sufficient for two months. On the 18th August 1907, 5.45 inches of rain fell and the Foysagar was overflowed and Ajmer was saved from the horrors of a water famine.

Later on various alternate sources were found to supply water for Ajmer City. The possible sources from which an adequate supply could be obtained were found to be Saraswati Valley (On the West) and Sagarmati Valley (On the South). The result of the investigations in the Saraswati valley revealed that on the eastern side it is enclosed by hills and in the ground a rock stratum was found at even level but no great depth of water. Also there was water in considerable quantities but not so much as in the Sagarmati valley and the distance was too much as compared to the proposed sites in the Sagarmati valley region.

The investigations in Sagarmati Valley reveled that there is approx. 1300 million gallons of water available per annum if supply is done at the maximum rate of 1,600,00 gallons per day. The bore holes which accompanies that report had the water level at 8 ft. This water level was recorded in January so the total fall in the water level from January to May was only 0.36 feet. It was noted that a good amount of irrigation was going on from the neighboring cultivators' wells at the same time. In addition to this the natural formation of the valley just west of the site of the water works makes it possible for this amount to be materially increased if necessary at a later date. If a concrete wall were built along this ridge to ground level it would prevent any sub soil water passing away from the valley and thus raise the groundwater level.

After various trials in different sites on Sagarmati valley and Saraswati valley, Bhanwta was selected as the major alternate source of water supply. The total cost of the scheme amounted to Rs. 1,05,621. The average income of the municipality at that time was 1,97,806 and annual expenditure Rs. 184,043. There was only one tax collected by the Municipality at that time and was called "Octroi Tax". To repay the loan taken for this project Municipality levied a new tax called as "House Tax" which must be paid by every house owner, whose house lied within 600 feet of a water pipeline. Its work started in 1907 and completed in 1909. At first, 13 wells were dig along with a pumping station and a steam power plant which runs on wood coal. The main supply pipeline is 18 inches broad which runs up to Kaiser ganj seven dials and from there splits into two mains one passing long the outside of the city by Madar Gate and then onto Kutchery Road to the Civil Lines. The other will go straight down the Srinagar Road to the Mayo College Gate with branches to the Nasirabad and Beawar Road.On this main a connection with the railway Mains will be made for supplying water to them as necessary. This was all done under the supervision of Municipal Engineer Mr. S.O. Heinemann.

The Bhanwta Water Supply Works had a maximum capacity of water supply at 1,600,000 Gallons per day. The probable consumption at the end of each decade was estimated as:

Table 3. Estimated quantity per day from Bhanwta wells

Year	Estimated Quantity Per Day
1911	1,000,000 Gallons
1921	1,200,000 Gallons
1931	1,400,000 Gallons
1941	1,600,000 Gallons

Source: Ajmer Municipality Report, 1945

Hence, there would be an increase of 200,000 Gallons per day in every 10 years. This means that infrastructure and facilities in 1911 was enough for 4 decades.

Present Status

Earlier 13 wells were dig during British time and another 15 were dug out by Indian Government after the independence. Bhanwta supplied water to Ajmer City for approx. 100 years till Bisalpur Water Supply came to the city in 2000. It's still in use but a very small amount of 1.5 MLD is being pumped daily which is 10% of what is possible.

Recently, there are 28 wells out of which 20 are in working conditions and well maintained. Interestingly all the wells are interconnected through pipelines with a tunnel of 6 ft. diameter which is used to maintain the transmission pipes. A steam power generating unit along with a water pumping station in which 100 HP motor imported from London is still present, it is capable of pumping water directly to the city without any delay.

Photo 1: Picture showing wells were numbered by engraved stones. On the right is the author taking coordinates of the wells; On the left is the caretaker.

Photo 2: Wells are still in good condition and have a good amount of water levels.

Photo 3: 100 HP motor still in place with **Made in London** engraving on top.

The site visits revealed that local family is doing their best to maintain this traditional water supply system. The government can restore this site quite easily as compared to finding out a new source.

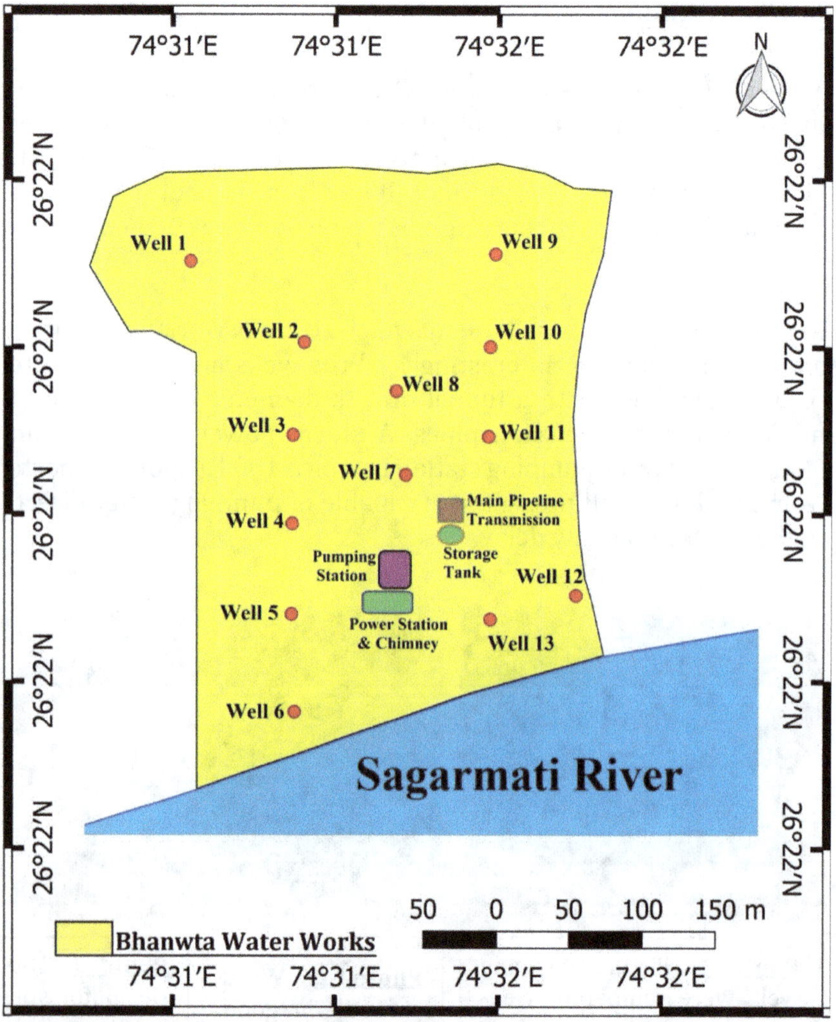

Figure 2. Map Showing Location of Wells in the Bhanwta Supply Works Area Adjacent to Sagarmati River.

Conclusion

The present study shows a large deficit of water supply in Ajmer City.

Therefore, an alternate, The Bhanwta Water Supply Works can be restored with new and modern technology as a major water supply for Ajmer City. The increasing population needs large amounts of water but due to fixed capacity of present water supply system it is getting inadequate. Hence, new water sources are needed to meet the requirements of the Ajmer city. Bhanwta water works is well established and could easily be brought in working condition and after its full capacity the deficit of 13 MLD at present will be fulfilled. If needed in future its supply could be increased because the surrounding has the maximum potential of groundwater nearest to Ajmer City.

References

Foster, Stephen, and Frank van Steenbergen (2011). Conjunctive groundwater use: a 'lost opportunity' for water management in the developing world? Hydrogeology Journal 19, no. 5: 959-962.

Fensham, R. J., J. L. Silcock, O. Powell, and M. A. Habermehl (2016). In search of lost springs: A protocol for locating active and inactive springs. Groundwater 54, no. 3: 374-383.

21

Remote Sensing Based Assessment of Mining Induced Subsidence Impact on Native Vegetation

Ashish Kumar Vishwakarma, Varun Narayan Mishra and Shruti Kanga

Introduction

Modern lives depends on minerals or mineral items, viz. base metals, valuable metals, coking coal, iron sands, limestone and industrial minerals. These are the fundamental and essential components utilized for structure and development, vehicle assembling and fuel, personal computers and various gadgets, transmissions, etc. (Rankin, 2011). The exploitation of underground natural resources often cause persistent land subsidence, ranging from small scale collapses to regional settlements. The subsidence rate varies with geological characteristics and the mechanical properties of rocks. Subsidence associated with abandoned mines can cause many problems, such as building collapse, road damage, environmental pollution, etc. (Jung *et al.,* 2007).Ground subsidence resulting from coal mining is a serious human geographical disaster (Guo-yun and Zhu-you, 2005)that not just harms much arable land, forest land, garden areas and urbanized areas but also changes the ecological landscape and physiognomy of the mining area (Qu *et al.,* 2006). Most of the researches focus on ecological restoration and land reclamation (Quanyuan *et al.,* 2009); however the investigation related to vegetation changes due to mining subsidence is relatively limited. Some of the studies have been performed in China (Hu and Gu, 1995; Donggan *et al.,* 2011), United Kingdom (Selman 1986), South Africa (van der Merwe and Peng 1992; Bell, Stacey, and Genske, 2000), Australia (Holla and Bailey, 1990), India (Tripathi, Singh, and Singh 2009; Vishwakarma *et al.,* 2018; Vishwakarma, Rai, and Shrivastva, 2019) and Illinois (Darmody *et al.,* 1989; Darmody *et al.,* 2014) showing various impacts such as soil erosion, declination in crop yields, disorder in the drainage of surface and subsurface levels, etc. Therefore, surface subsidence impact assessment on native vegetation needs to be enriched and innovated. Dynamism caused by mining subsidence resulted into various changes in local landscape like disturbances in drainage systems, generation of cracks, slope formations, etc., which ultimately results into soil erosion, mineral percolation, tree uprooting, soil and water contamination, inundation at the surfaces, etc. Thus the impact on native vegetation due to subsidence is the cumulative effects of the factors discussed above. Geographic information systems (GIS) and remote sensing (RS) techniques have shown evident advantage over conventional field inspection and lab estimations for evaluating any type of changes. Particularly for bigger areas, where ground leveling and surveying by global positioning system (GPS) are expensive, labor-intensive and time consuming, GIS and RS offer quick and effective information on geological variations and subsidence (Morfeld *et al.,* 2002).

This study aims to evaluate the effect of underground coal mining subsidence on the growth pattern and health condition of the native fauna through the analyses of long-term, i.e. eighteen year's landscape dynamics using RS and GIS techniques.

Materials and Methodology

Study Site

The study area is located in the Singareni Collieries Company Limited, Bhadradri Kothagudem district of the Indian state of Telangana at the longitude 80º40'E and latitude 17º30' N (Figure 1).The geology is nearly undulating plain landscape to slightly inclining towards the river Godavari in the south east with the average elevation differing from 119 to 157 m above mean sea level.

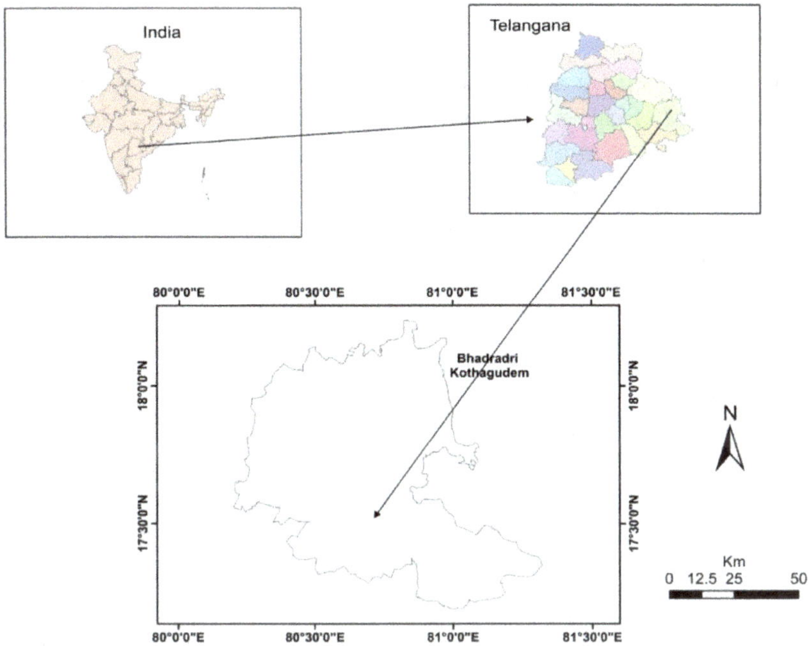

Figure 1. Location map of the study area.

Satellite Data

In this study, the vegetation change analysis of the study area was performed using four Landsat images covering the years 2000–2018, freely downloaded from USGS gateway (Vishwakarma *et al.,* 2018). To be reliable with seasonal fluctuations and vegetation health investigation, images with less or zero rainfall were chosen. The details of the images acquired for the study are as follows:

Table 1. Details of satellite data used

S.N.	Satellite-sensor	Path/Row	Date of acquisition
1	Landsat 8-OLI	142/048	22/10/2018
2	Landsat 5-TM	143/048	26/12/2010
3	Landsat 5-TM	143/048	26/11/2005
4	Landsat 7-ETM+	143/048	20/11/2000

Image Pre-processing

Standard procedure was followed for the image pre-processing of the acquired data for the assessment of changes occurred in the health condition of the native vegetation of the subsided mining area (Figure 2). Ortho-rectification, Geo-correction, Geo-referencing followed by Radiometric calibration and Atmospheric correction were performed for the said purpose (Mishra *et al.,* 2016; Mishra *et al.,* 2019).The image processing was performed using open source software QGIS (Vishwakarma *et al.,* 2018). Five natural breaks classes were prepared for the NDVI images of the study area. Multi-temporal NDVI images of the same area were prepared for the years 2000, 2005, 2010 and 2018.

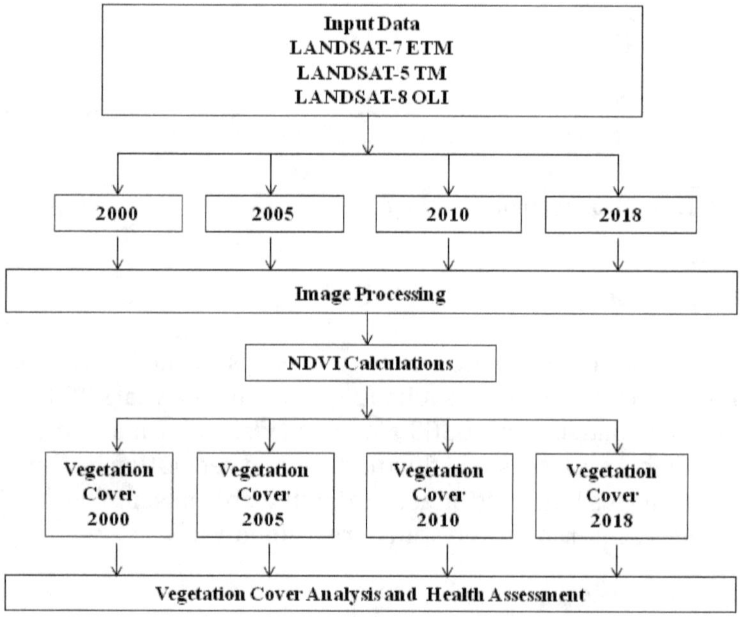

Figure 2. Methodology for the analysis of vegetation health.

Vegetation Health and Coverage

The vegetation change analysis was done using NDVI indices which were calculated by using the formula(Chander, Markham, and Helder 2009; Goslee 2011) as given below in equation (1)

$$NDVI= (NIR-Red)/(NIR+Red) \quad (1)$$

where, NIR = Near Infrared Band value, R = Red Band value recorded by Landsat-7 ETM+, Landsat-5 TM and Landsat-8 OLI imageries (Assal, Anderson, and Sibold 2015). The maximum and minimum value of NDVI ranges from +1 to -1, respectively (Walston, Cantwell, and Krummel, 2009; Dubovyk *et al.,* 2013).Good and healthy vegetation shows a high range of NDVI values, while poor and unhealthy vegetation show a low range of NDVI values. For example, NDVI values for forest ranges from 0.6 to 0.8 (Assal, Sibold, and Reich, 2014), while it is 0.2 to 0.3 for shrub and grassland (Díaz and Blackburn 2003; Bhowmik and Cabral, 2013), -1 for water and less than 0.1 for degraded barren areas like snow, rock and sand. In this study, vegetation indices were having the range of -0.19 to 0.72 indicating the study area to be a forest ecosystem with good and healthy vegetation as well as the presence of damaged and unhealthy vegetation. Based on comparative values of NDVI, slab with NDVI values 0.54-0.72 was considered as healthy, while slabs in between 0.47-0.31 of NDVI values were considered as damaged and unhealthy vegetation (Table 2). The changes in area coverage of vegetation for the years 2000, 2005, 2010 and 2018 were calculated to develop an understanding of the impact of mining subsidence on the native vegetation.

Results and Discussion

Area coverage of healthy vegetation with NDVI values 0.54-0.72 in the periods 2000-2005 and 2005-2010 had decreased by 36.22% and 15.95%, respectively which were followed by a significant increase in its area coverage by 267.38% in later period 2010-2018 (Figure 3, Table 2). In other words, periods 2000-2005 and 2005-2010 had seen a decrease in area coverage of healthy vegetation by 1.86 km^2 and 0.52 km^2, respectively followed by its increase in 2010-2018 by 7.38 km^2. In an associated manner, area coverage of unhealthy vegetation with NDVI slabs 0.47-0.54, 0.40-0.47 and 0.31-0.40 in the period 2000-2005 had increased by 74.17% (2.39 km^2), 32.72% (1.15 km^2) and 0.11% (0.005 km^2), respectively (Table 2). In periods 2005-2010 and 2010-2018, a continuous decrease in same NDVI slabs with unhealthy vegetation was observed (except the slab 0.47-0.54 in period 2005-2010). Vegetation of NDVI slab 0.47-0.54 were increased by 14.69% (0.82 km^2) in period 2005-2010, followed by a decrease in it by 34.91% (2.25 km^2) in period 2010-2018. Vegetation of NDVI slab 0.40-0.47 in periods 2005-2010 and 2010-2018 had decreased by 8.71% (0.41 km^2) and 40.53% (1.72 km^2), respectively, while vegetation of 0.31-0.40 slab had decreased by 9.89% (0.41 km^2) and 61.20% (2.28 km^2), respectively in the same time periods of 2005-2010 and 2010-2018.

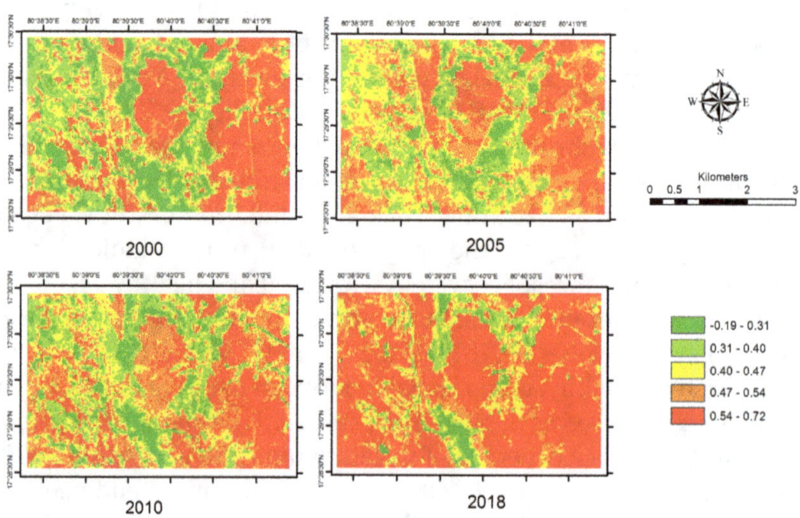

Figure 3. NDVI maps of the damaged area of the years 2000, 2005, 2010 and 2018.

It is clear from the above results that, during initial period of subsidence (2000-2005), a substantial decrease in area coverage of healthy vegetation had occurred with an increase in the area coverage of unhealthy vegetation, while in later period times after subsidence, area coverage of healthy vegetation had increased substantially with a decreasing pattern in the unhealthy vegetation. It can be inferred from the study that subsidence may have a negative impact on the native flora in near time of its occurrence, but in later periods of time, the negative effect gradually disappeared with well recovery of healthy vegetation. The reason behind this could be the gradual stabilization of the subsided disturbed surface with time.

Table 2. Vegetation health changes during 2000-2005, 2005-2010 and 2010–2018

NDVI Values	Vegetation area change from 2000 to 2005		Vegetation area change from 2005 to 2010		Vegetation area change from 2010 to 2018	
	(%)	(km^2)	(%)	(km^2)	(%)	(km^2)
0.54-0.72	Decreased by 36.22%	-1.86	Decreased by 15.95%	-0.52	Increased by 267.38%	7.38
0.47-0.54	Increased by 74.17%	2.39	Increased by 14.69%	0.82	Decreased by 34.91%	-2.25
0.40-0.47	Increased by 32.72%	1.15	Decreased by 8.71%	-0.41	Decreased by 40.53%	-1.72
0.31-0.40	Increased by 0.11%	0.005	Decreased by 9.89%	-0.41	Decreased by 61.2%	-2.28
0.19-0.31	Decreased by 58.15%	1.68	Increased by 42.51%	0.51	Decreased by 65.41%	-1.13

The health of plants depends on various factors like, soil moisture content, supply of soil available nutrients, soil textural composition, and various physicochemical, biological as well as environmental factors. Combinations of these factors aggregately have impact on overall growth and health conditions of native vegetation (Vishwakarma *et al.,* 2018). As there are various structural changes in land surface take place due to occurrence of land subsidence, various associated changes also come into existence in that particular affected land. Due to changes in textural composition, improper nutrient supply to the native plants occurs, which ultimately results into health degradation of the plants with improper leaf coloration. The principal of NDVI only depends upon the greenness of plants and as the plants health is directly related with the leaf colors; NDVI acts as a major tool to assess any change in the overall health conditions of the concerned fauna because of changes in anthropogenic or environmental factors.

Conclusions

The present study demonstrates the use of Landsat data to assess the effect of mine subsidence on health condition and growth patterns of the native flora of the Singareni Collieries Company Limited through one of the most popular methods of vegetation cover analysis using NDVI estimation. The investigation demonstrates that during initial time period of subsidence, native flora might be negatively affected due to subsidence, but the effect gradually disappeared with the time and vegetation recovered well in long run of time due to gradual stabilization of the land surface and absence of human interferences. Forest and grassland ecosystems could carry the partial destruction and ultimately stabilized the ecosystems by self-repair. Thus, in the long run of time, the exploitation of coal resources had no larger impact on the native fauna.

References

Assal, Timothy J, Patrick J Anderson, and Jason Sibold (2015). Mapping forest functional type in a forest-shrubland ecotone using SPOT imagery and predictive habitat distribution modelling. Remote sensing letters 6 (10):755-764.

Assal, Timothy J, Jason Sibold, and Robin Reich (2014). Modeling a historical mountain pine beetle outbreak using Landsat MSS and multiple lines of evidence. *Remote sensing of environment* 155:275-288.

Bell, FG, TR Stacey, and D.D. Genske (2000). Mining subsidence and its effect on the environment: some differing examples. *Environmental Geology* 40 (1-2):135-152.

Bhowmik, Avit Kumar, and Pedro Cabral (2013). Cyclone Sidr impacts on the Sundarbans floristic diversity. *Earth Science Research* 2 (2):62.

Chander, Gyanesh, Brian L Markham, and Dennis L Helder (2009). Summary of current radiometric calibration coefficients for Landsat MSS, TM, ETM+, and EO-1 ALI sensors. *Remote sensing of environment* 113 (5):893-903.

Darmody, RG, IJ Jansen, SG Carmer, and J.S. Steiner (1989). Agricultural impacts of coal mine subsidence: effects on corn yields. *Journal of environmental quality* 18 (3):265-267.

Darmody, Robert G, R Bauer, D Barkley, S Clarke, and D Hamilton (2014). Agricultural impacts of longwall mine subsidence: the experience in Illinois, USA and Queensland, Australia. *International Journal of Coal Science & Technology* 1 (2):207-212.

Díaz, B Meza, and George Alan Blackburn (2003). Remote sensing of mangrove biophysical properties: evidence from a laboratory simulation of the possible effects of background variation on spectral vegetation indices. *International Journal of Remote Sensing* 24 (1):53-73.

Donggan, Guo, Bai Zhongke, Shangguan Tieliang, Shao Hongbo, and Qiu Wen (2011). Impacts of coal mining on the aboveground vegetation and soil quality: a case study of Qinxin Coal Mine in Shanxi Province, China. *Clean–Soil, Air, Water* 39 (3):219-225.

Dubovyk, Olena, Gunter Menz, Christopher Conrad, Elena Kan, Miriam Machwitz, and Asia Khamzina (2013). Spatio-temporal analyses of cropland degradation in the irrigated lowlands of Uzbekistan using remote-sensing and logistic regression modeling. *Environmental monitoring and assessment* 185 (6):4775-4790.

Goslee, Sarah C. (2011). Analyzing remote sensing data in R: the landsat package. *Journal of Statistical Software* 43 (4):1-25.

Guo-yun, DU, and SUN Zhu-you. (2005). Research of Artificial Geologic Disasters of Coast Belt in East Bank of Laizhou Bay, Bohai Bay [J]. *Journal of Geological Hazards and Environment Preservation* 3.

Holla, L, and J G Bailey (1990). Land Subsidence Due to Coal Mining-a Look Back and a Look Forward. Paper read at 1990 International Coal Engineering Conference: Preprints of Papers.

Hu, Zhenqi, and Hehe Gu(1995). Reclamation planning for abandoned mining subsidence lands in Eastern China: A case study. *International Journal of Surface Mining and Reclamation* 9 (3):129-132.

Jung, Hahn Chul, Sang-Wan Kim, Hyung-Sup Jung, Kyung Duck Min, and Joong-Sun Won. (2007). Satellite observation of coal mining subsidence by persistent scatterer analysis. *Engineering Geology* 92 (1-2):1-13.

Mishra, Varun Narayan, Rajendra Prasad, Praveen Kumar Rai, Ajeet Kumar Vishwakarma, and Aman Arora (2019). Performance evaluation of textural features in improving land use/land cover classification accuracy of heterogeneous landscape using multi-sensor remote sensing data. *Earth Science Informatics* 12 (1):71-86.

Mishra, Varun Narayan, Praveen Kumar Rai, Pradeep Kumar, and Rajendra Prasad (2016). Evaluation of land use/land cover classification accuracy using multi-resolution remote sensing images. Paper read at Forum geografic.

Morfeld, P, J Ambrosy, U Bengtsson, H Bicker, B Kalkowsky, A Kösters, H Lenaerts, M Rüther, H-J Vautrin, and C Piekarski (2002). The Risk of Developing Coal Workers' Pneumoconiosis in German Coal Mining under Modern Mining Conditions. *Annals of Occupational Hygiene* 46 (suppl_1):251-253.

Qu, ZJ, H Cheng, YJ Yu, and XY Zou (2006). The effect of coal mine subsidence on the surface to landscape of vegetation. *J Ecol* 30 (3):414-420.

Quanyuan, Wu, Pang Jiewu, Qi Shanzhong, Li Yiping, Han Congcong, Liu Tingxiang, and Huang Limei (2009). Impacts of coal mining subsidence on the surface landscape in Longkou city, Shandong Province of China. *Environmental Earth Sciences* 59 (4):783.

Rankin, W John (2011). *Minerals, metals and sustainability: meeting future material needs*: CSIRO publishing.

Selman, PH. (1986). Coal mining and agriculture: A study in environmental impact assessment. *Journal of environmental management* 22 (2):157-186.

Tripathi, N, RS Singh, and JS Singh. (2009). Impact of post-mining subsidence on nitrogen transformation in southern tropical dry deciduous forest, India. *Environmental Research* 109 (3):258-266.

van der Merwe, JN, and SS Peng (1992). Experiences with undermining by coal in South Africa. Paper read at Proc. of third workshop on surface subsidence due to underground mining. Dept. of Mining Engr. West Virginia Univ. Morgantown, WV.

Vishwakarma, Ashish Kumar, Ashwani Kumar Agnihotri, Rajesh Rai, BK Shrivastva, and Sachin Mishra (2018). Impact Assessment of a Mine Subsidence on Native Vegetation of South Eastern Coalfields, India. *International Archives of the Photogrammetry, Remote Sensing and Spatial Information Sciences* 42:5.

Vishwakarma, Ashish Kumar, Rajesh Rai, and B K Shrivastva (2019). Distribution Characteristics of Micronutrients in Mining Induced Subsided Land of an Underground Coal Mine of South Eastern Coalfields, India. *Nature Environment and Pollution Technology* 18 (2):491-496.

Walston, Leroy J, Brian L Cantwell, and John R Krummel (2009). Quantifying spatiotemporal changes in a sagebrush ecosystem in relation to energy development. *Ecography* 32 (6):943-952.

22

Ecology based Integrated Pest Management for Low Rainfall Areas

Nisha Patel

Introduction

Agriculture in arid regions is fraught with risks and uncertainties due to extreme environmental conditions which lead to different types of abiotic stresses. To make matters worse, biotic stresses viz., insects and diseases sometimes cause huge losses to agricultural and horticultural crops in these regions. Various types of insects negatively affect the growth and productivity of crops, fruits and vegetables of arid zones. Insects adversely affect the yield and also quality of vegetables and fruits crops leading to deterioration of taste, attractiveness and palatability of fruits, which results in reduced market price and losses to the growers. Chemical pesticides are the most common methods by which insect pests are controlled by most of the farmers who spend thousands of rupees on pesticides. However many reports suggest that although the global production and use of pesticides in agriculture is increasing every year, the damage due to insects has not decreased rather it has increased, indicating towards a need for searching for better options. Indiscriminate and over use of synthetic pesticides in crop protection programs around the world has resulted in disturbances to the environment, pest resurgence, resistance to pesticides in many insects and lethal /sub-lethal effects on non-target organism's especially beneficial flora and fauna, including humans (Pimentel, 2005). The residues of pesticides in food and fruits are yet another cause of concern. Pesticides have been linked to many diseases especially cancer. Consignments of fruits and vegetables exported from India have sometimes been rejected due higher pesticide residues than the prescribed limit. The ill effects of too much dependence on synthetic pesticides in crop protection are becoming increasingly evident necessitating the need for a balanced approach for managing insect pests.

With a balanced approach towards insect management, the population and damage due to harmful insects can be reduced and enhanced crop productivity can be obtained with minimal use of pesticides and without adversely affecting the environment. Integrated Pest Management (IPM) is such an ecological approach for preventing or managing pests. It uses combination of all appropriate techniques such as biological control, habitat manipulation, modification of cultural practices, and use of resistant varieties in a compatible manner to keep pest populations below economically damaging levels while avoiding adverse effects to humans, wildlife and the environment. The goal in IPM is not to eliminate pests but to keep them at tolerable levels. Management decisions are made on the basis of the type and stage of pest, the stage of crop, climatic conditions, population of natural enemies etc. If monitoring indicates that the population of a particular pest is beyond acceptable level and a need for chemical pesticide spray they are used as per recommended guidelines.

Arid regions are endowed with considerable agro-ecological biodiversity. Proper manipulation of diversity and various components of production system viz. crop, animals, tree, grass etc. can also play important role in pest reduction by enhancing populations of natural enemies and suppressing pest insect populations (Gurr *et al.*, 2003; Landis *et al.*, 2000). In recent years habitat management has gained a lot of importance as sub discipline of pest management. It requires good understanding of biodiversity-ecosystem function relationships. Enhancing the population of natural enemies with shelter, nectar, alternative prey/hosts, and pollen (SNAP) has emerged as a major research topic and applied tactic (Gurr *et al.*, 2017).

Many cultural practices are helpful in reducing the risk of attack by insect pests. Cultural management manipulates the environment to make it more favourable for the plant and less favourable for the pest. Like all other control strategies, cultural control requires an understanding of the pest and the crop. Cultural control methods work by preventing the pest from colonizing the crop, creating adverse conditions that reduce survival of the pests and reducing the impact of pest injury. Proper water drainage and air circulation in crops , using insect and disease resistant or tolerant varieties or rootstocks, overwintering sites and alternate hosts of pest insects, managing irrigation schedules to avoid long periods of water stress or high relative humidity, avoiding excessive use of fertilizer are some practices which can reduce use of chemical pesticides and negative environmental impacts.

Major Insect Pests of Crops and their Management

Most of the insect pests associated with the arid crops are polyphagous in habits i.e they feed on many crops, although some insects are specific for certain crops. Pest insects are can be categorised according to the type of damage inflicted into defoliators, which feed on leaves ; sucking pests, which feed on the sap of plants from leaves, flowers seeds or pods; the borers, which enter and feed inside the plant tissue of stem, pod or fruit, or the root feeders. The control measures for protecting crops from these pests have to be planned according to the type of pests and nature of injury inflicted. Termites, white grubs, armyworm, locusts and grasshoppers are important polyphagous pests of the arid zone crops. Most of the crops cultivated in this area have many other common pests viz., whiteflies, jassids, grey weevils etc. (Singh *et al.,* 2009).

Termites

Termites (Isoptera) which feed primarily on cellulose are one of the most destructive insects. They are often referred to as "white ants" although they are unrelated to ants. They are social insects and most species live in huge underground colonies. Termites infest almost all components of arid production systems- crops, grasses, shrubs and trees. They cause damage to the stems and roots of young and mature plants which can result in significant plant mortality and yield losses.

Stressed plants are more prone to damage by termites. Healthy, well-grown crops and trees are rarely attacked even though termites may be present. Therefore improvement of the soil, especially with greater use of compost and green manure, although may not reduce the population of termites but would help in preventing crop damage by providing an alternative source of food. Better organic matter levels, moisture retention and nutrient maintenance in the soil improves root growth and plant production and thus minimises the risk of crop damage due to termites.

Application of neem, tumba (*Citrullus colocynths*) or castor cake has been demonstrated to have a suppressive effect on termites. Patel and Bhati, 2005 found reduced infestation of termites and thrips and increased yield of chilli with soil amendment of neem and tumba seed cakes @ 500 kg ha^{-1} before transplanting. Addition of chopped

Calotropis leaves to soil before sowing for reducing termite damage is a well-known ITK in various parts of the country (Mahapatro *et al.,* 2017). Providing alternative sources of cellulose at sites away from crops is sometimes used to lure the termites and prevent injury to crops. These could be mixed with insecticides or entomo-pathogens such as *Metarrhizium* and *Verticillium*. Seed treatment with thiamethoxam 30% FS @ 1.32 Kg per 40 Kg seeds or chlorpyrifos 20% EC @ 3 – 4 ml per kg or fipronil 5 SC @ 5ml per kg seed is helpful for preventing damage by termites during the initial stage of crop. Addition of insecticidal dust viz., chlorpyrifos or quinalphos in soil at the time of or before sowing is common method used for their control although it is costly. When standing crop is infested chlorpyrifos 20 EC @ 1kg a.i. per ha with irrigation water helps in controlling the problem.

White grubs

The larvae of beetles family Scarabaeidae (order Coleoptera) are commonly known as white grubs. The grubs of these insects are subterranean and feed on the roots of many cultivated and uncultivated plants. White grubs are polyphagous and cause damage to most of the crops cultivated in India during the kharif season viz., pearlmillet, groundnut, sorghum, cowpea, green gram and clusterbean.

The beetles feed on leaves of trees such as Neem (*Azadirachta indica*), Ber (*Zizyphus* spp.), Khejri (*Prosopis cineraria*) etc. are attracted to light in the night . The beetles can be collected while they are feeding on the trees, by vigorously jerking the branches of these trees. They can also be collected under sources of light and then destroyed mechanically or chemically by crushing, burning, burying, dipping in insecticidal solution or emulsion of spent oil or kerosene. The efficacy of these operations is better if they are carried out on a mass scale. The grubs are exposed to predators like birds during ploughing operations. Pheromone (synthetic pheromone-Anisole) dispenser traps or light traps @ 1 trap per ha in the evening after first monsoon showers can trap a large number of beetles. Deep summer ploughing exposes the pupae to scorching sun radiation and predators. Preventive measures like early sowing, using properly decomposed organic manures and crop rotation help in keeping the population of these insects to a low level.

Chemical treatment for the grubs involves high doses of insecticides incorporated in soil in dust formulation before or during crop sowing. Seed treatment with chlorpyriphos 20EC @ 6.5-12.0 ml per kg or imidacloprid 17.8SL @ 2.0 ml per kg can keep pest population low during the initial phases of the crop. Furrow application of insecticides such as, thiamethoxam 25 WS @ 1.9 lit per ha or fipronil 5FS @ 2.0 lit per ha or drenching the root zone of crop with chlorpyriphos 20 EC @ 4.0 lit per ha or quinalphos 25EC @ 3.2 lit per ha three weeks after the adult emergence is recommended for control of grubs (Nataraja *et al.*, 2015)

Grasshoppers and locusts

The grasshoppers damage crops by eating the leaves. Basically they are pests of grasses but often they shift to the cultivated crops. Each grasshopper can eat 30 to 100 mg of plant material (dry weight) each day. The stage of crop when these pests appear is important and more damage is caused if they appear during the early stage. Unless their population is large, these insects ordinarily do not cause serious injury to the crops. They have a fairly long life cycle, most species confining to a single generation in a year. The more common species of grasshoppers in dryland crops are *Chrotogonus* sp, *Hieroglyphus* sp, *Oxya* sp, *Atractomorpha* sp, *Acrida* sp and *Pyrgomorpha* sp.

The gradually dwindling populations of natural predators of grasshoppers viz., birds and reptilians which keep the population of grasshoppers under control and diminishing grasslands on the other hand are reducing availability of food resources has intensified problem of damage to crops by these insects. As the grasshoppers lay eggs on the field boundaries, these areas could be deep ploughed during summers to expose the eggs to sun and dehydration. Application of neem formulations or use of any contact insecticide in dust formulation has proved to be effective. Using control measures for grasshoppers when they are in early nymphal stages is a useful strategy for managing these pests.

Locusts also belong to the same family as grasshoppers. They exist in two types of phases, solitary and gregarious. In solitary phase, these insects feed like any other grasshopper species. In the gregarious phase they form swarms consisting of millions of individuals and migrate over large distances devouring all vegetation they come across.

Locust females lay eggs in egg pods in sandy soils at a depth of 10-15 centimeters below the surface. Each egg pod consist of 10-15 centimeters below the surface. A solitary female lays about 95-158 eggs whereas a gregarious female lays usually less than 80 eggs in an egg pod. The life cycle comprises three stages: egg, hopper and adult. Eggs hatch in about two weeks (the range is 10-65 days), hoppers develop in five to six stages over a period of about 30-40 days, and adults mature in about three weeks to nine months but more frequently from two to four months. At present the primary method of controlling desert locust swarms and hopper bands is with mainly organophosphate chemicals applied in small concentrated doses (referred to as ultra-low volume (ULV) formulation) by vehicle-mounted and aerial sprayers and to a lesser extent by knapsack and hand-held sprayers.

One of the mandates of the Food and Agriculture Organization (FAO) of the United Nations is to provide information on the general locust situation to all interested countries and to give timely warnings and forecasts to those countries in danger of invasion. Therefore, FAO operates a centralized Desert Locust information service within the locust Group at FAO Headquarters, Rome, Italy. All locust affected countries transmit locust data to FAO who in turn analyze this information in conjunction with weather and habitat data and satellite imagery in order to assess the current locust situation, provide forecasts up to six weeks in advance and issue warnings on an ad-hoc basis. FAO prepares monthly bulletins and periodic updates summarizing the locust situation and forecasting migration and breeding on a country by country basis (FAO, 1994).

Red hairy caterpillars

The red hairy caterpillar, *Amasacta moorei* also called *Katra* locally, is a polyphagous pest of sporadic occurrence. In recent years many farmers of Nagaur and Barmer districts have reported attack by this insect in crops especially in clusterbean and pearlmillet. The white coloured moths of this insect with a red lining on the forewings lay eggs on the wild vegetation around the field. After hatching, the larvae feed on the wild plants, before moving to the cultivated fields. The caterpillars move in large bands, devouring any vegetation coming their way, pulses being preferred food. They are voracious feeders and spoil far greater mass of plant material than is required to fulfil their appetite. The development of the hairy growth on their body provides protection from predators and also from direct contact of treated surface. In years of epidemic appearance, the entire field is devastated by the caterpillars.

The adult moth of the red hairy caterpillar often lays eggs on the uncultivated plants around the field. Therefore the weeds around the fields should be removed. Use of light traps is an effective measure against the adults of this pest. If the operations are carried out on community level, chances of success are higher. Moving bands of *Katra* during epidemics could be trapped in trenches. These populations could also be destroyed mechanically. For chemical control, the early instar larvae should be targeted, as the later instar larvae develop the hairy covering, hindering the contact of insecticide with body, and the dose requirement is also increased.

Insects of Crops

Pearl millet

The pearl millet crop generally faces less insect pest problems. In the early growth stage grasshoppers, crickets and weevils feed on leaves. An increase in grasshopper numbers can be controlled by application of dust formulation of any contact insecticide. Entomopathogens like *Metarhizium anisopliae* and *Beauveria bassiana* have been reported effective. Sporadic pests such as the red hairy caterpillar *Amsacta moorei* causes severe defoliation during the years of their appearance in some arid districts. Scarab grubs damage roots in the early crop stage, affecting normal growth of the plant, Termites affect the crop during the moisture stress periods at any stage. The ear head stage of pearl millet is susceptible to insect injury, besides bird feeding. The blister beetles *Cylindrothorax* (*Lytta*) spp feed on pollen and damage grains, while the *Rhinyptia laeviceps* and *R. meridionalis* var *puncticollis* beetles feed on milky and ripe grains. The beetles can be destroyed by dusting the crop with insecticides but trapping the beetles in light traps erected on mass scale is an effective means.

Arid pulses and legumes

The arid pulses and the legumes *viz.*, mung bean, moth bean, cowpea, pigeon pea and clusterbean are attacked by a number of regular and sporadic pests. The major pests are aphids (*Aphis craccivora*), weevils

Cyrtozemia spp., flower thrips (*Megalurothrips sjostedti*), white flies *Bemisia* spp, jassids *Empoasca* spp, the pod borer (*Maruca vitrata* syn. *Maruca testulalis*), blister beetles and bugs *Clavigralla, Nezara* and *Piezodorus* spp. The major Insect pests of clusterbean are leaf hopper, *Empoasca motti* Pruthi; whitefly, *Bemisia tabaci* (Genn.), *Acaudaleyrodes rachipora* (Singh); aphid, *Aphis craccivora* Koch; pod borer, *Helicoverpa armigera* (Hub.), leaf perforator, *Dichomeris inthes* Meyr and *Maruca testulalis* Geyer.

Jassids or leaf hoppers

Jassids are commonly found on all rainy season crops. Eggs are laid on petioles and veins on the leaf. The young nymphs and the adults feed on the underside of leaves sucking out the juice and injecting toxic saliva into the tissues. Infested plants look unhealthy and lack vigour. Under severe infestation, the leaves curl at the edges and turn brown, giving the field a scorched appearance called hopper burn. Seedling and pre-flowering stages of the plants are more susceptible to jassid attack.

White fly

The white fly *Bemisia tabaci* is a small insect, but often occurring in large numbers. It affects the pulses and castor crops, surviving on a wide range of host plants. Eggs are laid on the underside of leaves in a ring attached by a small stalk. The sessile nymphs feed by sucking sap from the leaves, devitalizing them. The adult white fly emerges in 3-4 weeks, depending on the temperature. Winged adults are highly active and like nymphs, feed on the underside of leaf. Severely infested plants show yellowing symptoms, wilting follows and the plant may dry ultimately, if infestation continues unchecked. The flies are vectors of yellow mosaic virus, which is the most injurious disease on moth bean, mung bean and other pulses. The injury inflicted through feeding by white flies, though significant, is considered less important in comparison to the act of YMV disease transmitted by this pest.

Aphids

The black aphid, *Aphis craccivora*, sometimes attacks the moth bean plants. Once settled on the host, the aphids reproduce both sexually and asexually (parthenogenetically), thus, multiplying rapidly. The colonies are often protected by the ants in return of the honeydew secreted by these insects. Colonies of aphids suck sap from the underside of leaves, top shoots and stem. The massive loss of sap from plants due to feeding by large aphid colonies results in wilting of plants. The vitality of the plants is greatly reduced. The leaves acquire a curly appearance. The yield of infested crop is greatly reduced. Humid conditions favour rapid multiplication of aphid colonies. When infestation occurs at flowering, the flowers fail to form pods and the developing pods do not produce healthy seeds. The cowpea aphid *A. craccivora* is reported to be a carrier of leaf crinkle virus.

Thrips

The thrips *Megaleurothrips distalis* are small insects with lacerating and licking mouthparts. They feed on leaves, buds and flowers. Plant tissue around the feeding site becomes discolored, subsequently turning brown and dying if feeding proceeds incessantly. Feeding on floral parts results in deformity of inflorescence and premature flower shedding, leading to reduced grain yields.

Mites

The mite is a minor pest of pulses, but occasionally it may cause serious losses. *Polyphagotarsonemus latus* is a small pale green mite, difficult to see without magnification. The symptoms of mite damage are leaf discoloration, curling or deformity in leaf and premature leaf drop. Sometimes pods are also affected.

Sesame

Til leaf and pod caterpillar *Antigastra catalaunalis* is the most destructive pest of sesame in arid regions. The pest attack results in a heavy loss (25 to 90%) in seed yield. It is about 2 cm in length when full grown, pale green in colour and with dark setae all over the body. The caterpillar rolls the leaves together, more frequently the top leaves and feeds on them from within. When pods are formed, it feeds on the developing grains inside, drastically reducing the yield. It may complete several generations before the crop matures. As soon as the damage of this caterpillar is observed, the affected parts should be removed and destroyed, to stop further multiplication of the pest. The crop may be dusted or sprayed with quinalphos. Spraying with

Neem seed kernels extract (5%) Neem oil (2%, two rounds) is effective during early stages of infestation. Spray of Spinosad 0.001% or profenofos 0.05%, is effective in reducing the population of this pest. Til hawk moth *Acherontia styx* is another damaging pest of sesame. The caterpillars of *A. styx* feed on leaves of sesame and sometimes on moth bean also. The full grown caterpillars are stout, greenish with dark green oblique lines on the sides, measuring up to 10 cm in length. Though few in numbers, they consume considerable foliage, in proportion to their body size. Handpicking of caterpillars and deep ploughing of soil in summers can reduce the problem of this pest. For chemical control quinalphos is effective.

Cumin

Cumin is an important seed spice of arid and semi-arid zones of Rajasthan. More than 80 % of cumin production of the country is from Rajasthan and Gujarat. Insect pests and diseases cause huge losses in cumin productivity. Technologies for insect and disease management can help in overcoming losses due to pests and improve productivity and production for arid zone farmers. Among insects aphids cause major losses by sucking sap from the plants thus devitalizing the plants and reducing productivity and yield of the crop. Two species of aphids *Hyadaphis coriandri* and *Myzus persicae* attack cumin crop in the arid region. Excess use of nitrogenous fertilizers leads to more attack of sucking pests therefore it should be avoided. Neem seed cake @ 2.5 ton/hectare prevents pest and disease attack. Indiscriminate spraying of pesticides should be avoided to encourage beneficial insects. Neem oil @ 3 ml per l water is effective for management of aphids .Spraying dimethoate 30 EC or malathion 50EC @ 1 ml per litre of water on crops or Imidacloprid 17.8 SL @ 1 ml per litre water is also suitable for bringing aphid population under control.

Major Pests of Fruit Crops and their Management

Insects can reduce productivity of fruit plants or vegetables in several ways. Insect pests may damage plants by eating leaves i.e. defoliation, by sucking plant juices or by feeding inside the stem fruit, roots or leaves. The more common insect pests of major importance in horticultural crops are the fruit flies, fruit borers, mealy bugs, caterpillars, whiteflies, aphids, bark eaters, and stem borers. In the arid region, ber or Indian Jujube (*Ziziphus mauritiana* Lam.), Pomegranate *Punica granatum*, citrus fruits and aonla Indian gooseberry (*Emblica officinalis* G.) are the commonly grown fruits. Many species of insect have been recorded on these fruit crops. However, only few species have attained the pest status and caused substantial economic damage.

Major insect pests of fruits in arid zone, symptoms and management:

Major insects	Symptoms and type of damage	Preventive and control measures
Fruit flies	Rotting and dropping of fruits	Use of resistant varieties, Raking of the orchard soil during summer and rainy season, destruction of pupa , chemicals
Fruit borer	Fruit rots and drop off from the tree, characteristic symptom is the bad smell and excreta coming out of the entry holes ultimately leading to fruit rot.	Affected fruits should be collected and destroyed, fruits should be bagged with butter paper. NSKE 5% or neem formulations 2 ml/1 should be sprayed at flowering stage. Chemicals ml/l or azadirachtin 1500 ppm @ 3.0 ml/l should be done till harvesting fruit borer. Soil under the tree or near the trees should be raked to destroy the maggots and pupae present in the soil.
Bark eating caterpillar	Several holes can be seen on the trunk and the trees lose productivity, wood dust and faecal matter hanging in the form of a web around the affected portion, breaking of branches.	The webs around the affected portion should be cleaned. Cotton swab soaked in petrol or kerosene or chemicals should be inserted in the holes and sealed with mud, cleanliness should be maintained in orchards
Mealy bugs	Sap sucking, loss of productivity	Preventive use of bio control agents, timely application of pesticides
Scale insects	Presence of small black swollen spots on the branches and fruits, drying of branches, secrete honey dew like substance which attracts black sooty mould. As a	Removal and destruction of alternate hosts, which harbor the scale insects. Spraying the affected patches with dimethoate or quinalphos at 15 days interval

	result, all the leaves and the branches turn black-ish affecting the growth of the plant.	
Aphids	Suck sap, devitalize plant, secrete sweet sticky substance, which attracts fungal growth, leaves show chlorotic patches viruses cause yellowing or, curling of leaves and stunt plant growth	Regular monitoring for aphids at least twice a week, when plants are growing , conserve natural enemies lady beetles adults and larvae, lacewing larvae and syrphid flies larvae, Manage ants, Spraying with chemicals

Year-long schedule of plant protection measures for managing insect pests on fruit crops in arid zone:

Month	Pest	Operation
Throughout the year	Mealy bugs	• Removal of weeds and alternate host plants harbouring the mealy bugs in and around the orchards , keeping orchard clean
June/ July	Chafer beetles Grey weevils	• Jerk plants, collect and destroy the beetles • Use light traps at different locations and kill the trapped beetles • Dust or spray quinalphos (1.5%)
August/ September	Bark eating caterpillar, Galls, lac insect Blue butterfly, tortoise beetle, Mealy bug	• Remove frassy web at junctions of branches and kill the caterpillar by inserting wire • Paint or spray the bark around holes with 0.025% quinalphos or 0.05% fenitrothion • Treat the live holes with 0.1% dichlorvos by injecting through used syringe or by dipping cotton swabs. Plug treated holes with mud • Remove galls, scratch lac • Spray 0.03% dimethoate. May have to be repeated if required • Apply quinalphos or other insecticide

		• Release the Australian ladybird beetle (*Cryptolaemus montrouzieri*) @ 5000/ha. in August- September
October	*Ber* fruit fly	• Spray quinalphos 25EC 1 ml/l water when first set fruits are pea sized, or when 75% fruit set is over. Repeat after three weeks
December/ January	Fruit fly	• Spread polythene sheet below plants to collect and destroy the pupae of fruit fly, or • Treat soil underneath with 1.5% quinalphos or other dust
May/June	Fruit fly Mealy bug	• Expose the pupae to sun through soil raking • Foliar spray with methomyl @ 1g/L after 30 days of soil drenching
April/May	Mealy bug	• Collection and destruction of all the pruned material from mealybug infested gardens in April/May. • Removal of loose bark and destruction of the debarked material • Locating the ant colonies and destroying them with drenching of chlorpyriphos 20 EC @ 2.5 ml/l or apply malathion 5% dust @ 25 kg/hectare • Swabbing/washing of trunk with 2 mL of dichlorvos 76 EC + 2 g of fish oil rosin soap in a litre of water • Soil drenching with imidacloprid 200 SL @ 1.50 ml/l/plant in the basins around the trunk or through drip irrigation @ 1 l/ ha in April-May.

Fruit flies Diptera, Tephritidae

The fruit flies such as *Bactrocera dorsalis, Dacus cucurbitae, Carpomyia vesuviana* etc. are one of the most destructive pests of fruits and vegetables. These flies look like house flies but are smaller in size. The body colour is variable from brown to black but usually there are prominent yellow and dark brown to black markings on the thorax. Generally, the abdomen exhibits two horizontal black stripes and a longitudinal median which may form a T-shaped pattern. The wings are clear. The female has a pointed slender ovipositor to deposit eggs under the skin of host fruit. The adult is a strong flyer and can fly long distances. This ability allows the fly to infest new areas very quickly.

The female lays eggs under the rind of the fruits by puncturing. After hatching the caterpillars feed on the pulp. Fruit that has been attacked may be unfit to eat as larvae tunnel through the flesh as they feed. Fungi and bacteria enter through these holes, leading to the rotting of the fruit. This makes the fruit unfit for consumption. The greatest enemies of the ber/ jujube in India are fruit flies, *Carpomyia vesuviana*. The adult of this pest is small fly with black spots on the thorax and dark bands on the wings. Some cultivars are more susceptible than others. Larger sized and sweeter fruits are preferred more sometimes all fruits of such varieties may be damaged by fruit flies, while trees with smaller, less-sweet type fruits may have only negligible damage. The female can lay more than 1,000 eggs in her lifetime. Time taken for development depends on the ambient temperature. Eggs are minute cylinders laid in batches. The maggots (larvae) are creamy-white, legless, and may attain a length of 10 mm inside host fruit. Maggots tunnel through the fruit feeding on the pulp, shed their skins twice, and emerge through exit holes in approximately 10 days. The larvae drop from the fruit and burrow two three cm into the soil to pupate. In 10 to 12 days, adults emerge from these pupa. The newly emerged adult females need eight to 12 days to mature sexually prior to egg laying. Breeding is continuous, with several generations in a year. Adults live 90 days on the average and feed on honeydew, decaying fruit, plant nectar, bird dung and other substances.

Treatment: Fruit fly larvae pupate in the soil and it has been found that treatment of the ground beneath the tree helps reduce the problem. Raking of the orchard soil during summer and rainy season destroys the hibernating pupae by exposing them to bright sunlight and birds and thus the extent of infestation is considerably reduced. Infested fruits having larvae should also be collected and destroyed.

Some ber varieties like Ilaichi, Mirchia, ZG-3 and 'Umran' are moderate to resistant to fruit fly damage. Early-maturing cultivars ('Gola', 'Jogia', 'Mundia' and 'Seb') attract higher fruit-fly infestation compared to those maturing late. The pest can be effectively controlled by two sprays of malathion. To prevent infestation, prophylactic sprays can be carried out with 0.03% oxydemeton or dimethoate starting from the stage when 70-80% fruits attain pea size and then repeating the spray at one month intervals. During the maturity of fruits, if necessary, spray with 0.5% Malathion+0.05% sugar solution at weekly intervals. Malathion has been observed to dissipate quickly in ber fruits decreasing well below the tolerance level of 3 ppm within two days.

Traps containing 1 ml methyl eugenol and 2 ml Malathion/ litre of water can be used for the control of other fruit flies. Traps made of plastic bottles with holes big enough for flies to enter, are hung at 10 traps per hectare. Solution (100 ml) of methyl eugenol and Malathion is put in each trap. Better control is obtained if all orchards of a particular area take up control measures collectively.

Fruit borers

Different types of caterpillars infest and destroy fruits. In Pomegranate the caterpillars of Pomegranate Butterfly (*Virachola isocrates*) is the most widespread, polyphagous and destructive pest of pomegranate. The adult female lays eggs on buds, flower or on young fruits. On hatching, the caterpillar bores into fruit and feed on the pulp. The infested fruit rots and drop off from the tree. The characteristic symptom is the bad smell and excreta of caterpillars coming out of the entry holes ultimately leading to fruit rot.

Management practices: The affected fruits should be collected and destroyed. Before maturity, the fruits should be bagged with butter paper. NSKE 5% or neem formulations 2 ml/l should be sprayed at flowering stage. From the stage of flowering to fruit development, regular sprays of malathion 50 EC 0.1% or methomyl 40 SP @ 1.0 ml/l or azadirachtin 1500 ppm @ 3.0 ml/l should be done till harvesting subjected to the presence of fruit borer.

In ber, the caterpillars of borer *Meridarches scyrodes* also feed on the ripe fruit and make it unfit for human consumption. The adult of this pest is small moth, dark brown in colour with fringed wings. Early instar larva is light yellowish and full-grown larva is red in colour. The larva bores into the fruits and feeds on the pulp around the seeds. The maggots bore into the ripe fruits and feed on the pulp. When full-grown, they come out of the fruit and drop to ground and pupate 6-

15 cm below the soil. Adult emerges from the soil. For controlling the fruit borer effectively, the attacked fallen fruits is needed to be collected regularly and destroyed. Wild ber trees present around or near to the orchard should be removed. Soil under the tree or near the trees should be raked to destroy the maggots and pupae present in the soil. Insecticides fenvelrate 0.4 per cent or chlorpyriphos 1.5 per cent dust should be mixed at 40 kg per hectare in the soil under the tree or near the trees to reduce the fruit borer incidence. Spray the crop three times, with malathion at 2 ml or fevelerate at 0.5 ml lit of water or neem seed kernel extract (NSKE) 5 per cent; first spray at pea-size stage, second spray at 15 days later and third spray at fruit ripening stage, by alternate use of insecticides. For the third spray add 10 g jaggery/ litre of spray mixture.

Bark-eating caterpillar *(Inderbela sp.)*

It damages stem and branches of grown up trees by eating bark and making tunnel into them. This insect is more common in neglected orchards. Several holes can be seen on the trunk and the trees lose productivity. Wood dust and faecal matter hanging in the form of a web around the affected portion is indication of the borer activity. Adult moths are active during May- June. Female moths deposit eggs under loose bark which hatch in 8-10 days. These larvae eat the bark and bore inside the tree. Larva remains hidden inside the tree during day time and becomes active at night. Severe infection reduces growth and productivity of the tree. The branches become weak and break from the point of attack by the larva.

Control: The webs around the affected portion should be cleaned. Cotton swab soaked in petrol or kerosene should be inserted in the holes and sealed with mud during September-October and February-March. Alternate sprays with carbaryl (2.5 g/litre of water) or quinalphos (2 ml/litre of water) or methomyl (3.5 g /litre of water) is effective in controlling the pest.

Mealy bugs (Homoptera: Pseudococcidae)

Mealybugs (Homoptera: Pseudococcidae) are polyphagous insect pests which cause severe losses in fruits trees. Mealybugs are small oval, wingless, soft-bodied sucking insects which appear like cotton. Most females can move slowly and are covered with whitish, mealy or cottony wax due to which it is difficult to eradicate them. The bugs lay eggs in the soil. When their population increases they form colonies on stems and leaves developing into dense, waxy, white masses.

Damage: Due to mealybug infestation the productivity of fruit trees declines considerably. Adult mealybugs feed on leaves, stems and fruits. They suck a large amount of sap from the plants with the help of piercing/sucking mouth parts, depriving it of essential nutrients. The excess sap is excreted as honeydew which attracts ants and develops sooty mould inhibiting the plant's ability to manufacture food. Infested fruits have uneven shapes, poor quality, and are susceptible to secondary infections by pathogens. High populations decrease plant growth and cause premature leaf or fruit drop.

Management: Management of adult mealy bugs is difficult because of its habit of hiding in crevices, and the waxy covering of its body. Unlike the adults, the crawlers are free from waxy coating and therefore the crawler stage is the more easily controlled by plant protection measures spraying pesticides. Successful control of mealy bug problem is possible with an integrated approach. For management of these pests, plants in the vicinity of the orchard serving as the alternate hosts for the mealy bugs should be destroyed. Ploughing of the orchard during November-December destroys the eggs in the soil. Flooding of orchard with water and raking of soil around tree trunk exposes the eggs to sun and natural enemies thereby destroying them.

Banding of tree trunk should be done with polythene sheets (400 gauge) 30 cm above ground level and just below the junction of branching to obstruct the ascent of the nymphs. Banding should be done well in advance before the hatching of eggs, i.e., around November - December. Pasting a grease band of 5 cm width on the main stem prevents the crawlers from reaching the branches. Early instar nymphs of the mealy bug can be controlled by spraying of 0.05 % carbaryl from January to March. Spraying of insecticides like Dichlorvos (0.02%), methyl demeton 25 EC 2 ml/l, chlorpyriphos 20 EC 2 ml/l, imidacloprid 200 SL 1ml/l or malathion 2.5ml/l of water at 15 days intervals has been found to control the insect population. Application of entomopathogenous fungi viz., *Beauveria bassiana* 5% NSKE around the tree trunk has also been found effective in reducing mealybug population. Encouraging biodiversity as well as beneficial insects in the orchards helps in increasing the natural predators such as ladybird populations in the orchard which helps in keeping the population of mealy bugs under control.

It is necessary to locate and destroy ant colonies with drenching of chlorpyriphos 20 EC @ 2.5 ml/l or apply 5% malathion dust @ 25 kg/ha. As the ants provide the mealy bugs protection from parasitoids and predators and also help in spreading the crawlers to non-infested plants.

Recently papaya mealy bug created havoc in the state of Tamil Nadu especially in papaya orchard. Because of its damage, papaya production came down by 60 – 80 per cent. Chemical pesticides could provide only partial and temporary relief and farmers went for repeated sprayings of pesticides. The NBAIR, Bangalore and TNAU, Coimbatore worked with three exotic papaya mealy bug parasitoids, standardized the mass production techniques of these parasitoids, and distributed them to farmers for field release, free of cost. All these parasitoids proved highly successful in Tamil Nadu. These parasitoids especially *Acerophagus papayae*, established very well and brought significant control of papaya mealy bug not only on papaya, but also on other crops.

Whitefly

Adult females lay eggs on the lower surface of apical leaves. Eggs hatch after a week. The crawlers dig their mouth parts into the leaf tissue for sucking the sap and remain static as "scales" throughout the remaining part of their larval and pupal period. Serious damage is caused by the excretion of honeydew secreted by the by whitefly, which runs down to the fruit and the upper surface of leaves. Under moist conditions, sooty molds can develop on the honeydew, reducing photosynthesis and hindering respiration of plants. The damage by whitefly also leads to yellowing of leaves and stunted growth, in severe cases leading to shedding of leaves.

Management: White flies can be trapped by hanging bright yellow sticky traps coated with polybutene adhesive at the height of the crop canopy. Spraying water with high volume sprayer by focussing the nozzle towards the under surface of leaves helps in washing out the honeydew, eggs, larvae, pupae and adult whitefly. This should be followed by spraying Triazophos 40 EC (1.5 ml/litre of water).

Aphids

Various species of aphids cause losses in fruit crops. Aphids have soft pear-shaped bodies with long legs and may be green, yellow, brown, red, or black depending on the species and the plants they feed on Aphids are have long slender mouthparts that they use to pierce stems, leaves, and other tender plant parts and suck out fluids.

Damage: They suck the cell sap from the lower surface of the leaves and devitalize the plant. They secrete sweet sticky substance, which attracts fungal growth. The affected leaves show chlorotic patches. High humidity favors the multiplication of aphids. Aphids may also transmit plant diseases. The viruses cause yellowing or, curling of leaves and stunt plant growth.

Management: Regular monitoring for aphids at least twice a week when plants are growing rapidly is important so that infestation can be caught early. Many aphid species prefer the underside of leaves, so turn leaves over when checking for aphids. On trees, clip off leaves from several areas of the tree. Also check for evidence of natural enemies such as lady beetles, lacewings, syrphid fly larvae, and the mummified skins of parasitized aphids. Look for disease-killed aphids as well; they may appear off color, bloated, flattened, or fuzzy. Substantial numbers of any of these natural control factors can mean the aphid population may be reduced rapidly without the need for treatment.

Managing ants is a key component of aphid management Ants are often associated with aphid populations, especially on trees and shrubs, and frequently are a clue that an aphid infestation is present. If you see large numbers of ants climbing your tree trunks, check higher up the tree for aphids or other honeydew-producing insects that might be on limbs and leaves. To protect their food source, ants ward off many predators and parasites of aphids

Biological control: Natural enemies can be very important for controlling aphids, especially in gardens not sprayed with broad-spectrum pesticides (e.g., organophosphates, carbamates, and pyrethroids) that kill the natural enemy species as well as pests. Usually natural enemy populations don't appear in significant numbers until aphids begin to be numerous.

Many predators also feed on aphids. The most well-known are lady beetles adults and larvae, lacewing larvae and syrphid flies larvae. Naturally occurring predators work best, especially in garden and landscape situations. Creating favorable conditions for these predators by enhancing plant diversity is an important tool in reducing population of this pest.

Chemical control: Spraying with Dimethoate (0.03%) or Malathion (0.1%) at 15 days interval effectively controls the aphid population.

Scale insects

The scale insects can be identified by presence of small black swollen spots on the branches and fruits. Adults and pupa suck the cell sap from the fruit and tender shoots causing drying of branches. In case of severe infestation, the whole tree dries up. The insects secret honey dew like substance which attracts black sooty mould. As a result, all the leaves and the branches turn blackish affecting the growth of the plant.

Control: Removal and destruction of alternate hosts, which harbor the scale insects. Spraying the affected patches with Dimethoate or Quinalphos at 15 days interval helps to control the pest.

Beneficial Insects

Contrary to popular perception that all insects are harmful and cause damage to our plants animals and to human beings, many of the insects are beneficial and aid in food production by way of pollination and controlling insects which are pests. All insects that are on the plant are not are "pests"! Actually, some insects are needed to keep the natural enemy population alive. Natural enemies (predators, parasitoids, pathogens and nematodes) reduce pest insect populations. Natural enemies either can live naturally in and around the field, or be reared in large numbers to be released into the field. Rearing and releasing natural enemies is becoming an important IPM option in many crops.

Pollination

Pollinators are essential for fruit and seed production in many vegetables and fruits crops. For human nutrition, the benefits of pollination include not just abundance of fruits, nuts and seeds, but also their variety and quality.

Among insects bees (Hymenoptera: Apidae) are the most important and effective pollinators, but other insects such as wasps, flies, moths, butterflies, and beetles also have a major contribution as pollinating species. Vertebrate such as bats, birds, hummingbirds and rodents, squirrels birds are also important pollinators. In an agro- ecosystem with diversity of flowering plants, there is abundance and diversity of pollinators too. At the organic farm in CAZRI there was a large diversity of pollinators which included different types of honeybees wasp, syrphids and many other dipterans and hymenopterans due to different types of flowering plants.

Natural Enemies

The use of natural enemies to maintain pest populations below damaging levels is known as Biological control. Natural enemies of insects fall into three major categories: predators, parasitoids, and pathogens. Predators catch and eat their prey. Some common predatory arthropods include ladybird beetles, lacewings, syrphid flies, carabid (ground) beetles, big-eyed bugs, and spiders. Parasitoids (sometimes called parasites) do not usually eat their hosts directly. Adult parasitoids lay their eggs in, on, or near their host insect. When the eggs hatch, the immature parasitoids use the host as food. Pathogens are organisms that cause diseases in insects. The main groups of insect disease-causing organisms are insect-parasitic bacteria, fungi, protozoa, viruses, and nematodes. The bacterium *Bacillus thuringiensis* (Bt) is a well-known microbial control agent that is available commercially. Several insect-pathogenic fungi are used as microbial control agents, including *Beauveria*, *Metarhizium*, and *Paecilomyces*

Protecting Pollinators and Beneficial Insects from Pesticides

Pesticides which are sprayed for control of pest insects also affect the pollinators and other beneficial insects if used improperly, therefore before the use of pesticides the other options for management of pest insects in orchards should be explored. Insecticides which have least toxicity to bees and minimum length of time it persists in the environment should be considered for using in orchards viz., Bt, chlorantranilipole whereas those with high toxicity to bees such as Malathion, imidacloprid, acephate etc. should be avoided. Avoid applying pesticides to plants during flowering. Pesticides if necessary should be applied in evenings when the honeybees are not foraging. Under the environmental hazards section of the label, note any bee hazard and other environmental warnings. Make targeted applications. Scout for pests regularly and selectively treat pest problems, avoiding blanket applications to the landscape. Take action while pest numbers are low and manageable in order to use the least amount of pesticide. Do not over apply the pesticide, since more is not necessarily better. Follow label directions and use only the amount indicated. The instructions given on labels especially with regard to safety to pollinators should be strictly adhered to. If insecticide is necessary, choose the least toxic product and plan to apply it well before or after the plants flower. Consider using an alternate control tactic such as removing pests manually.

Enhancing Biodiversity

The larval or young stages of many beneficial insects feed on other insects but the adults of most of the predators and parasitoids of pest insects require nectar and pollen at certain stages of their life for growth and reproduction. A diversified ecosystem with many types of trees, shrubs and small plants provides microhabitats, food sources (prey, nectar, pollen), alternative hosts and shelter for natural enemies and thus encourages colonization and population build-up of natural enemies. The plant species chosen for this purpose should be such that they themselves do not attract insect pest of the crops chose for the farm and also the flowering should be in succession so that food is available for beneficial round the year.

Plants that have flowers with a good amount of nectar/ pollen, like *Compositae* and *Brassicaceae* are especially attractive for insects. Flowers which are attractive in colors, small shallow flowers, extra floral nectaries promote or encourage beneficial insects on the farm. Senna plant in the fields also supported parasitoid population as they have extra floral nectaries which provided food to beneficial insect during off season.

Shrubs

Under arid conditions many plants like ber and henna in the organic farm provide pollen and nectar to variety of pollinators, predators and parasitoids besides providing shelter for insects like lacewing. Predatory wasps were observed in good numbers on ber flowers. Some wild plants which encourage beneficial insects like coccinellids syrphids can be left on the farm. Moreover *Calotropis* and henna also supported a population of prey insects aphids and whiteflies which provided food to predators when the crops were not in field. *Acacia senegal* trees was also a good source of pollen and nectar and it attracted supported many types of pollinators and parasites, in the orchard.

Soil Management

Soil management and plant health have a crucial role in prevention of insect damage. Healthier soils produce plants that are less damaged by pests. Practices that promote soil health help in keeping plants healthy and it can better express its inherent abilities to resist pests. Some soil-management practices boost plant-defence mechanisms, making plants more resistant and/or less attractive to pests. Healthier soils also harbour more diverse and active populations of the soil organisms that compete with, antagonize and ultimately reduce soil-borne pests. Application of nutrients only through organic source of fertilizers such as farm yard manure and oilseed cakes has long-term benefits in terms of building up of soil organic matter which favour multiplication of microorganisms besides improving the physical properties of soil. Oilseed cakes are a source of valuable major and micro-nutrients essential for optimal crop growth and yield and enrich the arid zone soils which are highly deficient in micronutrients. Moreover some oil seed cake viz., neem works both as a fertilizer as well as a pesticide.

Precautions

The waiting period needs to be strictly observed so that the fruit is free from harmful residues. The chemical pesticides also affect the beneficial insects- pollinators, predators and prasitoids so spraying should be done in the morning and evening hours when these insects are not active.

References

FAO. (1994). Desert Locust Guidelines (five volumes). Rome: FAO.

Geoff, M., Gurr, Steve, D., Wratten, Douglas A. Landis, and Minsheng You (2017). Habitat management to suppress pest populations: Progress and prospects. *Annual Review of Entomology*. Vol. 62:91-109.

Gurr, G.M., Wratten, S.D., and Luna J. Michael (2003). Multi-function agricultural biodiversity: Pest management and other benefits. *Basic and Applied Ecology* 4:107-116.

Landis, D.A., Wratten, S.D., and Gurr, G.M. (2000). Habitat management to conserve natural enemies of arthropod pests in agriculture. *Annual Review of Entomology* 45:175-201.

Mahapatro ,G.K., Debajyoti, C., and Gautam, R.D. (2017). Indian Indigenous Traditional Knowledge (ITK) on termites: Eco-friendly approaches to sustainable management. *Indian Journal of Traditional Knowledge*. Vol. 16 (2), April 2017. 333-340

Nataraja, M.V., Jadon, K. S., Ram Dutta and Savalia, S.D. (2015). White grubs and their management in groundnut ICAR-Directorate of Groundnut Research pp5

Patel, Nisha and Bhati, T.K. (2005). Organic soil amendments for pest management in chilli crop in the arid region. In *National Conference on Applied Entomology*. Entomological Research Association. September 26-28, MPUAT, Udaipur. 288-289.

Pimentel, D. (2005). Environmental and economic costs of the application of pesticides primarily in the United States. Environ Dev Sustain 7, 229–252. https://doi.org/10.1007/s10668-005-7314-2

Singh, M.P., Satyavir, Patel Nisha, Lodha S, Bhansali R. R., Kumar Arun, Tripathi R. S, Soni B.K., Kaul R.K., Singh Raj, Rajpurohit T. S., Rathore B. S., Ahmed S.I., Srivastava K. K., Sharma Meeta, Verma Neelam and Singh Sangeeta (2009). Management of pests, diseases and weeds in arid production systems. In: *Trends in Arid Zone Research in India* (Eds. Amal Kar, Garg B K, Singh M.P and Kathju S.) Central Arid Zone Research Institute. Jodhpur. 411-457.